Biochemistry of
Selenium

BIOCHEMISTRY OF THE ELEMENTS

Series Editor: **Earl Frieden**
Florida State University
Tallahassee, Florida

A Continuation Order Plan is available for this series. A continuation order will bring delivery of each new volume immediately upon publication. Volumes are billed only upon actual shipment. For further information please contact the publisher.

Biochemistry of Selenium

Raymond J. Shamberger

The Cleveland Clinic Foundation
Cleveland, Ohio

PLENUM PRESS • NEW YORK AND LONDON

Library of Congress Cataloging in Publication Data

Shamberger, Raymond J., 1934–
  Biochemistry of selenium.

  (Biochemistry of the elements, v. 2)
  Bibliography: p.
  Includes index.
  1. Selenium—Physiological effect. 2. Selenium—Metabolism. I. Title. II. Series.
[DNLM: 1. Selenium. QU 130 B6144 1980 v. 2]
QP535.S5S45 1983   574.19'214   82-22316
ISBN 0-306-41090-7

© 1983 Plenum Press, New York
A Division of Plenum Publishing Corporation
233 Spring Street, New York, N.Y. 10013

Printed in the United States of America

Preface

In recent years many exciting research results have indicated that selenium, depending on its concentration, can influence mammalian metabolism. It has been estimated that in selenium-deficient areas, selenium or selenium–vitamin E combinations added to animal feed can prevent annual losses to beef and dairy cattle and sheep valued at 545 million dollars and poultry and swine losses valued at 82 million dollars.

Some animal diseases that can be prevented by a selenium-supplemented diet include liver necrosis, nutritional muscular dystrophy, exudative diathesis, pancreatic degeneration, mulberry heart disease, infertility, growth impairment, periodontal disease, and encephalomalacia. Selenium intake levels are dependent on the plant or animal feed concentrations, which, in turn, are dependent on the pH of the soil and the types of rocks from which the soils are derived.

At normal metabolic levels selenium possesses an antioxidant affect manifested through glutathione peroxidase, and selenium also has an effect on cytochrome P-450 and heme metabolism. Comparisons are made between metabolism of selenium and sulfur in plants, animals, and humans. At greater selenium intake levels acute poisoning occurs when high-selenium-content (10,000 ppm Se) plants are consumed in large quantities. The toxic reactions were first manifested in cavalry horses near Fort Randall, Nebraska, in the 1860s.

Selenium has been found to be a potent anticarcinogen for a variety of chemically-induced cancers, and some inverse relationships between selenium occurrence and human cancer mortality have been demonstrated. The anticarcinogenicity of selenium holds much promise in regard to finding the body's anticancer mechanism. If this mechanism is elucidated, perhaps cancer control can be achieved. The potential of selenium may even surpass that of interferon. In addition, selenium has been shown to prevent Keshan disease, a severe myocardiopathy which occurs in young Chinese children. Inverse epidemiological relationships have also been observed between coronary heart disease and environmental selenium.

v

Much evidence indicates that selenium or selenium–vitamin E combinations counteract the toxic effects of mercury, methylmercury, cadmium, lead, silver, thallium, and arsenic in animals. Selenium–vitamin E combinations may offer a workable antidote to metal toxicity in humans exposed to these environmental hazards.

The forms of naturally occurring selenium are classified according to whether they are of low or high molecular weight. The synthetic forms of selenium along with their chemotherapeutic effects are also outlined. Because the levels of selenium are important in tissues, numerous nondestructive as well as destructive methods of selenium analyses are outlined.

In order to reduce the cost of the textbook, lists of several references were reduced to the most recent reference. The author wishes to thank Kathryn Risko, Phyllis Pittman, and Helen Brewster for their excellent typing, and Barbara DeWitt, Cindy Kopf, Ann McHugh, and Andrea Yartin for their careful help in proofreading.

Contents

5. Biological Interactions of Selenium with Other Substances

6. Environmental Occurrence of Selenium

7. Toxicity of Selenium

8. Selenium in Health and Disease

Forms of Selenium

1

The forms of selenium that occur in living systems (Table 1-1) and the forms that are biologically available to the organism depend on the form of selenium supplied to the organism, the amount of selenium supplied, and the species of plant or animal. For the purpose of this chapter it is most convenient to distinguish the forms of selenium which are present naturally. These include compounds of low molecular weight, which occur in the free form, and also natural forms of selenium which are present in high-molecular-weight compounds. Forms which are synthesized and have biological activity are listed in Chapter 9.

1.1 Low-Molecular-Weight Compounds

1.1.1 Selenocysteine

Smith (1949) has subjected protein hydrolysates of corn and wheat seeds to one-dimensional paper chromatography. He then analyzed the paper for selenium and selenium locations and compared them to positions of synthetic seleno-amino acids. He found the R_f migrations to be similar to that of selenocysteine (Table 1-1). McConnell and Wabnitz (1957) have injected dogs with ^{75}Se and have analyzed the protein hydrolysates of liver by column and two-dimensional paper chromatography. Radioactive spots were close but not necessarily coincident with synthetic seleno-amino acids. The *Escherichia coli* L-methionine requiring mutant has been grown with ^{75}Se-selenite (Cowie and Cohen, 1957). The protein hydrolysate was analyzed by two-dimensional paper chromatography. There was a radioactive spot in the region of cysteine, but synthetic seleno-cysteine was unavailable for comparison. McConnell *et al.* (1959) injected dogs with $H_2{}^{75}SeO_3$ and noted an increase in urinary ^{75}Se after the animals were fed bromobenzene. The urinary ^{75}Se radioactivity was mainly associated with the mercapturic acid fraction. Wool hydrolysates

Table 1-1. Naturally Occurring Organic Selenium Compounds

Compound	Formula
Selenocysteine	$HSe-CH_2CH(NH_2)COOH$
Selenocystine	$HOOCCHNH_2CH_2-Se-Se-CH_2CH(NH_2)COOH$
Selenohomocystine	$HOOCCH(NH_2)CH_2CH_2-Se-Se-CH_2CH_2-CH(NH_2)COOH$
Se-methylselenocysteine	$CH_3-Se-CH_2CH(NH_2)COOH$
Selenocystathionine	$HOOCCH(NH_2)CH_2-Se-CH_2CH_2CH(NH_2)COOH$
Selenomethionine	$CH_3-Se-CH_2CH_2CH(NH_2)COOH$
Se-methylselenomethionine	$(CH_3)_2-Se-CH_2CH_2-CH(NH_2)COOH$
Dimethylselenide	$CH_3-Se-CH_3$
Dimethyldiselenide	$CH_3-Se-Se-CH_3$
Trimethyl selenonium	$(CH_3)_3Se^+$
Selenotaurine	$H_2NCH_2CH_2-SeO_3H$

from sheep injected with H_2 $^{75}SeO_3$ were analyzed by column and paper chromatography. Peaks of standard seleno-cysteine were similar to those found in the wool hydrolysates.

1.1.2 Selenocystine

Peterson and Butler (1962) have fractionated the 80% ethanol extracts of red clover, white clover, and rye grass which were given radioactive selenite. Radioactive tracings of paper chromatograms showed substantial amounts of selenocystine (Table 1-1), selenomethionine, and their oxidation products in the soluble fraction. Onions (*Allium cepa*) contain a large variety of sulfur compounds and were used in an attempt to isolate seleno-amino acids after the injection of radioactive selenite into the bulb (Spare and Virtanen, 1964). After extraction with 70% ethanol, the amino acid fraction was isolated by ion exchange chromatography. Two-dimensional paper chromatograms showed that two of the five radioactive spots observed were similar to selenocystine and selenomethionine. The authors also believed that one of the five major radioactive spots was propenyl β-amino-β-carboxyethyl selenoxide, the selenium analog of the lachrymotory precursor.

1.1.3 Selenohomocystine

Leaves from Astragalus *crotalariae* metabolize ^{75}Se-selenomethionine to selenohomocystine (Table 1-1) (Virupaksha et al, 1966). Dowex-50 column chromatography followed by further fractionation by paper chromatography and electrophoresis gave a compound which yielded selenomethionine after reduction and methylation.

1.1.4 Se-methylselenocysteine

Aqueous extracts (Trelease *et al.*, 1960) were analyzed from soil grown plants of *Astragalus bisulcatus*. Eighty percent of the total selenium was present as Se-methylselenocysteine (Table 1-1), which was isolated as a crystalline solid with small amounts of S-methylcysteine.

1.1.5 Selenocystathionine

Hot water extracts of soil grown *Astragalus pectinatus* were analyzed and selenocystathionine (Table 1-1) was isolated as a crystalline solid (Horn and Jones, 1941). Elemental analysis yielded the empirical formula $C_{21}H_{42}N_6Se_2SO_{12}$, and this was interpreted to be a 2:1 mixture of $C_7H_{14}N_2O_4Se$ and $C_7H_{14}N_2O_4S$. Selenocystathionine was biosynthesized from radioactive selenite in the selenium accumulator *S. pinnata* by Virupaksha and Shrift (1963). After chromatography of a trichloroacetic acid extract on Dowex 50, selenocystathionine comprised about 10% of the radioactive selenium in the extract. High-voltage electrophoresis further purified the compound, which was finally identified by paper chromatography by comparing it to the unmodified form and to derivatives obtained by oxidation with hydrogen peroxide or reduction with Raney nickel.

Selenocystathionine has been identified as the cytotoxic compound in seeds of coco de mono (*Lecythis ollaria*) or monkey nut tree, which is distributed in Central and South America. Aqueous extracts of these nuts were subjected to ion exchange chromatography, which yielded a crystalline material. The structure was established by infrared and nuclear magnetic resonance spectra, along with elemental analysis and paper chromatography (Kerdel-Vegas, 1965). Tissue culture studies showed that almost all of the cytotoxic activity present in water extracts of coco de mono could be accounted for by selenocystathionine (Aronow and Kerdel-Vegas, 1965). Selenocystathionine also has been identified in the seeds of the selenium accumulator (*Neptunia amplexicaulis*) (Peterson and Butler, 1967). The compound comprised about one third of the radioactivity in aqueous ethanol extracts of plants grown on ^{75}Se-labeled selenite.

1.1.6 Selenomethionine

Protein hydrolysates from livers of dogs injected with ^{75}Se were analyzed by column and two-dimensional paper chromatography (McConnell and Wabnitz, 1957). The radioactive spots were close but not coincident with synthetic selenomethionine (Table 1-1), possibly becuase of the large volumes of hydrolysates used. Selenomethionine was also found in the wool of sheep that were injected with $H_2{}^{75}SeO_3$. The wool hydrolysates were analyzed by column and paper chromatography (Rosenfeld, 1961). Tuve and Williams (1961) have grown *E. coli* with $Na_2{}^{75}SeO_3$ on a sulfur-deficient medium and analyzed the protein hydrolysates by Dowex-50 and Dowex-2 columns and two-dimensional paper chromatography. Radioactive spots were eluted from paper chromatogram and rechromatographed with synthetic selenomethionine. Other workers have also found chromatographic evidence for radioactive selenomethionine in protein hydrolysates of microorganisms grown in the presence of radioactive selenite (Weiss *et al.*, 1965). Replacement of methionine by selenomethionine in β-galactosidase isolated from a strain of *E. coli* grown on a medium high in selenate and low in sulfate has been reported by Huber and Criddle (1967a). About 80 of its 150 methionine residues have been replaced by selenomethionine, but there was no apparent replacement of cystine by selenocystine. Rumen microorganisms also can incorporate inorganic selenium into selenoamino acids. Paulson *et al.* (1968) compared the incorporation of selenate and selenite to that of selenomethionine and sulfate in rumen fluid. Selenate appeared to undergo rapid reduction to selenite. Selenite was rapidly and extensively bound to protein by a nonenzymic process because a substantial portion of the protein-bound selenium was removed by dialysis in the presence of glutathione, indicating incorporation of selenium into protein S–Se–S cross-linkages. Pronase digestion released 13% of the added selenite or selenate from reduced glutathione-treated acids that were observed after chromatography. A time-dependent incorporation of selenomethionine into protein was observed. This incorporation could also be inhibited by excess methionine. pronase released most of the incorporated selenium, and paper and ion exchange chromatography showed that the predominant form was the unchanged selenomethionine. Hidiroglou *et al.* (1968) have also reported that selenite was incorporated firmly into proteins of rumen bacteria. Selenomethionine was demonstrated autoradiographically on paper chromatograms of the acid-hydrolyzed proteins, but percentages of the total protein-bound selenium were not reported. Selenocystine was not detected, but this compound in the free form is easily destroyed under the usual conditions of protein hydrolysis (Huber and Criddle, 1967b).

The protein hydrolysates from corn and wheat seeds of soil-grown plants were subjected to one-dimensional paper chromatography. The paper, analyzed for selenium and selenium locations, compared with positions of synthetic seleno-amino acids (Smith, 1949).

Peterson and Butler (1962) used paper chromatography and paper electrophoresis to examine the forms of selenium in enzymic digests of proteins from plants grown in solutions containing radioactive selenite and from aqueous ethanol extracts of various plants. Extensive incorporation of selenium into selenomethionine was observed in the protein portion of ryegrass, wheat, red clover, and white clover. Much smaller incorporation of selenium into proteins was observed in a selenium accumulator plant, *Neptunia amplexicaulis*. In the small aquatic plant *Spirodela oligorrhiza* Butler and Peterson (1967) found 90% of the radioactive selenium from enzymatically hydrolyzed plant protein was present as selenomethionine. Small amounts of selenocystine were found in protein, but in the aqueous ethanol extracts of the same plant, selenocystine and oxidation products were always in excess of selenomethionine. The reason for this difference is not known but could be due to a greater instability of selenocysteine or to enzymic discrimination against its incorporation into protein.

Jenkins and Hidiroglou (1967) have found 59% of the selenite sprayed directly into brome grass was present as selenomethionine and selenocystine after one week. They were present primarily in the polypeptide form. The amount of selenocystine was $2\frac{1}{2}$ times greater than the amount of selenomethionine. This work stands in contrast to the work of Peterson and Butler (1962), who fractionated their proteins by similar procedures but added the selenite to the leaf portion of the plant; this may be the reason for the different metabolic products.

1.1.7 Se-methylselenomethionine

Peterson and Butler (1962) first detected a radioactive compound of high electrophoretic mobility in aqueous ethanol extract of clover and ryegrass roots after growth on radioactive selenite. The compound was identified by paper chromatography as Se-methylselenomethionine (Table 1-1). Virupaksha and Shrift (1965) found this compound to be the predominant soluble organoselenium compound synthesized from selenite by several species of Astragalus which lack the ability to accumulate selenium. The compound was eluted from a Dowex-50 column with 4 N HCl and identified by paper chromatography and electrophoresis. The occurrence of this substance in nonaccumulators may distinguish these species from the accumulator species, where Se-methylselenocysteine is the predominant compound.

1.1.8 Dimethyl Selenide

McConnell and Portman (1952) demonstrated that rats injected with radioactive selenate exhaled a radioactive compound that could be isolated with carrier dimethyl selenide and recrystallized to a constant specific activity as $(CH_3)_2SeHgCl_2$, the mercuric chloride derivative. Dimethyl selenide (Table 1-1) has also been isolated from *Scopulariopis brevicaulis* and *Aspergillus niger* grown with Na_2SeO_4 or Na_2SeO_3 (Dransfield and Challanger, 1955). The methyl groups originated from methionine. Klug and Froom (1965) by gas chromatography reported the presence of dimethyl selenide as a respiratory product in rats given selenite and in Astragalus plants and seeds. Ganther (1962) studied the enzymic synthesis of dimethyl selenide in liver extracts and found that a double-labeled product containing ^{14}C and ^{75}Se was formed in a ratio close to 2:1 after incubation with ^{75}Se-labeled selenite and ^{14}C-labeled S-adenosylmethionine. Even though the double-labeled study is consistent with CH_3SeCH_3, CH_3SeSCH_3 is not excluded.

1.1.9 Dimethyl Diselenide

Four volatile compounds from *Astragalus racemosus* were obtained by Evans *et al.* (1968). The compounds were trapped on active carbon followed by extraction from the carbon in various solvents. One compound was identified by gas chromatography as dimethyl diselenide (Table 1-1), $CH_3SeSeCH_3$. Dimethyl selenide was not obtained.

1.1.10 Trimethyl Selenonium

Byard (1969) identified one of the urinary selenium compounds excreted by rats as the trimethyl selenonium ions (Table 1-1) $(CH_3)_3Se^+$. This identification was confirmed by Palmer *et al.* (1969). Both studies isolated the trimethyl selenonium ion by adsorption on Dowex 50, followed by elution and precipitation as the Reineckate derivative. Synthetic trimethyl selenonium ion was compared to the unknown using paper and ion exchange chromatography, nuclear magnetic resonance, infrared, cocrystallization, and mass spectrometry. Since the compound was not formed when normal rat urine was mixed with selenite (Palmer *et al.*, 1969) and could be detected in rat tissues (Byard, 1969), it appears to be an actual metabolite and not an artifact of the isolation procedure.

1.1.11 Elemental Selenium

Microorganisms have been reported to reduce selenium salts to elemental selenium (Woolfolk and Whitcley, 1962). McCready *et al.* (1966) have identified amorphous selenium as the end product of selenite reduction by *Salmonella heidelberg*. The red precipitate from a 48-hr culture which was grown in 0.1% Na_2SeO_3 was isolated and analyzed by X-ray fluorescence. The red precipitate was heated for one hour and yielded a crystaline black powder, which was identified as metallic selenium by X-ray diffraction. Elemental selenium has been reported to be deposited in the roots of plants given toxic selenium levels. In addition, a high proportion of selenium, presumably elemental selenium, was extracted from the roots of a selenium accumulator with bromine water (Peterson and Butler, 1962).

1.1.12 Selenotaurine

After sheep were injected with $H_2{}^{75}SeO_3$, the selenotaurine derivative of bile cholic acid separated on a celite column (Rosenfeld, 1961).

1.1.13 Selenocoenzyme A

After rats were injected with $Na_2{}^{75}SeO_3$, selenocoenzyme was identified chromatographically (Lam *et al.*, 1961). The formula for selenocoenzyme A is listed in Chapter 9, structure number **LXXVI**.

1.1.14 Other Compounds

McColloch *et al.* (1963) obtained evidence for a seleniferous wax

$$R-\overset{\displaystyle O}{\overset{\displaystyle \|}{C}}-Se-R$$

containing selenoesters of the type R—C—Se—R after *Stanleya bipinnati* was grown in the presence of ^{75}Se-selenite. About 5%–10% of the total radioactivity was extracted from the leaves by a 30-sec rinse in Skelly F, suggesting the presence of seleniferous wax on the external leaf surfaces. The infrared spectrum of the leaf extract had an absorption pattern expected for waxes. Silica gel chromatography provided further evidence of the presence of selenium in the hydrocarbon–vegetable wax fraction.

Virupaksha and Shrift (1965) detected in plants appreciable amounts of an acidic ninhydrin-positive selenium compound in *Astragalus vasei* (a nonaccumulator) when this plant was given selenite or selenate. The compound was not found in species of *Astragalus* that accumulated selenium. The compound had a mobility slightly lower than that of oxidized glutathione, and glutamic acid was found after acid hydrolysis. The compound appears to be a glutamyl peptide containing a seleno-amino acid which decomposes during hydrolysis. After milder hydrolysis, a spot in the area of serine was detected as well. This compound may be the same as that found on the amino acid chromatograms of *Astragalus pectinatus* seed extracts (Nigam and McConnell, 1972). The compounds were identified as two isomers of glutamylselenocystathionine. Both isomers were contaminated to the extent of about 10% with the corresponding sulfur analogs (Nigam and McConnell, 1976). Stewart *et al.* (1974) studied the metabolism of $Na_2{}^{75}SeO_4$ in horseradish leaves and seedlings and the glucosinolate sinigrin was isolated. The isolated sinigrin fraction contained selenium, and the authors suggested that the selenium in this fraction was present as selenosinigrin. On paper chromatograms radioactivity was also present in the position of selenomethionine, which is similar to the results of several other workers who have observed incorporation of $^{75}SeO_4^{2-}$ and $^{75}SeO_3^{2-}$ into selenomethionine in higher plants. Dimethyl selenone $(CH_3)_2SeO_2$ was produced by soil and sewage sludge samples when they were incubated with sodium selenite (Na_2SeO_3) or elemental selenium (Reamer and Zoller, 1980). Dimethyl selenone was confirmed by gas chromatography–mass spectrometry using known compounds for comparison. Dimethyl selenide and dimethyl diselenide were also produced by the soil and sewage sludge samples. The identification of $(CH_3)_2SeO_2$ as a volatile metabolite gives some support to the methylation reaction proposed by Challenger (1951), who has postulated that dimethyl selenone is the final intermediate prior to its reduction to $(CH_3)_2Se$. The results of this study suggest that microbial methylation of selenium commonly occurs in various media and under different conditions. Thus, biomethylation could be a pathway for the mobilization of selenium to the atmosphere.

1.2 Macromolecular Forms of Selenium

1.2.1 Formate Dehydrogenase

One of the first reports of a "nutritional" effect of selenium was the demonstration by Pinsent (1954). In a purified culture medium, traces of selenite were needed for the production of formate dehydrogenase in *E.*

coli. Molybdenum and iron were also shown to be required for the activity of this enzyme. Lester and DeMoss (1971) have demonstrated that protein biosynthesis apparently is essential to show the response to selenium. Selenite and selenocystine were shown to be about equal in stimulating the synthesis of formate dehydrogenase, but selenomethionine was much less active (Enoch and Lester, 1972). The latter investigators suggested that the selenium might be present in the enzyme as nonheme iron selenide. Shum and Murphy (1972) have also substantiated the relative differences in the potencies of different forms of selenium in inducing formate dehydrogenase. They also reported a coincidence of ^{75}Se incorporation and formate dehydrogenase activity after a partially purified enzyme preparation was subjected to sucrose gradient centrifugation. The formate dehydrogenase isolated was somewhat unstable. Although these studies were carried out with *E. coli,* similar effects were observed of selenite and molybdate on the activity of the formate dehydrogenase of *Clostridium thermoaceticum* (Andressen and Ljungdahl, 1973). Homogenous preparations of ^{75}Se-labeled enzyme have been isolated from this organism, but the number of selenium atoms per mole of protein and the chemical form of the selenium which is incorporated are unknown. Studies on the formate dehydrogenases of *C. sticklandii* and *Methanococcus vannielii* (Stadtman, 1974a,b) indicate that selenium is an essential component of the enzyme from these sources as well. Molybdenum is likely to be present in some or all of the formate dehydrogenases. In addition, nonheme iron is known to be a component of the *C. thermoaceticum* enzyme. During the catalytic reaction in which formate is oxidized to carbon dioxide, the iron, selenium, and molybdenum may participate in the electron transfer process.

Studies on the formate dehydrogenase activity of *M. vannielii* revealed that this organism produces both a selenium-dependent and a selenium-independent formate dehydrogenase (Jones *et al.*, 1979). *M. vannielii* is strictly an anaerobic microorganism that grows on formate as the sole organic substrate and converts it into carbon dioxide and methane:

$$4HCOOH \rightarrow 3CO_2 + CH_4 \qquad (1\text{-}1)$$

In a selenium-deficient formate–mineral salts medium the organisms grow very slowly, and under these conditions a 100,000-dalton formate dehydrogenase that contains 1 g-atom of molybdenum and ten each of iron and acid-labile sulfide is present in the cells. However, there is no selenium in the enzyme. When the medium is supplemented with 1 μm of selenite, the rate of growth is markedly stimulated. These cultures contain both a high-molecular-weight (>500,000) selenium-containing formate dehydrogenase and the 100,000-dalton enzyme. Further supplementing the culture medium with tungstate additionally stimulates growth, and using these conditions the large selenium-containing formate dehydrogenase is the predominant form. In the presence of radioactive tungstate growth of *M. vannielii* results in incorporation of ^{185}W into the enzyme and a simultanoues decrease in the amount of molybdenum. This suggests the partial replacement of molybdenum with tungsten in the enzyme. This replacement may favor subunit interaction in the high-molecular-weight enzyme complex and thus may tend to stabilize this form. Partial replacement of molybdenum by tungsten also has been observed in the selenium-dependent formate dehydrogenase of *C. thermoaceticum* (Ljungdahl and Andreesen, 1975).

Easy dissociation by treatment with 0.25% sodium dodecylsulfate into a 100,000-dalton molybdo–iron sulfur protein subunit and a selenoprotein subunit has been observed with the selenium-dependent formate dehydrogenase complex of *M. vannielii*. This molybdo–iron sulfur protein subunit may be identical to the 100,000-dalton formate dehydrogenase which was isolated from selenium-deficient cells. The precise selenium-dependent formate dehydrogenase complex subunit structure has not been determined. Both the 100,000-dalton molybdo–iron sulfur protein and the high-molecular-weight complex are extremely oxygen labile. Therefore, fully active forms of the two enzymes have not been available in amounts sufficient for accurate assessment of relative catalytic activities. Because there is a correlation between the rapid growth of *M. vannielii* and the selenium-containing enzyme complex cells, it is possible that this is the more active formate dehydrogenase. The immediate electron acceptor for formate oxidation by both formate dehydrogenases of *M. vannielii* is an 8-hydroxy-5-deazaflavin which is an abundant cofactor in methane bacteria. Reduction of FAD, FMN, and tetrazolium dyes but not pyridine nucleotides was observed with the isolated enzymes. A second protein, and 8-hydrozy-5-deazaflavin-dependent $NADP^+$ reductase, is required to link formate oxidation to $NADP^+$ reduction.

The chemical form of selenium in the selenium-containing formate dehydrogenase of *M. vanniellii* is selenocysteine (Jones *et al.*, 1979). This is also likely true for the *E. coli* formate dehydrogenase and thus there should be four selenocysteine residues in the *E. coli* enzymes or one per 110,000 $\propto$ subunit.

In *E. coli* the concentration of selenium-dependent formate dehydrogenase is elevated when the organism is cultivated anaerobically with nitrate as the terminal electron acceptor (Enoch and Lester, 1975). In this organism the nitrate reductase complexes and the formate dehydrogenase both possess multiple redox centers and are bound to membranes. Electron transfer between these two complexes is mediated by ubiquinone (Enoch and Lester, 1975) (Figure 1-1).

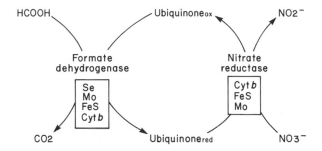

Figure 1-1. Formate dehydrogenase–nitrate reductase system of *Escherichia coli*. Reproduced with permission, from the *Ann. Rev. Biochem.* V. 49, 1980, by Annual Reviews, Inc.

Composition studies were done on a highly purified *E. coli* formate dehydrogenase (Enoch and Lester, 1975). They showed that the approximately 600,0000-dalton enzyme consists of three types of subunits in an $\propto 4$, $\alpha 4$, $\beta 2$ or, $\gamma 4$ structure. The 110,000-dalton $\propto$ subunits contain the 4 g-atoms of selenium present in the enzyme. The enzyme also contains, in equivalents per mole (equiv/mol): molybdenium, 4; β-type heme, 4; acid-labile sulfide, 52; and nonheme iron, 56. The distribution of the various components among the subunits was not determined. The two smaller subunits, β and γ, are 32,000 and 20,000 daltons, respectively. The isolated enzyme had electron acceptors which included phenazine methosulfate, dichlorophenolindophenol, viologens, ferricyanide, and tetrazolum dyes in addition to the normal acceptor, ubiquinone. On the other hand, flavins and pyridine nucleotides were not reduced by the purified enzyme.

1.2.2 Glycine Reductase

1.2.2.1 Protein A

Turner and Stadtman (1973) reported that their low-molecular-weight "protein A" of the clostridial glycine reductase system is a selenoprotein. At first, these workers noted that the yield of protein A was strongly influenced by the age of the cells at the time of harvest. Young cultures were good sources of protein A when the cells were harvested. On the other hand, cells from old cultures were poor sources of this protein. Very young cultures of *C. thermoaceticum* also contain detectable levels of formate dehydrogenase under the usual growth conditions. This microbe continued to synthesize this enzyme throughout the log phase of growth when selenite and molybdate were added to the culture medium. The latter observation suggested that inorganic micronutrients might be important in glycine reductase, because when selenite was added to the medium used for *C. sticklandii*, the activity of the enzyme was markedly increased. With molybdate there was no effect when it was added to the medium. This suggested that there was already an ample amount of molybdenum in the growth medium or that molybdate was not required for the synthesis of glycine reductase. The precise chemical nature of the selenium at the active site of protein A is not known, but an aromatic substance or heterocyclic compound may contain the selenium. A selenol group is produced after borohydride reduction (Stadtman, 1974a,b).

1.2.2.2 Reaction Catalyzed

The clostridial glycine reductase system participates in a specific electron transfer process that is coupled to the esterification of orthophosphate and synthesis of ATP:

$$NH_2CH_2COOH + R(SH)_2 + Pi + ADP \rightarrow CH_3COOH + NH_3 + RS_2 + ATP \quad (1\text{-}2)$$

Support for the stoichiometry of the glycine reductase indicated in equation (1-2) is supported by numerous chemical balance experiments (Stadtman, 1958). For each two reducing equivalents utilized in the conversion of glycine to acetate and ammonia there is generated 1 equiv of ATP. Arsenate can replace phosphate in the enzyme system *in vitro* without affecting the reaction rate. However, with arsenate an adenylate acceptor is not required. This may indicate that substitution of arsenate for phosphate results in the formation of an unstable arsenate intermediate that spontaneously hydrolyzes. Monovalent cations such as K^+, Rb^+, or NH_4^+ may be related to conformational effects required for interaction of the protein components of the system. A dimercaptan such as 1,4-dithiothreitol or 1,3-dimercaptopropanol serves as electron donor in the purified enzyme system *in vitro*. They substitute for a more complex series of electron carriers that normally transfer reducing equivalents from reduced pyridine nucleotides (NADH and NADPH). Glycine is the terminal electron acceptor and in both cases is reduced to acetate and ammonia. The reductive deamination of glycine serves as a terminal electron acceptor process for several amino acid-fermenting anaerobic bacteria, but for many of these microorganisms there are alternate acceptors present in the complex culture media in which they are normally cultivated. For example, *C. sticklandii* and *C. malenominatum* grow luxuriantly in rich media that contains mixtures of amino acids and peptides, even though the amount of selenium present is too low to allow synthesis of the glycine reductase protein (Turner and Stadtman, 1973). When a selenium is added to these media the selenoprotein appears but has little, if any, effect on growth even though glycine reductase activity appears. *Clostridium sporogenes*, in contrast, can be cultivated in a partially defined medium as the principal electron acceptor, and under these conditions both growth and glycine reductase activity are dependent on the availability of selenium (Venugopolan, 1980). The soluble glycine reductase system that catalyzes the reaction in equation (1-2) consists of at least three or four different proteins. Two of these have been isolated in homogenous form and a third has been extensively purified (Turner and Stadtman, 1973).

1.2.2.3 Properties

Protein A, the selenoprotein, is a heat-stable, acidic protein with a low molecular weight of about 12,000. It can be obtained in purified form if a reducing agent such as dithiothreitol is included in all solutions to protect the protein from oxidation and aggregation during the isolation procedure. Growth of *C. sticklandii* in media containing ^{75}Se-labeled selenite (1 μm) results in the synthesis of highly radioactive protein A. In contrast, ferredoxin, a 6000-molecular-weight protein, contains approximately eight iron sulfide residues and eight cysteine residues. Ferredoxin is produced in large amounts by *C. sticklandii*, but becomes only slightly radioactive. This demonstrates that little substitution of selenium for sulfur occurs under the growth conditions used and the incorporation of selenium in protein A is a highly specific process.

When protein A was isolated from extracts of bacteria grown on ^{75}Se-labeled selenite, the glycine reductase activity and the ^{75}Se content of the fractions were enriched in parallel. Addition of a few micrograms of highly purified protein A restores full glycine reductase activity to a protein fraction which contains the remaining enzyme complex components. This biological assay is accurate, precise, and sensitive (Turner and Stadtman, 1973). When *C. sticklandii* is grown on media deficient in selenium, protein A synthesis occurs only during the early stages of growth, but ceases when the selenium is exhausted. The other glycine reductase proteins continue to be formed and, if extracts of the deficient cells are supplemented with protein A, full glycine reductase activity is restored. There is a possibility that an inactive form of protein A is produced by the cell in the absence of selenium. On the other hand, synthesis of the entire polypeptide chain may cease. It is thought that protein A is a sulfhydryl protein since, after reduction with dithiothreitol, treatment with iodoacetamide destroys its biological activity in the glycine reductase reaction. It is possible that an SH group may be modified by the alkylating agent. Chemical and enzymatic degradation procedures of ^{75}Se-labeled protein A, following alkylation of the protein, allows the isolation of the labeled material in good yield in a molecular weight fraction of about 200–300 (Stadtman, 1974a). It is possible that the selenium-containing moiety of protein A may be an aromatic substance or a heterocyclic compound with aromatic properties. The exact chemical role of the selenoprotein in the glycine reductase reaction is likely a carrier function rather than a truly catalytic function. The greater reactivity and lower oxidation–reduction potential (Fredga, 1972; Huber and Criddle, 1967b) of some organoselenium compounds compared to their sulfur counterparts makes the selenoprotein A an especially attractive candidate to be an electron carrier that may also participate in the reduction process.

The glycine reductase enzyme complex is composed of selenoprotein A, a 12,000-dalton, acidic, heat-stabile protein; protein B, approximately 200,000 daltons, which contains one or more essential carboxyl groups; and a fraction C protein of approximately 250,000 daltons. Because the 250,000-dalton fraction C protein copurifies with iron, glycine reductase has similarities to another iron-containing enzyme complex, namely, *E. coli* ribonucleotide, reductase (Reichard, 1978). Both enzyme complexes utilize dithiols as electron donors and catalyze reactions that are chemically comparable. In one reaction a hydrogen atom replaces an amino group and in the other a hydrogen atom replaces a hydroxyl group. Since a tyrosine radical was generated on the iron-containing ribonucleotide reductase protein, one could postulate that a radical mechanism may be involved in glycine reduction (Reichard, 1978).

The glycine reductase selenoprotein A contains 1 g-atom of selenium per mole, which is present within the polypeptide chain as a single selenocysteine residue. This is the first selenium-dependent protein in which the chemical form of the selenium-containing molecule has been identified (Cone *et al.*, 1976). In addition to selenocysteine, two cysteine residues are also contained in the 12,000-dalton selenoprotein A. All three of these residues in the reduced form of the protein upon exposure to oxygen are rapidly reoxidized. The electronic absorption spectrum of the oxidized form of selenoprotein A is typical of a protein that contains five or six residues of phenylalanine, one of tyrosine, and no tryptophan. The acidity of the protein may be explained by the high content of aspartate and glutamate residues. Glycosyl groups are also present and may explain the blocked amino groups. Reduction of the protein A with borohydride at neutral pH gives rise to an additional chromophore, which is thought to be the selenocysteine residue in the ionized selenol form.

Using antibodies prepared to the native selenoprotein A, no precursor cross-reacting protein was detected in selenium-deficient cells of *C. sticklandii* (Stadtman, 1980). Inhibition of cell multiplication by several antibiotics that block protein synthesis as well as by ribampicin, an inhibitor of DNA-dependent RNA synthesis, also prevents the formation of selenoprotein A by selenium-supplemented cells. In addition, selenium supplied in the form of [^{75}Se] selenocysteine is used preferentially for synthesis of [^{75}Se] protein A as compared to H_2^{75}Se (added as H_2^{75}SeO$_3$ and reduced by excess thiols in the medium). There were other possible interpretations than the possibility that selenocysteine is incorporated into protein A during protein-synthesis. Unfortunately, the results of parallel experiments in which ^{14}C-labeled or ^{3}H-labeled selenocysteine was tested as substrate cannot be interpreted because of the extremely rapid cleavage of selenium from the amino acid catalyzed by *C. sticklandii* (Stadtman, 1980). Possibly current studies with selenoproteins produced by microorganisms that do not rapidly break down selene-cysteine will provide more information as to whether direct incorporation of the intact seleno-amino acid into these proteins can occur.

1.2.3 Nicotinic Acid Hydroxylase

Imhoff and Andreesen (1979) have reported about a 16-fold elevation in nicotinic acid hydroxylase when supplemental selenium at about 10^{-7} M is added to the media in which *Clostridium barkeri* and *Clostridium acidi-urici* were cultured. Nicotinic acid hydroxylase catalyzes an anaerobic reaction which is depicted formally as the addition of water to nicotinic acid followed by the removal of two hydrogen atoms and generation of NADPH (Figure 1-2). This formation of 6-oxonicotinic acid by nicotinic acid hydroxylase is the first step in the fermentation of nicotinic acid to propionic acid, acetic acid, carbon dioxide, and ammonia. Earlier compositional studies had shown when the enzyme was purified that the

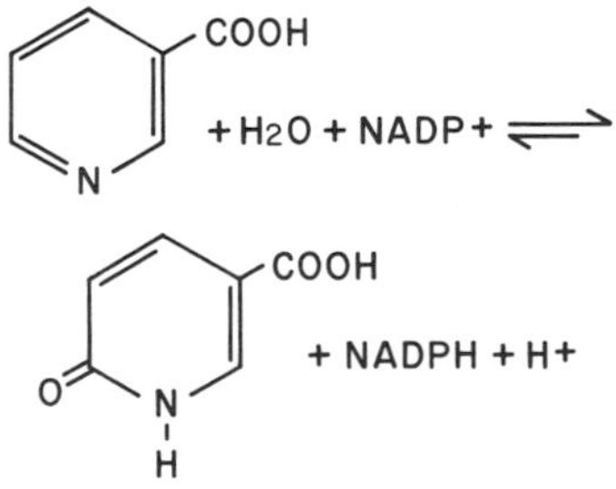

Figure 1-2. Reaction catalyzed by nicotinic acid hydrolase.

300,000-dalton enzyme contained 6 equiv/mol of acid-labile sulfide, 1.5 of FAD, and 11 of iron. When nicotinic acid hydroxylase was purified from *C. barkeri* cells labeled with ^{75}Se, both the enzyme activity and ^{75}Se were enriched in parallel (Stadtman, 1980). These results indicated that selenium might indeed be an essential cofactor for the enzyme with the possibility of serving as yet another redox center in addition to the bound flavin and the iron sulfur centers.

1.2.4 Xanthine Dehydrogenase

Another clostridial enzyme, xanthine dehydrogenase, also increased in activity in response to selenite supplementation of growth media (Wagner and Andreesen, 1979). Xanthine dehydrogenase also contains multiple redox centers. FAD and molybdenum have been found in the relative molar amounts of 8.5, 3.3, and 1. The enzyme may also contain acid-labile sulfide. The reduction of uric acid to xanthine is considered to be the major role of xanthine dehydrogenase in the purine-fermenting anaerobic bacteria:

$$\text{uric acid + reduced acceptor} \rightarrow \text{xanthine + H}_2\text{O + acceptor} \qquad (1\text{-}3)$$

The purified enzyme can oxidize a variety of purines, substituted purines, and aldehydes using dyes, ferricyanide, or oxygen as electron acceptor. A highly purified enzyme preparation was subjected to neutron activation analysis to determine the presence of selenium (Wagner and Andreesen, 1980). Even though selenium was found, there was much less selenium than the expected minimal value of 1 equiv/mol of enzyme. Although this may be due to a very labile selenium-containing moiety being lost from the enzyme during isolation or more than one type of xanthine dehydrogenase being present in the preparation, e.g., selenoenzyme as well as a selenium-independent enzyme, has not been determined. The relatively low substrate specificity of the enzyme preparations might also be an indication of the presence of a family of related enzymes.

1.2.5 Thiolase

The first selenium-dependent enzymes to be recognized were observed to function as redox catalysts. This suggested that other selenoenzymes might also fit this pattern. One exception has already been discovered; this enzyme is known as thiolase. Thiolase is known to catalyze the coenzyme A-dependent cleavage of acetyl-CoA to form 2 acetyl-

CoA:

$$\text{acetoacetyl-CoA} + \text{CoASH} \leftrightarrow 2 \text{ acetyl-CoA} \qquad (1\text{-}4)$$

Clostridium kluyveri grows on ethanol and lactate and anaerobically produces large amounts of butyric and caproic acids. Catalysis of reaction (1-4) appears to occur in the unfavorable direction of the synthesis of acetoacetyl-CoA (Goldman *et al.*, 1961). Hartmanis (1980) has found an extremely active form of thiolase in *Clostridium kluyveri* that contains covalently bound selenium which is interesting from the chemical point of view. The condensation reaction should be considerably facilitated if an enzyme-bound selenolacyl ester rather than a thiolacyl ester, took part as an intermediate, since a selenol group is more reactive than a thiol group. The catalytic properties of the two *C. kluyveri* thiolases should be compared in detail. Both thiolases catalyse the CoA-dependent cleavage of acetyl-CoA, but only one contains selenium. This comparison should provide clues as to their normal biochemical roles in the strictly anaerobic microorganism.

1.2.6 Glutathione Peroxidase

1.2.6.1 Discovery

Glutathione peroxidase is the first selenium-dependent enzyme that has been extensively studied in mammals. Even though the enzyme was first discovered in 1957, the same year selenium was found to be essential, it was not until 1971 that the two trials of research came together. The investigations of Rotruck *et al.* (1971) that led to this finding grew out of the nutritional relationship between selenium and vitamin E. The workers reasoned that because several of the effects of selenium deficiency in animals could be prevented by vitamin E or antioxidants as well as by selenium, selenium might play a role in preventing oxidative damage. Since the membrane lipids of the red blood cell can be damaged by oxidant stressors, resulting in rupture of the cell and release of its hemoglobin through hemolysis, Rotruck *et al.* (1971) chose erythrocytes as a model system for investigation. The hemolysis is prevented by vitamin E and has long been used as an *in vitro* test for vitamin E deficiency.

Their most important observation was that selenium added to the diet of vitamin E-deficient animals prevented the hemolysis of their red blood cells *in vitro*, but only if glucose was present in the incubation medium (Rotruck *et al.*, 1971). The glucose-dependent effect of selenium implicated some step in the sequence of enzymic reactions that link glucose oxidation to the destruction of peroxides by reduced glutathione.

Subsequently, it was shown that the GSH concentration was effectively maintained during an *in vitro* incubation, and that the defect was clearly not in the generation of GSH, but in its utilization to protect the cell, possibly by means of glutathione peroxidase. Subsequent experiments confirmed this (Hafeman *et al.*, 1974) showing that erythrocytes as well as other tissues from animals deficient in selenium had much lower levels of glutathione peroxidase. The glutathione peroxidase from rats given ^{75}Se-labeled selenite contained a large portion of the total erythrocyte ^{75}Se (Rotruck *et al.*, 1973). When the enzyme was purified from sheep erythrocytes, at each stage of the purification the selenium content rose progressively and reached a value of 0.34% in the pure enzyme, equivalent to nearly 4 g-atoms/mole (Oh *et al.*, 1974). The same value has been obtained from several other laboratories for erythrocyte glutathione peroxidase isolated in crystalline form from cows (Flohe *et al.*, 1973), humans (Awasthi *et al.*, 1975), and the rat liver enzyme (Nakamura *et al.*, 1974). Based on the earlier studies it had been deduced that, in mammals, this enzyme is mainly responsible for the protection of the membranes from red blood cells and other tissues that must function in a highly aerobic environment, from damage due to organic peroxides (Mills and Randall, 1958). Flohe (1976) pointed out the important role of glutathione peroxidase in a number of defense mechanisms essential for the survival of animals as well as several possible regulatory functions of the enzyme. Glutathione peroxidase is capable of reducing a wide variety of organic peroxides in addition to H_2O_2, but reduced glutathione (GSH) is the only reductant of physiologic significance (Flohe, 1976; Günzler *et al.*, 1972):

$$2GSH + H_2O_2 \rightarrow GSSG + 2H_2O \qquad (1\text{-}5)$$
$$2GSH + ROOH \rightarrow GSSG + ROH + H_2O \qquad (1\text{-}6)$$

Some thiol compounds are also accepted as donor substrates, if they show structural similarities to GSH. An extensive specificity study revealed that both carboxylic groups of the GSH molecule are essential for substrate binding (Flohe *et al.*, 1979).

With regard to the hydroperoxide substrate, GSH peroxidase is unspecific. With only a few exceptions all hydroperoxides studied were reduced by GSH peroxidase (Table 1-2). The rate constants of the enzyme reaction with the hydroperoxides may well reflect the chemical reactivity of the substrates. This broad specificity range of GSH peroxidase provides a basis for the interaction with metabolic pathways.

1.2.6.2 Distribution

By conventional tissue fractionation of rat liver, glutathione peroxidase can be characterized as a soluble enzyme present in the cytosol and the

Table 1-2. Acceptor Substrates for GSH Peroxidase[a]

Substrate	Source of enzyme	Reference
H_2O_2	Bovine red blood cells	Flohe et al., 1972
Allopregnanolone 17α-hydroperoxide	Pig red blood cells	Little, 1972
Cholesterol 7β-hydroperoxide	Pig red blood cells	Little, 1972
Cholesteryl hydroperoxyoctadecadienoates	Pig aorta	Smith et al., 1973
Cumene hydroperoxide	Rat liver supernatant	Little and O'Brien, 1968
	Bovine lens	Holmberg, 1968
	Bovine red blood cells	Günzler et al., 1972
	Pig red blood cells	Little, 1972
Ethyl hydroperoxide	Bovine red blood cells	Günzler et al., 1972
Ethyl hydroperoxyactadecatrienoate	Rat liver supernatant	Little and O'Brien, 1968
Glyceryl 1-hydroperoxyoctadecadienoates	Pig aorta	Smith et al., 1973
15-Hydroperoxyprostaglandine E_1	Bovine red blood cells	Christ-Hazelhof, 1976
Linoleic acid hydroperoxide	Rat liver supernatant	Little and O'Brien, 1968
	Rat liver supernatant	Christophersen, 1968
	Bovine lens	Holmberg, 1968
	Pig red blood cells	Little, 1972
	Pig aorta	Smith et al., 1973
Linolenic acid hydroperoxide	Rat liver supernatant	Christophersen, 1969a
Methyl hydroperoxyoctadecadienoates	Pig aorta	Smith et al., 1973
Peroxidized DNA	Rat liver supernatant	Christophersen, 1969b
Pregnenolone 17α-hydroperoxide	Pig red blood cells	Little, 1972
Pregnenolone 17α-hydroperoxide	Pig red blood cells	Little, 1972
Prostaglandin G_2	Bovine red blood cells	Nugteren and Hazelhof, 1973
Tert-butyl hydroperoxide	Rat liver supernatant	Little and O'Brien, 1968
	Bovine lens	Holmberg, 1968
	Bovine red blood cells	Günzler et al., 1972
Thymine hydroperoxide	Rat liver supernatant	Christophersen, 1969b

[a] From Flohe et al., The glutathione peroxidase reaction: A key to understand the selenium requirement of mammals, in Trace Metals in Health and Disease, N. Kharasch (ed.), Raven Press, New York, 1979.

mitochondrial matrix. Flohe and Schlegel (1971) reported on cell fractionation that 25.9% of the enzyme activity in the high- and low-density mitochondria and about 73.3% of the enzyme activity was found in the soluble fraction (Table 1-3). Almost no activity was measured in the nuclei, liposomes, peroxisomes, and the high- and low-density microsomes. The mitochondria also have been fractionated into its components by Parsons et al.. (1967). No glutathione peroxidase was found in the inner or outer membrane. The matrix space contains 92% of the activity

Table 1-3. Subcellular Distribution of Glutathione
Peroxidase in Rat Liver (Flohe and Schlegel, 1971)[a]

Fraction	Specific activity U/mg protein	Percent of total activity
Homogenate	0.58	100
Liposomes	0.00	0
Microsomes, high density	0.06	0
Microsomes, low density	0.04	0
Mitochondria, high density	0.48	25
Mitochondria, low density	0.04	0
Nuclei	0.01	0
Peroxisomes	0.09	0
Soluble fraction	1.38	73.3

[a] From Flohe et al., The glutathione peroxidase reaction: A key to understand the selenium requirement of mammals, in *Trace Metals in Health and Disease*, N. Kharasch (ed.), Raven Press, New York, 1979.

and the intermembrane space contains 4% of the activity. These results, however, do not rule out the possibility that GSH peroxidase might be loosely associated with biomembranes *in vivo*. Zakowski and Tappel (1978) have reported that the mitochondrial enzyme is also selenium-dependent and therefore, despite minor differences, could be considered homologous to cytoplasmatic GSH peroxidase. It is interesting that the distribution of selenium described by Diplock *et al.* (1973) does not exactly reflect the subcellular localization of glutathione peroxidase. This difference suggests that the glutathione peroxidase reaction is possibly not the only selenium-dependent metabolic pathway in mammals. Flohe *et al.* (1972) studied kinetically glutathione peroxidase from bovine blood and concluded that the enzyme is present almost exclusively in the reduced form in the presence of physiologic concentrations of reduced glutathione and concentrations of peroxides. This observation may be important in view of the lack of agreement in published reports concerning the chemical composition and stability properties of the enzyme, which in many cases have been measured using preparations allowed to undergo spontaneous oxidation under conditions which were not controlled when it was found that the reduced enzyme (reduced with GSH, borohydride, or its substrate) contained four selenocysteine residues in the selenol form (Forstrom *et al.*, 1978). It is theoretically possible that the exposure of the protein to air might lead to the formation of unstable oxidized derivatives that tend to undergo elimination reactions and deposit elemental selenium. *In vivo*, after a selenol group on a protein interacts with a peroxide, the resulting oxidized enzyme form is likely to undergo reduction rapidly by excess GSH present normally in the cell (Flohe *et al.*, 1972). This cyclic oxidation reduction process is different from the uncontrolled air oxidation of the protein *in vitro*, as the latter process often involves exposure to light.

1.2.6.3 Properties

Glutathione peroxidase has been studied in several mammalian and avian species and has been obtained in highly purified or crystalline form from bovine erythrocytes (Flohe *et al.*, 1971), ovine erythrocytes (Oh *et al.*, 1974), human erythrocytes (Awasthi *et al.*, 1975), bovine lens (Holmberg, 1968, and rat liver Nakamura *et al.*, 1974). The molecular weight values reported for the enzymes were between 76,000 and 92,000. The enzyme is composed of four almost identical 19,000–23,000–dalton subunits (Flohe *et al.*, 1971; Oh *et al.*, 1974; Awasthi *et al.*, 1975). Each subunit contains selenium in the form of a single selenocysteine residue (Forstrom *et al.*, 1978). A small tryptic peptide (molecular weight 2000) that contained the selenocysteine residue was partially purified from rat liver glutathione peroxidase which had been carboxymethylated (Zakowski *et al.*, 1978). When the peptide mixture was subjected to a sequential Edman degradation, it was found that the seleno-amino acid was located within the polypeptide chain of a ^{75}Se-labeled tryptic amino acid composition of homogeneous rat liver glutathione peroxidase. The residues found were calculated per 19,000-dalton subunit and were 1 tryptophan (determined spectrophotometrically), 2 half-cystine (determined as cysteic acid after performic acid oxidation of native protein), 5 tyrosine, 7 phenlyalanine, 3 methionine, 11 lysine, 4 histidine, 7 arginine, 14 aspartic acid, and 15 glutamic acid. In addition, all other amino acids normally found in proteins were also present. After digestion of the protein with nitric acid and perchloric acid, the selenium content of the enzyme was measured by fluorimetric analysis. Nakamura *et al.* (1974) found 4 g-atoms/mol, which is in agreement with the results of other investigations. If the enzyme was dialyzed for two days against five changes of 50 mM sodium phosphate buffer, pH 7 containing *p*-chloromercuribenzoate, titration showed the presence of 8 equiv of SH per mole of protein, which accounts for the two cysteine residues per subunit. Under these conditions the selenocysteine residues of the protein would not have been in the titratable selenol form.

The selenocysteine and the cysteine residues of the reduced form of native glutathione peroxidase (reduced with GSH and its substrate) have not been shown to be fully alkylated by any one alkylating agent tested (Forstrom *et al.*, 1978). The four selenocysteine residues of the native enzyme react quantitatively with chloroacetic acid (Stadtman, 1980) or iodoacetic acid (Forstrom *et al.*, 1978) and the resulting derivatized enzyme is completely inactivated. On the other hand, treatment of the reduced native enzyme with iodoacetamide (Wendel and Kerner, 1977) or N-ethylmaleimide (Stadtman, 1980) was reported to result in no inactivation. The findings that the selenocysteine residues are unreactive with neutral reagents, but react readily with acidic alkylating reagents, are indicative of the steric hindrance effects of neighboring groups of the native protein. X-ray analysis have shown that a histidine and an arginine residue as well as numerous aromatic side chains are near each selenocysteine residue in the active sites. Even though it is implied that the cysteine residues in glutathione peroxidase may react with neutral alkylating agents under nondenaturing conditions and therefore are not essential for enzymic activity, there seems to be no quantitative evidence to support this claim. Under comparable experimental conditions glutathione peroxidase from human erythrocytes was reported to be inactivated 51% by 10 mM N-ethylmaleimide and 73% by 10 mM iodoacetamide (Awasthi *et al.*, 1975). However, the precise form of enzyme that was treated was not specified and the alkylated groups were not identified. After denaturation of the protein with guanidinium hydrochloride (Forstrom *et al.*, 1978), the selenocysteine residues of reduced rat liver [^{75}Se] glutathione peroxidase were derivatized with ethyleneimine. The reactivity of the —SH or the —SeH groups of native protein with this reagent has not been reported. Further studies using pure glutathione peroxidase and labeled alkylating agents appear to be needed to resolve the question of a possible catalytic role of one or both of the two cysteine residues present in each subunit of the enzyme.

Substantial progress has been made in the elucidation of the three-dimensional structure of bovine erythrocyte glutathione peroxidase using X-ray analysis at 2.8-Å resolution (Stadtman, 1980). Crystals of the reduced form of the enzyme were so thin that they were unsuitable for X-ray analysis (Ladenstein and Wendel, 1976). After treatment of the reduced enzyme with H_2O_2 thick or crystals which were suitable for x-ray analysis were obtained. A cell unit of this form of the enzyme is composed of two tetramers. About 178 of the amino acids of the 180 and 183 amino acid residues determined by amino acid analysis to be present have been detected in the crystals. The catalytically active selenocysteines are located in flat depressions on the molecular surface of the N-terminal ends of long α helices. The minimum distance between selenocysteine has been calculated to be 20 Å. It is likely that the intramolecular diselenide bonds cannot be present in the oxidized form of the enzyme. Treatment of glutathione peroxidase with cyanide at pH 7.5–9.5 resulted in extensive loss of enzyme activity and a simultaneous release of selenium from the protein (Prohaska *et al.*, 1977). Cyanide inactivation only affected the oxidized form of the enzyme, which suggests that during reaction with its peroxide substrate the enzyme is converted to its selenic form (Enz–RSeOH). Despite these indications there is no direct evidence concerning the oxidation state of the selenocysteine residues after reaction of the reduced enzyme with a peroxide. Evidence in general shows that the selenocysteine residues are in the selenol (RSeH) form in the reduced enzyme and that they undergo changes in valence upon reaction with a peroxide substrate, but direct proof is still lacking that this oxidation step of the reaction actually converts the selenocysteine residues to oxygen-containing derivatives (e.g., selenenic acids).

Assay of glutathione peroxidase levels in blood is often used as a way of detecting selenium deficiency because of the frequently observed correlation between level of this enzyme and the selenium nutritional status of humans and animals. In some cases, however, assays using organic peroxides (e.g., cumene hydroperoxide) as substrate may give misleading results, especially if tissues such as liver were analyzed. The lack of correlation in certain deficiency conditions was attributed to the presence of a non-selenium-dependent glutathione peroxidase (Lawrence and Burke, 1976; Prohaska and Ganther, 1976), but it became clear later that glutathione transferase (Jakoby, 1977), particularly glutathione transferase B (ligandin) was the responsible enzyme (Stadtman, 1979). Because detoxification enzymes of this class may account for as much as 5% or more of the soluble protein of liver, their reactivity with organic peroxides, even though much less when compared to glutathione peroxidase, may be an important factor in protecting liver from damage due to peroxides in selenium-deficient animals.

1.2.7 Miscellaneous Selenoproteins

There were several selenium-containing proteins whose biochemical roles are still unknown, but they have been found in various biological samples. In some instances these are probably the missing entities in certain deficiency syndromes. One such example is about a 10,000-dalton-molecular-weight selenoprotein that has been isolated from normal semitentinous muscles and the hearts of lambs (Pedersen *et al.*, 1972), but is lacking from selenium-deficient animals. White muscle disease of lambs and calves, gizzard myopathy of domestic birds, and pancreatic fibrosis of chicks are known selenium-deficiency syndromes which appear to be a type of nutritional dystrophy in which collagen replaces damaged tissue. It is not known whether the muscle small-molecular-weight selenoprotein which has been reported to possess a cytochrome c chromophore (Whanger *et al.*, 1973) is the component that is limiting in all of these conditions.

The spectral properties of this protein are similar to cytochrome c, even though the molecular weight and amino acid composition are different (Whanger *et al.*, 1973), resembling those of cytochrome b_5 (Strittmatter and Ozols, 1966). The spectral characteristics, however, are different from those reported for cytochrome b_5 (Strittmatter and Ozols, 1966). This selenium-containing protein appears to possess the same protein composition as cytochrome b_5, but contains a heme group identical to that found in cytochrome c. Based on the extinction coefficient derived for reduced cytochrome c at 550 nm (Massey, 1959), this selenoprotein was calculated to contain 1.2 mol of heme per 10,000 g of protein, which suggests one heme group per mole protein. This 10,000-dalton selenoprotein contains no glutathione peroxidase activity.

Farmers and animal nutritionists concerned with the efficiency of production in domestic animals have come to realize that fertility, especially in males, is directly related to the availability of selenium in the diet. McConnell and coworkers (1979) have discovered that a selenium-containing protein of about 15,000 daltons appears in the testis of the rat concomitant with the onset of sexual maturity and may be of special importance in regard to fertility. A 17,000-dalton selenoprotein isolated from rat sperm tail (Calvin, 1978) may be identical.

Burk and Gregory (1982) have separated and identified fractions of glutathione peroxidase and another selenoprotein, called ^{75}Se-P, from rat liver. Both selenoproteins were also found in the plasma. Selenium deficiency decreased ^{75}Se incorporation by glutathione peroxidase at 3 and 72 hr after ^{75}Se injection but increased ^{75}Se incorporation by ^{75}Se-P. Apparent molecular weights of ^{75}Se-P from liver and plasma were determined by gel filtration and were found to be 83,000 and 79,000. ^{75}Se-P may account for some of the physiological effects of selenium.

Another bacterial selenoprotein that has not yet been identified is produced by *C. kluyveri* (Stadtman, 1980). This protein, which has been isolated and almost completely purified, appears to be a component of a high-molecular-weight iron-containing enzyme complex.

1.2.8 Seleno-tRNA's

An example of the natural occurrence of a different type of macromolecule which contains selenium is provided by the discovery (Chen and Stadtman, 1980) that a few bacterial aminoacyl transfer nucleic acids (tRNA's) are specifically modified with selenium in the polynucleotide portions of the molecules. *C. sticklandi*, which has been studied in the greatest detail, has been shown to synthesize three readily separable ^{75}Se-labeled tRNA's when incubated with ^{75}Se-selenite or ^{75}Se-selenocysteine. Seleno-tRNA formation in this system is highly specific for selenium and is not decreased in the presence of 1000-fold molar excesses of sulfur analogs. One of the three seleno-tRNA's isolated from *C. sticklandii* copurified with an *L-prolyl-tRNA species*. *M. vannielii* and several other anaerobic bacteria synthesize selenium-containing tRNA's, but in each of these only a few of the tRNA's contain selenium (Chen and Stadtman, 1980). In several cases a specifically modified base in a particular tRNA appears to play a role in some type of regulatory process. One could also assume that modification of the selenium may serve a similar role. In *E. coli* it is known that selenium can be incorporated into tRNA (Saelinger *et al.*, 1972; Hoffman and McConnell, 1974), and a ^{75}Se-labeled nucleoside that chromatographed with 4-selenouridine was isolated from the enzymic digests of ^{75}Se-labeled tRNA. In this case it was thought that selenium incorporation had occurred by the known pathway of sulfur transfer to a specific uracil residue of *E. coli* tRNA, resulting in the formation of 4-selenouracil instead of the normal 4-thiouracil. However, when varying ratios of sulfur to selenium were utilized in the medium, no inhibition effects were reported. The sulfur transferase that catalyzes this reaction utilized either selenocysteine or cysteine as donor with similar efficiency (Chen and Stadtman, 1980). Thus if the observed ^{75}Se labeling had occurred by this process it should have been markedly decreased by the addition of high sulfur levels. Transfer RNA has also been isolated from the selenium indicator plant *A. bisulcatus*. (Young and Kaiser, 1979). This material had a high guanosine to cytidine ratio and showed a major and modified nucleoside composition characteristic of plant transfer RNA's, and exhibits chromatographic and electrophoretic properties similar to transfer RNA's from other well-studied plant and bacterial systems. RNA's isolated from *A. bisulcatus* seedlings incubated in the presence of ^{75}Se indicated some radioactive incorporation, but at extremely low levels. The isolated transfer RNA's were active in accepting amino acids, but their overall levels of activity seemed to be low when compared with those from a homologous *E. coli* aminoacylation reaction system.

References

Aronow, L., and Kerdel-Vegas, F., 1965. Seleno-cystathionine, a pharmacologically active factor in the seeds of *Lecythis ollaria, Nature* 205:1185–1186.
Andressen, J. R., and Ljungdahl, L. G., 1973. Formate dehydrogenase of *Clostridium*

thermoaceticum: Incorporation of selenium-75, and the effects of selenite, molybdate, and tungstate on the enzyme, *J. Bacteriol.* 116–867–873.

Awasthi, Y. C., Beutler, E., and Srivastava, S. K., 1975. Purification and properties of human erythrocyte glutathione peroxidase, *J. Biol. Chem.* 250:5144–4149.

Blau, M., 1961. Biosynthesis of {^{75}Se} selenomethionine and {^{75}Se} selenocystine. *Biochim. Biophys. Acta* 49:389–390.

Burk, R. F., and Gregory, P. E., 1982, Some characteristics of ^{75}Se-P, a selenoprotein found in rat liver and plasma, and comparison of it with selenoprotein peroxidase, *Arch. Biochem. Biophys.* 213:73–82.

Butler, G. W., and Peterson, P. J., 1967. Uptake and metabolism of inorganic forms of selenium-75 by *Spirondela oligorrhiza, Aust. J. Biol. Sci.* 20:77–86.

Byard, J. L., 1969, Trimethyl selenide. A urinary metabolite of selenite, *Arch. Biochem. Biophys.* 130;556–560.

Calvin, H. I., 1978. Selective incorporation of selenium-75 into a polypeptide of the rat sperm tail, *J. Exp. Zool.* 204:445–452.

Challenger, R., 1951. Biological methylation, *Adv. Enzymol.* 12:429–491.

Chen, C. S., and Stadtman, T. C., 1980. Selenium-containing tRNA's from *Clostridium sticklandii*: Cochromatography of one species with L-prolyl-tRNA, *Proc. Natl. Acad. Sci.* 77:1403–1407.

Christ-Hazelhof, E., Nugteren, D. H., and Van Dorp, D. A., 1976. Conversions of Prostaglandin endoperoxides by glutathione-S-transferases and serum albumins, *Biochim. Biophys. Acta* 450:450–461.

Christophersen, B. O., 1968. Formation of monohydroxypolyenic fatty acids from lipid peroxides by a glutathione peroxidase, *Biochim. Biophys. Acta* 164:35–46.

Christophersen, B. O., 1969a. Reduction of linolenic acid hydroperoxide by a glutathione peroxidase, *Biochim. Biophys. Acta* 176:463–470.

Christophersen, B. O., 1969b. Reduction of x-ray-induced DNA and thymine hydroperoxides by rat liver glutathione peroxidase, *Biochim. Biophys. Acta* 186:387–389.

Cone, J. E., Martin del Rio, R., Davis, J. N., and Stadtman, T. C., 1976. Chemical characterization of the selenoprotein component of Clostridial glycine reductase: Identification of selenocysteine as the organoselenium moiety, *Proc. Natl. Acad. Sci.* 73:2659–2663.

Cowie, D. B., and Cohen, G. N., 1957. Biosynthesis by *Escherichia coli* of active altered proteins containing selenium instead of sulfur, *Biochim. Biophys. Acta* 26:252–261.

Diplock, A. T., Caygill, C. P. J., Jeffery, E. H., and Thomas, C., 1973. The nature of the acid-volatile selenium in the liver of the male rat. *Biochem. J.* 134:283–293.

Dransfield, P. B., and Challenger, F., 1955. Studies on biological methylation. Part XV. The formation of dimethyl selenide in mould cultures in presence of D- and L-methionine, or of thetins, all containing the $^{14}CH_3$ group, *J. Chem. Soc.* 1955:1153–1160.

Enoch, H. G., and Lester, R. L., 1972. Effects of molybdate, tungstate and selenium compounds on formate dehydrogenase and other enzyme systems in *Escherichia coli, J. Bacteriol.* 110:1032–1039.

Enoch, H. G., and Lester, R. L., 1975. The purification and properties of formate dehydrogenase and nitrate reductase from *Escherichia coli, J. Biol. Chem.* 250:6693–6705.

Evans, C. S., Asher, C. J., and Johnson, C. M., 1968. Isolation of dimethyldiselenide and other volatile selenium compounds from *Astragalus racemosus* (Pursh), *Aust. J. Biol. Sci.* 21:13–20.

Flohe, L., 1976. Role of selenium in hydroperoxide metabolism, in *Proceedings of the Symposium on Selenium—Tellurium Environment*, Industrial Health Found., Inc., Pittsburgh, pp. 138–157.

Flohe, L., and Schlegel, W., 1971. Glutathion-peroxidase, IV. Intrazellulare verteilung des glutathion-peroxidase-systems in der rattenleber, *Hoppe Seylers Z. Physiol. Chem.* 352:1401–1410.

Flohe, L., Eisele, B., and Wendel, A., 1971. Glutathion-peroxidase. I. Reindarstellung and Molekulargewichtsbestimmungen, *Hoppe-Seylers Z. Physiol. Chem.* 352:151–158.

Flohe, L., Loschen, G., Günzler, W. A., and Eichele, E., 1972. Glutathione peroxidase. V. The kinetic mechanism, *Hoppe-Seylers Z. Physiol. Chem.* 353:987–999.

Flohe, L., Günzler, W. A., and Schock, H. H., 1973. Glutathione peroxidase: A selenoenzyme, *FEBS Lett.* 32:132–134.

Flohe, L., Günzler, A., and Loschen, G., 1979. The glutathione peroxidase reaction: A key to understand the selenium requirement of mammals, in Khrasch, N., *Trace Metals in Health and Disease*, Raven Press, New York, pp. 263–286.

Forstrom, J. W., Zakowski, J. J., and Tappel, A. L., 1978. Identification of the catalytic site of rat liver glutathione peroxidase, *Biochemistry* 17:2639–2644.

Fredga, A., 1972, Organic selenium chemistry, *Ann. N.Y. Acad. Sci.* 192:1–9.

Ganther, H. E., and Baumann, C. A., 1962. Selenium metabolism. II. Effects of diet, arsenic, and cadmium, *J. Nutr.* 77:210–216.

Goldman, P., Alberts, A. W., and Vagelos, P. R., 1961. Requirement for a malonyl CoA-CO_2 exchange reaction in long chain but not short chain fatty acid synthesis in *Clostridium kluvveri*, *Biochem. Biophys. Res. Commun.* 5:280–285.

Günzler, W. A., Vergin, H., Müller, I., and Flohe, L., 1972. Glutathione–peroxidase. VI. Die reaktion der glutathione peroxidase mit verschiedenen hydroperoxiden, *Hoppe-Seylers Z. Physiol. Chem.* 353:1001–1004.

Hafeman, D. G., Sunde, R. A., and Hoekstra, W. G., 1974. Effect of dietary selenium on erythrocyte and liver glutathione peroxidase in the rat, *J. Nutr.* 104:580–587.

Hartmanis, M., 1980. A new selenoprotein from *Clostridium Bluyveri* that copurifies with thiolase, *Fed. Proc. Fed. Am. Soc. Exp. Biol.* 39:1772.

Hedegaard, J., Falcone, G., and Calabro, S., 1963. Incorporation of selenium into analogs of sulfurated amino acids in *Candida albicans*, *Compt. Rend. Soc. Biol.* 187:280–284.

Hidiroglou, M., Heaney, D. P., and Jenkins, K. J., 1968. Metabolism of inorganic selenium in rumen bacteria, *Can. J. Physiol. Pharmacol.* 46:229–232.

Hoffman, J. L., and McConnell, K. P., 1974, The presence of 4-selenouridine in *Escherichia coli* tRNA, *Biochim. Biophys. Acta* 366:109–113.

Holmberg, N. J., 1968, Purification and properties of glutathione peroxidase from bovine lens, *Exp. Eye Res.* 7:570–580.

Horn, M. J., and Jones, D. B., 1941. Isolation from *Astragalus pectinatus* of a crystalline amino acid complex containing selenium and sulfur, *J. Biol. Chem.* 139:649–660.

Huber, R. E., and Criddle, R. S., 1967b. Comparison of the chemical properties of selenocysteine and selenocystine with their sulfur analogs, *Arch. Biochem. Biophys.* 122:164–173.

Huber, R. E. and Criddle, R. S., 1967a. The isolation and properties of beta-galactosidase from *Escherichia coli* grown on sodium selenite, *Biochim. Biophys. Acta* 141;587–599.

Imhoff, D., and Andreesen, J. R., 1979, Nicotinic acid hydroxylase from *Clostridium barkeri*: Selenium-dependent formation of active enzyme, *FEMS Microbiol. Lett.* 5:155–158.

Jackby, W. B., 1977. The glutathione S-transferases: A group of multifunctional detoxification proteins, *Adv. Enzymol.* 46:383–414.

Jenkins, K. J., and Hidiroglou, M., 1967. The incorporation of 75-Se-selenite into dystrophogenic pasture grass. The chemical nature of the seleno compounds formed and their availability to young ovine, *Can. J. Biochem.* 45:1027–1039.

Jones, J. B., Dilworth, G. L., and Stadtman, T. C., 1979. Occurrence of selenocysteine in the selenium-dependent formate dehydrogenase of *Methanococcus vannielii*, *Arch. Biochem. Biophys.* 195:255–260.

Kerdel-Vegas, F., Wagner, F., Russell, P. B., Grant, N. H., Alburn, H. E., Clark, D. E., and Miller, J. A., 1965, Structure of the pharmacologically active factor in the seeds of *Lecythis ollaria*, *Nature* 205:1186–1187.

Klug, H. L., and Froom, J. D., 1965. Identification of dimethyl selenide as a respiratory product from rats administered sodium selenite, *S. Dakota Acad. Sci. Proc.* 64:247.

Ladenstein, R., and Wendel, A., 1976. Crystallographic data of the selenoenzyme glutathione peroxidase, *J. Mol. Biol.* 104:877–882.

Lam, K. W., Riegl, M., and Olson, R. E., 1961. Biosynthesis of selenocoenzyme A in the rat, *Federation Proc.* 20:229.

Lawrence, R. A., and Burk, R. F., 1976. Glutathione peroxidase activity in selenium deficient rat liver, *Biochem. Biophys. Res. Commun.* 71:952–958.

Lester, R. L., and DeMoss, J. A., 1971. Effect of molybdate and selenite on formate and nitrate metabolism in *Escherichia coli*, *J. Bacteriol.* 105:1006–1014.

Little, C., 1972. Steroid hydroperoxides as substrates for glutathione peroxidase, *Biochim, Biophys. Acta* 284:375–381.

Little, C. and O'Brien, P. J., 1968. An intracellular GSH-peroxidase with a lipid peroxide substrate, *Biochem. Biophys. Res. Commun.* 31:145–150.

Ljungdahl, L. G., and Andreesen, J. R., 1975. Tungsten, A component of active formate dehydrogenase of *Clostridium thermoaceticum*, *FEBS Lett.* 54:279–282.

Massey, V., 1959. The microestimation of succinate and the extinction coefficient of cytochrome C, Biochim. Biophys. Acta. 34:255–256.

McCollach, R. J., Hamilton, J. W., and Brown, S. K., 1963. An apparent seleniferous leaf wax from *stanleya bipinnata*. *Biochem. Biophys. Res. Commun.* 11:7–13.

McConnell, K. P., and Portman, O. W., 1952. Excretion of dimethyl selenide by the rat, *J. Biol. Chem.* 195:277–282.

McConnell, K. P., and Wabnitz, C. H., 1957. Studies on the fixation of radioselenium in proteins, *J. Biol. Chem.* 226:765–776.

McConnell, K. P., Kraemer, A. E., and Roth, D. M., 1959. Presence of selenium-75 in the mercapturic acid fraction of dog urine, *J. Biol. Chem.* 234:2932–2934.

McConnell, K. P., Burton, R. M., Kute, T., and Higgins, T., 1979. Selenoproteins from testis cytosol, *Biochim. Biophys Acta* 588:113–119.

McCready, R. G. L., Campbell, J. N., and Payne, J. I., 1966. Selenite reduction by *Salmonella heidelberg*, *Can. J. Microbiol.* 12:703–714.

Mills, G. C., and Randall, H. P., 1958. Hemoglobin catabolism. II. The protection of hemoglobin from oxidative breakdown in the intact erythrocyte, *J. Biol. Chem.* 232:589–598.

Nakamura, W., Hosoda, S., and Hayashi, K., 1974. Purification and properties of rat liver glutathione peroxidase, *Biochim. Biophys. Acta* 358:251–261.

Nigam, S. N., and McConnell, W. B., 1972. Isolation and identification of L-cystathionine and L-selenocystathionine from the foliage of *Astragalus pectinatus*, *Phytochemistry* 11:377–380.

Nigam, S. N., and McConnell, W. B., 1976. Isolation and identification of two isomeric glutamylselenocystathionines from the seeds of *Astragalus pectinalus*. *Biochim. Biophys. Acta* 437:116–121.

Nugteren, D. H., and Hazelhof, E., 1973. Isolation and properties of intermediates in prostaglandin biosynthesis, *Biochim. Biophys. Acta* 326:448–461.

Oh, S. H., Ganther, H. E., and Hoekstra, W. G., 1974. Selenium as a component of glutathione peroxidase isolated from ovine erythrocytes, *Biochemistry* 13:1825–1829.

Palmer, I. S., Fischer, D. D., Halverson, A. W., and Olson, O. E., 1969. Identification of a major selenium excretory product in rat urine, *Biochim. Biophys. Acta* 177:336–342.

Parsons, D. F., Williams, G. R., Thompson, W., Wilson, D., and Chance, B., 1967. Improvements in the procedure for purification of mitochrondrial outer and inner membrane. Comparison of the outer membrane with smooth endoplasmatic reticulum, in *Mitochondrial Structure and Compartmentation*, E. Quagliariello, S. Papa, E. C. Slater, and J. M. Tager (eds.) Adriatica Editrice, Bari, Italy, pp. 29–70.

Paulson, G. D., Baumann, C. A., and Pope, A. L., 1968. Metabolism of ^{75}Se-selenite, ^{75}Se-selenite, ^{75}Se-selenomethionine and ^{35}S-sulfate by rumen microorganisms *in vitro, J. Anim. Sci.* 27:497–503.

Pedersen, N. D., Whanger, P. D., Weswig, P. H., and Muth, O. H., 1972. Selenium binding proteins in tissues of normal and selenium responsive myopathic lambs, *Bioinorg. Chem.* 2:33–45.

Peterson, P. J., and Butler, G. W., 1962. The uptake and assimilation of selenite by higher plants, *Aust. J. Biol. Sci.* 15:126–146.

Peterson, P. J., and Butler, G. W., 1967. Significance of selenocystathionine in an Australian selenium-accumulating plant, *Neptunia amplexicaulis, Nature* 213:599:600.

Pinsent, J., 1954. The need for selenite and molybdate in the formation of formic dehydrogenase by members of the coli-aerogenes group of bacteria, *Biochem. J.* 57:10–16.

Prohaska, J. R., and Ganther, H. E. 1976. Association with GSH: organic hydroperoxide oxidoreductase activity with glutathione S-transferse A activity in testicular cytosol, *Fed. Proc. Fed. Am. Soc. Exp. Biol.* 36:1094.

Prohaska, J. R., Oh, S. H., Hoekstra, W. G., and Ganther, H. E., 1977. Glutathione peroxidase: Inhibition by cyanide and release of selenium, *Biochem. Biophys. Res. Commun.* 75:64–71.

Reamer, D. C., and Zollar, W. H., 1980. Selenium biomethylation products from soil and sewage sludge, *Science* 208:500–502.

Reichard, P., 1978. From deoxynucleotides to DNA synthesis, *Fed. Proc. Fed. Am. Soc. Exp. Biol.* 37:9–14.

Rosenfeld, I., 1961. Biosynthesis of seleno-compounds from inorganic selenium by sheep, *Fed. Proc. Fed. Am. Soc. Exp. Biol.* 20:10.

Rotruck, J. T., Hoekstra, W. G., and Pope, A. L., 1971. Glucose dependent protection by dietary selenium against haemolysis of rat erythrocytes *in vitro, Nature* (London) *New Biol.* 231:223–224.

Rotruck, J. T., Pope, A. L., Ganther, H. E., Swanson, A. B., Hafeman, D. G., and Hoefstra, W. G., 1973. Selenium: Biochemical role as a component of glutathione peroxidase, *Science* 179:588–590.

Saelinger, D. A., Hoffman, J. L., and McConnell, K. P., 1972. Biosynthesis of selenobases in tRNA by *Escherichia coli, J. Mol. Biol.* 69:9–17.

Shum, A. C., and Murphy, J. C., 1972. Effects of selenium compounds on formate metabolism and coicidence of selenium-75 incorporation and formic dehydrogenase activity in cell-free preparation of *Escherichia coli, J. Bacteriol.* 110:447–449.

Smith, A. G., Harland, W. A., and Brooks, C. J. W., 1973. Glutathione peroxidase in human and animal aorta, *Steroids Lipids Res.* 4:122–128.

Smith, A. L., 1949. M. S. thesis, South Dakota State College of Agriculture and Mechanical Arts, Brookings, South Dakota, cited by A. Shrift, 1961. Biochemical interrelations between selenium and sulfur in plants and microorganisms, *Fed. Proc. Fed. Am. Soc. Exp. Biol.* 20:695–702.

Spare, C. G., and Virtanen, A. I., 1964. Occurrence of free selenium-containing amino acids in onion (Allium cepa), *Acta Chem. Scand.* 18:280–282.

Stadtman, T. C., Elliott, P., and Tiemann, J., 1958. Studies on the enzymic reduction of amino acids, *J. Biol. Chem.* 231:961–973.

Stadtman, T. C., 1974a. Selenium biochemistry, *Science* 183:915–922.

Stadtman, T. C., 1974b. Composition and some properties of the selenoprotein of glycine reductase, *Fed. Proc. Fed. Am. Soc. Exp. Biol.* 33:1291.

Stadtman, T. C., 1979. Some selenium-dependent biochemical processes, *Adv. Enzymol.* 48:1–28.

Stadtman, T. C., 1980. Selenium-dependent enzymes, *Ann. Rev. Biochem.* 49:93–110.

Stewart, J. M., Nigam, S. N., and McConnell, W. B., 1974. Metabolism of Na_2 $^{75}SeO_4$ in horseradish: Formation of Selenosinigrin, *Can. J. Biochem.* 52:144–145.

Strittmatter, P., and Ozols, J., 1966. The restricted tryptic cleavage of cytochrome b_5, *J. Biol. Chem.* 241:4787–4792.

Trelease, S. F., DiSomma, A. A., and Jacobs, A. L., 1960. Seleno-amino acid found in *Astragalus bisulcatus*, *Science* 132:618.

Turner, D. C., and Stadtman, T. C., 1973. Purification of protein components of the clostridial glycine reductase system and characterization of protein A as a selenoprotein, *Arch. Biochem. Biophys.* 154:366–381.

Tuve, T., and Williams, H. H., 1961. Metabolism of selenium by *Escherichia coli*: Biosynthesis of selenomethionine, *J. Biol. Chem.* 236:597–601.

Venugopolan, V., 1980. Influence of growth conditions on glycine reductase of *Clostridium sporogenes*, *J. Bacteriol.* 141:386–388.

Virupaksha, T. K., and Shrift, A., 1963. Biosynthesis of selenocystathionine from selenate in *Stanleya pinnata*, *Biochim. Biophys. Acta* 74:791–793.

Virupaksha, T. K., and Shrift, A., 1965. Biochemical differences between selenium accumulator and non-accumulator *Astragalus* species, *Biochim. Biophys. Acta* 107:69–80.

Virupaksha, T. K., Shrift, A., and Tarrer, H., 1966. Biochemical differences between selenium accumulator and non-accumulator *Astragalus* species, *Biochim. Biophys Acta*. 130:45–55.

Wagner, R., and Andreesen, J. R., 1979. Selenium requirement for active xanthine dehydrogenase from *Clostridium acidiurici*, *Arch. Microbiol.* 121:255–260.

Wagner, R., and Andeesen, J. R., 1980. Selenium requirement for active xanthine dehydrogenase from *Clostridium acidiurici* and *Clostridium cylindrosporum*, *Arch. Microbiol.* 121:255–260.

Weiss, K. F., Ayres, J. C., and Kraft, A. A., 1965. Inhibitory action of selenite on *Escherichia coli*, *Proteus vulgaris* and *Salmonella thompson*, *J. Bact.* 90:857–862.

Wendel, A., and Kerner, B., 1977. Modification of red cell glutathione peroxidase by alkyl halides, *Hoppe-Seyler's Z. Physiol. Chem.* 358:1296.

Whanger, P. D., Pedersen, N. D., and Weswig, P. H., 1973. Selenium proteins in ovine tissues. II. Spectral properties of a 10,000 molecular weight selenium protein, *Biochem. Biophys. Res. Commun.* 53:1031–1035.

Woolfolk, C. A., and Whiteley, H. R., 1962, Reduction of inorganic compounds with molecular hydrogen by *Micrococcus lactilyticus*, *J. Bacteriol.* 84:647–658.

Young, P. A., and Kaiser, I. I., 1979. Isolation and partial characterization of transfer RNAs from *Astragalus bisculcatus*, *Plant Physiol.* 63:511–517.

Zakowski, J. J., and Tappel, A. L., 1978. Purification and properties of rat liver mitochondrial glutathione peroxidase, *Biochim. Biophys. Acta* 526:65–76.

Selenium Deficiency
Diseases in Animals

2

2.1 Introduction

Biological function is thought to depend on the tissue concentration or the intake of a nutrient. The severity of deficiency signs and the effects of resupplementation depend on the degree of deficiency. This dependency has been formulated mathemetically by Bertrand (1912). According to Bertrand's rule, a function for which a nutrient is essential and the nutrient is low or absent results in a theoretical deficiency, but the function increases with increasing exposure to the essential nutrient. This increase in function is followed by a plateau representing the maintenance of optimal function through homeostatic regulation, and a decline of the function toward zero as the regulatory mechanisms are overcome by increasing concentrations that become toxic. Bertrand's work has been graphically interpreted by Mertz (1981) in a review article (Figure 2-1). This type of classification of function should help in more completely understanding the complete picture in regard to trace elements. In the past, certain trace elements have been labeled as either toxic, essential, or carcinogenic, etc. by various groups without regard to its other equally important functions. Recent work has demonstrated that several of the trace elements including selenium possess the characteristics outlined by Bertrand and Mertz. It is likely that each essential nutrient has its own specific curve which differs from that of other nutrients, i.e., the width of the plateau.

The principle of Bertrand's model is probably applicable as well to other essential nutrients, including the bulk elements, water, and oxygen. One can reach two conclusions from this model which are relevant to the understanding of trace element research: (i) for each element there is a range of safe and adequate exposures, within which range the tissue is able to maintain optimal tissue concentration; (ii) every trace element is potentially toxic when the range of safe and adequate exposure has been exceeded. The next three chapters follow this pattern of increasing concentration: essentiality, normal metabolic function, and toxicity.

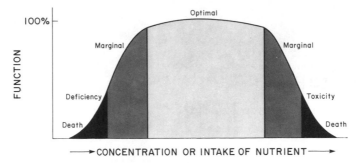

Figure 2-1. Dependence of biological function on tissue concentration or intake of a nutrient. Permission Walter Mertz, *Science*, 213:1331–1338, 1981, by the American Association for Advancement of Science.

Several nutritionally related pathological conditions occur in animals of economic importance. Some of these disorders respond to selenium treatment and some of these disorders are responsive to selenium and vitamin E. The disorders and which of the two nutrients they respond to are listed in Table 2-1. Marked variability in the deficiency symptoms and conditions between species do exist. The types of deficiency diseases and the descriptive features of some of the more important conditions are discussed.

2.2 Dietary Liver Necrosis and Factor 3

2.2.1 Discovery

The lesion of dietary liver necrosis was first described by Weichsel-baum (1935), who attempted to maintain these animals on a cystine-free diet. This is a rapidly developing condition which can cause an apparently healthy rat to become ill and die within a day or two. Upon necropsy, massive hepatic necrosis is found. A similar necrotic degeneration of the liver was produced by dietary means in the rat (Schwarz, 1948). Schwarz was using diets based on extremely pure casein to determine whether a then new bacterial growth factor, H′, was required as a vitamin by the rat. Factor H′ was later identified as *p*-aminobenzoic acid and eventually turned out to be nonessential. In these initial experiments almost all rats died within an average of 32 days from an acute necrosis of the liver.

Relatively early it became apparent that the sulfur amino acids also exerted some protective effect (Daft *et al.*, 1942). The role of vitamin E as an agent which prevented dietary liver necrosis was established between 1939 and 1944 (Schwarz, 1944). By using protection against liver necrosis as an assay of purification, α-tocopherol was practically reisolated from wheat germ by Schwarz. At this time it becomes clear that another unidentified dietary factor occurred in nutrients which independently prevented this disease. This dietary factor was designated Factor 3 and was present in crude casein, kidney, and liver powder, and also in brewer's yeast. However, it was absent from Torula yeasts. Because of its absence from Torula yeast the latter has been used for years as the sole source of protein in standard diets for producing dietary liver necrosis and related diseases (Schwarz, 1951). Factor 3 at first appeared to be a new vitamin. The discovery of organically bound selenium as the essential constituent of Factor 3 was made later after about seven years to isolate this elusive agent (Schwarz and Foltz, 1957). Factor 3 was also isolated from American brewer's yeast and liver. Schwarz and Foltz obtained two chemically similar substances with Factor 3 activity which were designated α and β Factors 3. The α Factor 3 was found to be water soluble, strongly ionic, and stable aganist oxidation, but sensitive to reducing agents. Factor 3 activity was entirely eliminated by dry ashing. Fractionation of α Factor 3 led to a semicrystalline preparation which developed a characteristic garliclike odor upon alkali addition. This observation led to the discovery that selenium is an integral part of Factor 3, since it has been reported that the breath of cattle consuming high-selenium plants from seleniferous soils has a garliclike odor.

Selenium was detected as an integral constituent of highly concentrated Factor 3 fractions which were prepared by a complex fractionation scheme from kidney powder. The very best Factor 3 preparations obtained were approximately 10,000-fold as active as the initial material. Quantitative analyses showed that the purified preparation contained about 7% of the element. If one could assume the molecular weight of Factor 3 to be around 300, these fractions would be roughly 25% pure. Efforts to isolate the active compound from kidney powder as well as other natural sources were halted because of very low yields and extreme instability of the purified agent.

In natural source materials and food, selenium always appears to be organically bound. However, not only organic but also inorganic selenium compounds (selenite, selenate, and selenocyanate) were found effective. These inorganic compounds were one third as potent as the naturally occurring form of Factor 3 from kidney powder. It became apparent that not all organic selenium derivatives were equally active, and that a chemical specificity prevailed to some extent. Although some organoselenium derivatives were less active than Factor 3, they were approximately as active as selenite, others were less potent, and some were not at all

Table 2-1. Vitamin E and Selenium Deficiency Disease of Animals

Disease	Animal	Tissue affected	Vitamin E	SE	Antioxidants
Reproductive failure					
Fetal death, resorption	Rat	Embryonic vascular system	X		X
	Cow, ewe		X	X	
Testicular degeneration	Rooster, rat, rabbit, hamster, dog, pig, monkey	Germinal epithelium	X	X	
Nutritional Myopathies					
Nutritional muscular dystrophy (NMD)	Chick,[a] rat, guinea pig, rabbit, dog, monkey	Striated muscle	X	[b]	
"Mulberry heart" disease	Pig	Cardiac muscle	X		
NMD	Mink	Striated cardiac muscle	X	X	
Gizzard myopathy	Turkey	Gizzard muscle	X	X	
	Duck		X		
"Stiff lamb" disease	Newborn lamb	Striated muscle	[c]	X	
NMD	Sheep, goat, calf	Striated muscle		X	
Creatinuria	Rat, rabbit, guinea pig, monkey	Plasma	X		
Erythrocyte hemolysis	Chick, rat, rabbit	Erythrocyte	X	X	X

Disorder	Species	Site			
Incisor depigmentation	Rat	Incisor enamel	[c]		
Systemic disorders					
Liver necrosis	Rat, mouse, pig	Liver	X	X	X
Membrane lipid peroxidation	Chick, rat	Hepatic mitochondria and microsomes	X	X	X
Accumulation of ceroid[d]	Rat, mink, calf, lamb, dog, chick, turkey	Adipose	X	[c]	[e]
Plasma protein loss	Chick, turkey	Serum albumin	X	X	
Kidney degeneration[d]	Mouse, rat, pig	Renal tubule contorti	X	X	X
Anemia	Monkey, pig	Bone marrow	X		
Encephalomalacia[d]	Chick	Cerebellum	X		X
Exudative diathesis	Chick	Capillary walls	X	X	[f]
Pancreatic fibrosis	Chick	Pancreas	X	X	
Lack of growth	Rat, monkey	Body size, mass	X	X	
Lack of hair growth	Rat, monkey	Hair	X		
Lack of feather growth	Turkey	Feathers	X		

[a] Responsive to sulfur-containing amino acids.
[b] May partially reduce severity.
[c] Syndrome not easily produced in absence of dietary polyunsaturated fatty acids.
[d] Accelerated by polyunsaturated fatty acids.
[e] Involvement proposed but not conformed.
[f] Active only in presence of selenium.

effective. Schwarz *et al.* (1972) have tested several hundred selenium compounds in an effort to match the potency of Factor 3. From these observations they tried to gain some insight into the actual structure of Factor 3. Monoselenodicarboxylic acids with five, seven, and 11 carbon atoms had potency equal to or greater than selenite. The even-carboned monoselenodicarboxylic acids had little activity. The diselenodicarboxylic acids with 2–12 carbon atoms had about the same activity as selenite. The even-carboned benzylseleno-*n*-carboxylic, 5–11 had almost double the activity of selenite. The even-carboned compounds 2–10 had about the same potency as selenite. Although this approach yielded some interesting comparisons and insight into the structure of Factor 3, the structure of Factor 3 has not been established.

The purified and concentrated Factor 3 preparation from pork kidney powder, as well as the most active synthetic selenium compounds, will afford 50% protection against liver necrosis when added to the diet at levels supplying only seven parts per billion (0.007 ppm) of selenium. Inorganic selenium compounds such as selenite show an effective dose of (ED_{50}) of 0.02 ppm selenium in the diet. Factor 3 selenium is approximately 1000-fold as effective as vitamin E, when compared on a molar basis (Schwarz and Pathak, 1975).

2.2.2 Pathology

Studies of the pathology of this dietary liver necrosis have been done (Porta *et al.*, 1968). Focal necrosis of the centrilobular hepatocytes are the earliest abnormality found with light microscopy. Eventually, a massive necrosis occurs when the lesion heals after selenium or vitamin E supplementation, and focal scarring results. However, fat accumulation is not prominent in these livers. Cirrhosis has not been reported after dietary liver necrosis. This observation indicates that the process of cirrhosis may not be related to selenium or vitamin E deficiency. Daft *et al.* (1942) have demonstrated that choline can completely prevent cirrhosis of the liver, but that choline had no effect on necrotic liver degeneration. However, in Daft *et al.*'s experiment cystine completely prevented liver necrosis but did not prevent cirrhosis. These results also suggest that cirrhosis and dietary liver necrosis are two separate and distinct entities.

Schwarz (1965) has postulated that there are three important phases in the development of dietary liver necrosis. These phases may also be important in many other deficiencies. First of all, there is likely an induction period, during which the animal becomes deficient. During this period the stores of the two primary protective agents vitamin E and selenium are being depleted gradually, but are nonetheless in adequate amount to maintain normal function. Next there could be a latent phase during which no gross pathological lesions are detectable. Marked disturbances of intermediary metabolism as well as the subcellular, electron microscopic structure are detectable. In the last phase there are severe anatomical macroscopic changes. Changes are often irreversible in this acute phase. The gross macroscopic changes of the liver develop extremely fast, usually over a period of only a few hours. These changes are often fatal within a few hours or days in the majority of animals.

2.2.3 Biochemical Defect

The exact biochemical lesion of dietary liver necrosis is uncertain, but it is likely that this condition may be due to a lipid peroxidation. When the rats are dying of apparent liver necrosis, large amounts of ethane are exhaled. Ethane is an established end product of lipid peroxidation (Burk, 1976). One could postulate that in the absence of vitamin E and of selenium-dependent glutathione peroxidase, the defensive mechanisms against lipid peroxidation are inadequate to hold the process in check. This could result in the uncontrolled oxidation of polyunsaturated fatty acids and the production of free radicals as well as other toxic breakdown products which ultimately result in cell death. There also could be another as yet unknown destructive process which occurs at the same time as the lipid peroxidation, with this latter process being primarily responsible for the necrosis.

Death from liver necrosis in rats is usually preceded by a latent phase of about seven days' duration (Schwarz and Corwin, 1960). A characteristic impairment of energy metabolism can be observed in liver slices and homogenates during this last phase. This disorder has been designated "respiratory decline" and is characterized by the sudden failure of oxygen consumption which follows an apparently normal initial period of respiration which lasts for approximately 30 min. Several aspects of the respiratory decline have been thoroughly investigated (Schwarz, 1976). Electronmicroscopy showed that the respiratory decline is related to swelling of the mitochondria as well as degenerative cystic degeneration of the elongated profiles of the ergastroplasmic reticulum. These studies were among the first to establish a connection between electronmicroscopic subcellular histophathic changes and a distinct metabolic dysfunction. These findings can now be interpreted because it was found that selenium is the active site of glutathione peroxidase, an enzyme which protects against the swelling of mitochondria.

2.2.4 Hepatosis Dietetica

When the liver necrosis occurs in pigs it is called hepatosis dietetica. There are two clinical patterns of hepatosis dietetica in pigs: In one more severe type, apparently healthy, rapidly growing three-month-old piglets become ill and die within hours after the start of the disease. Upon autopsy massive liver necrosis is observed (Obel, 1953). This first pattern markedly resembles dietary liver necrosis in rats. In other animals, on the other hand, acute liver failure may not occur. Instead, the disease runs a subacute course with the development of jaundice and ascites (Trapp *et al.*, 1970). When the liver is examined it shows areas of massive necrosis over the capsular and cut surfaces of the liver and collapse as well as fibrosis (Obel, 1953; Trapp *et al.*, 1970). The affected lobules may be swollen with hemorrhagic necrosis or show nonhemorrhagic coagulative necrosis. Many of the lobules may be totally unaffected. Lesions of hepatosis dietetica are frequently found in pigs that have died from mulberry heart disease. Other conditions frequently found accompanying hepatosis dietetica are effusions of fluids into the body cavities, skeletal muscle degeneration, and esophagel ulcers. Hepatosis dietetica is less common than mulberry heart disease, but when present usually involves a very high proportion of pigs in a herd and numerous mortalities.

In addition to rats and pigs, hepatic necrosis can also occur in monkeys. Seven adult squirrel monkeys were fed a low-selenium semipurified feed with *Candida utilis* as the protein source and adequate vitamin E (Muth *et al.*, 1971). After nine months, the monkeys developed a variety of problems such as alopecia, loss of body weight, and listlessness. At that point, three of the monkeys were given 40 μg of selenium as selenite by injection at two-week intervals. The three monkeys given selenium became moribund and died. Upon necropsy several lesions were noted including hepatic necrosis, skeletal muscle degeneration, myocardial degeneration, and nephrosis.

2.3 Nutritional Muscular Dystrophy

2.3.1 Pathology

Several important sheep-raising areas have either a low soil selenium content or selenium which is not available to the plants. A high incidence of selenium and vitamin E deficiency in sheep have been observed both in New Zealand and Western Oregon (Lannek and Lindberg, 1975; Muth, 1970). These deficiencies of vitamin E and selenium bring about a nutritional muscular dystrophy (NMD). This disease has also been called white muscle disease and stiff lamb disease. Occurrences of NMD in the United States have led to a highly useful study by Allaway and his associates of the U.S. Plant, Soil and Nutrition Laboratory at Ithaca, New York. These workers completely mapped the seleniferous and selenium-deficient areas of the United States and showed that NMD occurs only in areas in which the soil and, therefore, the forage crops are severely deficient in selenium. This disorder has been observed mostly in growing animals. Lambs and calves three months old are the ones usually affected. The lambs gradually lose strength and become weak. They then develop a stiff appearance which results in an abnormal gait. Their weakness may prevent foraging or suckling and lead to death from starvation. In addition, secondary infections such as pneumonia are frequently found. However, if cardiac involvement occurs sudden death could result. Upon necropsy the heart shows a mild focal to an extensive diffuse grayish white discoloration of the subendocardial myocardium. This lesion extends for only about 1 mm into the myocardium and most commonly affects the right ventricle, even though other cavities may be involved. Even though this condition occurs mostly in the age range of one to three months, occasionally NMD occurs in newborns and in adults. The disorder appears frequently after the dams and offspring have been released to pasture and seems to be exacerbated by muscular activity (Muth, 1970). In mature animals diagnosis may be missed or confused with heavy metal poisoning or conditions involving the central nervous system. Muth (1963) has listed the areas of NMD occurrence in the United States (Figure 2-2).

If the muscles are examined on necropsy, they reveal symmetrical involvement with the pale areas of several skeletal muscles indicating the areas of greatest degeneration. The skeletal muscles commonly affected are those most active, e.g., diaphragm, intercostal, myocardium, and the pelvic and hindleg muscles. The deep muscles overlying the cervical vertabrae are particularly affected with typical chalky-white lesions. The lesions are white striations which are characteristically bilateral and symmetrical. These lesions are particularly characteristic of chronic dystrophy. Microscopically the lesions range from hyaline degeneration to a coagulation necrosis of the muscle fibers. There is also fragmentation and disappearance of a varying number of muscle fibers. Fibrous proliferation occurs to a moderate degree where the muscle fibers have disappeared. This process is accompanied by the appearance of macrophages and lymphocytes. In some cases calcification and attempts at regeneration are often present.

In experimental animals NMD has been produced experimentally. The experimental NMD has been shown to respond to either selenium

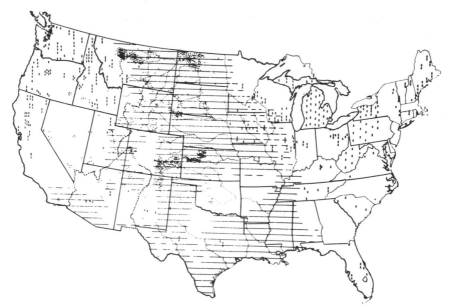

Figure 2-2. The relationship of white muscle disease to the distribution of naturally occuring selenium in the United States. \\\\ White muscle disease; ⁙ locations of plants in excess of 50 ppm Se; ══ areas where seleniferous formations contribute to soil parent material; ⌒⌒ rivers draining Se areas. Permission *Am. Vet. Assoc.* 142:1379–1384, 1963.

(Lannek and Lindberg, 1975) or vitamin E administration (Whanger *et al.*, 1976). Measurement of blood selenium levels has been a predictor of NMD's occurrence (Andrews *et al.*, 1968). In addition, blood glutathione peroxidase activity also correlates well with the level of blood selenium (Thompson, 1976). Abnormally elevated levels of serum glutamic oxal-acetice transaminase as well as lactic acid dehydrogenase and creatinine were also observed with NMD.

2.3.2 Prevention of NMD

Prevention of NMD has been successful using several different meth-ods. These include administration of selenium by drench, injection, and also by placing heavy pellets in the rumen (Kuchel and Godwin, 1976). Injections, however, are time consuming and expensive. For these rea-sons, their use has been primarily as therapy rather than as a preventative measure. Australian workers have investigated the use of intraruminal selenium pellets in grazing sheep. Heavy pellets were made out of 9 g iron and 1 g elemental selenium. These pellets were well retained and free of calcium phosphate coating (Kuchel and Buckley, 1969). The pellets worked as selenium was released to the sheep. Concentrations of the element in edible tissues were similar to those in normal animals after 6 or 12 months. When Handreck and Godwin (1970) labeled the pellets with ^{75}Se, the pellets were calculated on the basis of tissue and excretory analysis, to release about 1 mg selenium daily, and blood and tissue selenium were within normal limits. Godwin's group has also observed that lambs fed a selenium-deficient ration until NMD appeared recovered after introduction of ruminal selenium pellets. With grazing beef cows on selenium-deficient pastures, 30-g pellets (10% Se, 90% Fe) increased blood selenium levels. Within five weeks, the blood selenium levels were increased to levels about equal to another group of animals which had received a selenite drench. When as many as three heavy pellets (Se, Co, and steel grinder) were in the rumen at the same time, the selenium pellets were well retained with the apparent loss of only one selenium pellet per 40 or 50 animals in one year. When steel grinder pellets were present, the release of selenium was increased.

Canadian workers (Jenkins and Hidiroglou, 1972) have investigated the use of slow-release vitamin E and selenium pellets in ewes and lambs for control of NMD. Pellets containing 20 mg selenium as sodium selenite with stearic acid and silastic glue were implanted in the loose connective tissue behind the shoulder in pregnant ewes or in lambs which were fed dystrophy-producing hay. These pellets were of the slow release type and were implanted in ewes 40 to 78 days (mid-gestation) prior to lambing. Another group of ewes were implanted with pellets containing 10 mg selenium as sodium selenite with hydrogenated peanut oil and magnesium stearate. These pellets were placed in ewes 1–15 days (late gestation) before lambing (Hidiroglou *et al.*, 1971). Hay containing about 15 ppb selenium and 15 ppm tocopherol was sprayed with a urea–sucrose mixture and fed to all of the animals, including an untreated control group. Both of the implantation treatments increased blood selenium in the ewes and lambs two to four times that of untreated animals. Milk selenium was measured at 1, 60, and 90 days of lactation and was 9.5, 7.2, and 8.0 ppb, respectively, for untreated animals. The treated ewes averaged 25.0, 15.4, and 10.6 ppb, respectively. The abnormalities of the lamb muscle pathology were reduced with the selenium treatment of the ewes. In an experiment done in a similar way (Hidiroglou *et al.*, 1972a), NMD was prevented without additional selenium in vitamin E-implanted ewes or lambs, suggesting that most of the lamb's requirement for selenium was reduced with adequate vitamin E. Selenium implantation was completely effective even when lambs were receiving relatively low amounts of vitamin E in the milk.

Vitamin E plus 10 mg selenium administered as a drench every two weeks to gestating ewes did not increase tissue selenium levels in the lambs to concentrations above those for lambs from ewes receiving only selenium. Blood selenium was maintained at normal levels in the ewe by a 10-mg drench every two weeks. A single drench of 50 mg selenium before mating raised blood selenium for two to three months but did not increase selenium concentrations in milk. Levels of selenium were normal in edible tissues of lambs from implanted ewes (Jenkins and Hidiroglou, 1972).

NMD was also prevented in neonatal beef calves born to cows fed selenium-deficient rations (Hidiroglou *et al.*, 1972b) by the implantation of selenium pellets. The pellets contained 15 mg selenium as sodium selenite with stearic acid and silastic acid as vehicles and were placed in the subcutaneous tissue at the base of the ear. After one month, no trace of the pellet was observed and a hole was present in the conchal cartilage. The selenium in the plasma of implanted calves was higher than the controls, but within the normal range. NMD as well as an elevated serum glutamic oxalacetic transaminase content resulted in several of the un-implanted calves but not in the implanted calves.

The use of selenium-fortified salt, blocks, or animal mixtures appears to be the most promising procedure for the prevention of ruminant NMD resulting from a deficiency of selenium and/or vitamin E. Selenium-deficient cows were fed a selenium- and vitamin E-fortified mineral supplement during the latter 2/3 of gestation as well as to lactating cows through the first month (Jenkins *et al.*, 1974). The mineral mixture contained 2700 IU/kg of vitamin E and about 14.8 ppm of selenium. These amounts were calculated to provide 0.07 ppm of selenium and 13.2 IU vitamin E/kg of diet. The blood selenium levels of the dams and offspring were within normal limits during supplementation, while milk selenium was significantly increased by remained extremely low, averaging less than 18 ppb during the first month of lactation. When the supplement was withdrawn at one month postpartum, the selenium milk concentration dropped to that of the control animals by the end of the second month of lactation. At four weeks of age, the tissue selenium in kidney and liver of calves was not significantly different for treated and untreated animals. For untreated and treated calves the selenium of the muscle dry matter was significantly increased at 0.050 and 0.062 ppm, respectively.

In pigs, NMD often occurs when hepatosis dietetica and mulberry heart disease are also present. Although NMD in farm animals is considered to be a disease of the young, in swine the disease occurs in piglets, sows, and fattening pigs. NMD has been reported in Sweden to be associated with the feeding of grain containing rancid fat (Swahn and

Thafvelin, 1962) or with rations containing low levels of selenium (Lindberg and Siren, 1965) or vitamin E (Thafvelin, 1960). In field cases in the United States it has been reported that pigs which have NMD are frequently weak and unsteady. Lesions usually are not grossly apparent, but in some cases pale areas are grossly visible in skeletal muscle. The most commonly affected muscles with NMD are the longissimus, diaphragm, and the adductor muscles. In some cases the lesions observed are loss of striations, and fragmentation of fibers with some mineral deposition. Zenker's necrosis may be present as well as infiltration with phagocytes. NMD lesions which are not always grossly apparent can be observed microscopically. The degeneration of the muscle fibers can be observed microscopically in all cases (Michael *et al.*, 1969; Van Vleet, 1970; Van Vleet *et al.*, 1976).

In Canada, congenital NMD in piglets has been occasionally diagnosed. This condition is usually observed in piglets shortly after birth and affects most of the litter. If the hindlegs are affected, this produces a condition which is termed *splay-leg* or *spraddle-legged* pigs. In Sweden, Lannek *et al.* (1962) have observed an interrelationship between Vitamin E and iron. Pigs born from sows consuming diets low in vitamin E were found to die or develop NMD after iron was injected. Tollerz and Lannek (1964) have observed that vitamin E, and not selenium, when administered 8 hr before the iron injection prevented the effect. Most studies indicate that, in general, NMD is preventable by vitamin E but not by selenium (Swahn and Thafvelin, 1962; Michel *et al.*, 1969). While there is experimental evidence that vitamin E is effective against muscular dystrophy, better results have been obtained when both vitamin E and selenium were administered. In general, NMD does not assume the clinical importance in pigs that it does in sheep because the pigs are more susceptible to severe liver or heart disease. The animals frequently die from these problems before NMD can develop to a life-threatening extent.

In chickens, selenium spares vitamin E in preventing nutritional muscular dystrophy. The disease in chickens is primarily due to vitamin E and sulfur-containing amino acid deficiencies (Scott, 1970). NMD in chicks is characterized by white striations in the fiber bundles of the breast and in some cases the leg muscles (Scott *et al.*, 1967). Necropsy studies revealed histological changes in the skeletal muscle typical of Zenker's degeneration. The order of prominance of the "selenium-responsive" diseases of the young poult appears to be first, myopathy of the smooth muscle (gizzard); second, myocardial myopathy; and third, myopathy of the skeletal muscle.

Walter and Jensen (1964) have found the levels of serum glutamic–oxalacetic transaminase to be related to the incidence of muscular dystrophy in chicks and poults which have been maintained on selenium and vitamin E-deficient diets. The sulfur amino acids and vitamin E alone, or selenium along with vitamin E at levels too low to in themselves prevent this dystrophy, have all been reported to be effective preventatives.

2.4 Exudative Diathesis

Exudative diathesis in chickens and turkeys is characterized by an edema of body tissues, with large amounts of fluid accumulation under the skin of the abdomen and breast (Dam and Glavind, 1938). Most outbreaks occur in chicks between three and six weeks of age. The chicks become dejected, lose condition, show leg weakness, and some become prostrate and die. Swelling of the subcutaneous tissue, particularly under the wings and down the thighs is frequently observed. There is a reddish color of the skin due to hemorrhages in subcutaneous tissues near the areas of fluid accumulation. The fluid is often a blue-green color and when subjected to electrophoresis, its protein composition is similar to that of plasma (Dam et al., 1957). Trypan blue permeability of the capillaries is increased in chickens with exudative diathesis (Dam and Glavind, 1940). The increase of permeability of the capillaries may account for the edema. In the chicks there is a low blood serum protein, especially the albumin, which probably contributes to the edema. Anemia has been described as occurring with exudative diathesis.

Once animals develop this condition, they will usually die within a few days. By adding as little as 0.1 ppm of selenium as selenite to a vitamin E-deficient diet, exudative diathesis can be prevented in chicks. Vitamin E alone has been reported to prevent this disease (Scott et al., 1955). In this report, it appeared that vitamin E completely spared selenium so that selenium would not be classified as essential. On the other hand, Thompson and Scott (1969) have reported that chicks on diets prepared with crystalline amino acids containing less than 0.005 ppm of selenium, and d-α-tocopherol at levels as high as 200 ppm in the diet, did not prevent poor growth and mortality. The investigators conclude that the chicks have a requirement for selenium. In other experiments as high as 50 IU of vitamin E per pound of diet failed to completely prevent exudative diathesis, whereas 0.1 ppm of selenium prevented this disease. Mathias and Hogue (1971) have reported that while ethoxyquin and N,N'-diphenylphenylene diamine can partially prevent exudative diathesis in the presence of low levels of selenium and vitamin E, the two artificial antioxidants were unable to prevent death from selenium and vitamin E deficiency. These reports also support the essentiality of selenium. The differences in some of the experimental results may be due to slight impurities of selenium in some of the diets. Apparently only when crystalline diets are used can be true assessment of essentiality be made.

Several controlled treatment trials have been carried out in areas of New Zealand experiencing outbreaks of exudative diathesis. In every case, 15 mg of selenium added to one gallon of drinking water for one day, immediately controlled the disease and many of the affected birds made a rapid recovery (Hartley and Grant, 1961).

2.5 Pancreatic Degeneration

Chicks fed a severely selenium-deficient diet but supplemented with vitamin E and bile salts develop a severe atrophy of the pancreas (Thompson and Scott, 1970). If selenium is not added to the deficient diet, the animals will die subsequently of pancreatic insufficiency. This disease has not been observed in other than experimental animals. After a marked decrease in absorption of lipids including vitamin E, death usually results. After the pancreatic degeneration begins, the pancreatic and intestinal lipase decrease. The decrease in lipase results in a failure in fat digestion and also causes a failure in fat absorption. As a result, there is an absence of bile and monoglycerides in the intestinal lumen. The bile and monoglycerides are needed for micelle formation, and when they are decreased, the vitamin E absorption is impaired. Unhydrolyzed fat appears in the feces. If bile salts were added to the diet, fat digestion was not returned to normal levels, and enhancement of vitamin E absorption was observed only temporarily. During the experimental period of four weeks, the absorption of vitamin E and survival of the animals improved if free fatty acids, monoglycerides, and bile salts were added to the basal diet. Like many of the other deficiency diseases reviewed, the selenium requirement for prevention of pancreatic degeneration was found to depend on the vitamin E level in the diet. As little as 0.01 mg of selenium as sodium selenite per kilogram of diet completely prevented pancreatic degeneration even when vitamin E levels were very high (100 IU/kg or more). However, when the vitamin E of the diet was at more nearly normal levels (10–15 IU/kg), a level of 0.02 mg Se/kg of diet was required.

Chicks produced from hens fed a low-selenium, low-vitamin-E practical diet had low activities of glutathione peroxidase in plasma and pancreas when they hatched (Bunk and Combs, 1981). Low levels of glutathione peroxidase were observed during all stages after the onset of nutritional pancreatic atrophy. Selenium supplementation of the diet prevented nutritional pancreatic atrophy and resulted in significant elevations of glutathione peroxidase activity. Even though the early stages of nutritional pancreatic atrophy are believed to involve mitochondrial swelling, no significant differences were observed in rate of oxygen uptake, respiratory control index, or adenosine diphosphate-to-oxygen ratio between pancreatic mitrochondria isolated from either selenium-deficient or selenium-adequate chicks. If the maternal selenium status was improved, there was significantly increased chick pancreatic glutathione peroxidase activity at hatching. The onset of nutritional pancreatic atrophy was delayed in the chicks were fed a selenium-deficient diet. On the other hand, a significant proportion of the second-generation selenium- and vitamin E-depleted chicks used in these studies was found to grow at nearly normal rates when fed the selenium-deficient diet. These chicks, which were designated as refractory to the growth-depressing effect of severe selenium deficiency, were biochemically deficient with very low glutathione peroxidase activities and showed nutritional pancreatic atrophy. Bunk and Combs (1981) concluded that the selenium-responsive lesion, which results in nutritional pancreatic atrophy in the chick, is different from the lesion which results in depressed growth.

2.6 Mulberry Heart Disease

Mulberry heart disease often occurs together with hepatosis dietetica in pigs, but sometimes will itself cause sudden death (Van Vleet *et al.*, 1970). The clinical patterns are usually those of a rapidly growing healthy feeder animal suddenly developing heart failure and dying within hours. Before death, the animals may also show dyspnea and severe muscle weakness. The heart at necropsy may have pale streaks of necrosis as well as patches of congestion and hemorrhage. The capillary endothelium and the arteries of the heart muscle walls may also show thickening. The heart muscle itself may become thin and the heart lesions may vary from obvious extensive subepicardial, myocardial and subendocardial ecchymotic and suffusive hemorrhage to a subtle mottling of the subepicardial myocardium. The lesions usually involve the atria and ventricles bilaterally, but the right side is most often affected. Excessive amounts of fluids are frequently found in the pericardial sac and the thoracic and abdominal cavities, and the lungs are frequently congested.

The disease has been called "mulberry heart disease" because the affected animals frequently have such extensive cardiac hemorrhage that the organ has a reddish-purple gross appearance comparable to that of a mulberry. Mulberry heart disease has been shown in several studies to occur in animals deprived of selenium and vitamin E (Piper *et al.*, 1975). Selenium levels were low in affected animals. However, administration of selenium increased tissue selenium levels and the illness was prevented (Van Vleet *et al.*, 1973). Blood selenium concentrations in pigs, in contrast to sheep and cattle, did not correlate well with blood glutathione peroxidase activity (Thompson, 1976).

2.7 Reproductive Problems

The embryonic mortality which has occurred in ewes in the South Island of New Zealand for decades often runs as high as 75% (Andrews *et al.*, 1968). This infertility tends to occur in those areas where congenital white muscle disease is found. The affected ewes may be of any age and are usually in fair to good condition at mating. Although many of the ewes conceive, many fail to produce a lamb. The cause of the infertility is embryonic death at about three–four weeks postconception (Hartley, 1963). Administration of selenium to the ewe before mating prevents this infertility in ewes. Lambing percentages were increased from 25% to 90%. The increase in lambing percentages was likely due to decreased embryonic mortality, but improvement in the ova fertilization rate is also possible (Segerson and Ganapathy, 1980), or the improvement may be due to a combination of these factors. Segerson and Ganapathy (1980) suggested that the improvement in ova fertility in selenium- and vitamin E-treated ewes was due to an increase in the number of uterine contractions migrating toward the oviduct at mating. Presumably sperm would be transported to the oviduct. Segerson *et al.* (1981) also found that selenium is more important than vitamin E in influencing uterine motility and contraction velocity but does not appear to influence the number of electrical spikes per minute or the mean amplitude of the spikes. This infertility due to selenium deficiency apparently does not affect cattle. In some area calving percentages appear satisfactory for cattle grazing side by side with sheep who are experiencing selenium-responsive infertility.

Improved egg production, hatchability, and growth of chicks have been demonstrated when selenium has been added to low-selenium, low-vitamin-E diets (Cantor and Scott, 1974). However, in another experiment, selenium failed to improve egg hatchability when a Torula yeast diet was used, but in this experiment tocopherol was effective (Cregar *et al.*, 1960). On the other hand, different results were observed in Japanese quail. When these birds were raised from one day of age on a diet of low-selenium and low-vitamin-E content, additional selenium was about equal to vitamin E supplementation in maintaining hatchability of the eggs (Jensen, 1968). The young quail which were able to hatch from the eggs laid by the hens on the selenium-deficient diets were observed to have a high incidence of paralysis of the legs and degeneration of the muscle in the gizzard and legs. Eggs from chickens fed practical feedstuffs had about an equal selenium content of dried egg white and yolk. However, when selenite was fed, the selenium content of dried yolk was higher (Latshaw and Osman, 1975).

In rats fed Se-deficient (<0.02 ppm) diets, sperm morphology and motility appeared to be normal at four months, but 50% of the animals in the 11–12-month interval produced sperm with impaired motility and a characteristic midpiece breakage (Wu et $al.$, 1979).

Although active spermatogenesis was observed in some of the seminiferous tubules of selenium-deficient rats born to females on a selenium-deficient diet, poor motility of spermatazoa from the cauda epididymus of these males was observed with most of the sperm showing breakage of the fibrils in the axial filaments. Neither vitamin E nor other antioxidants could counteract these effects (Wu et $al.$, 1973). In second-generation selenium-deficient male rats, testicular atrophy with aspermatogenesis is regularly seen (Sprinker et $al.$, 1971). Aspermatogenesis has also been observed in many first generation rats fed selenium-deficient diets for a year or longer (Burk, 1978).

Calvin et $al.$ (1981) have produced evidence that the selenium in rat sperm is associated with a cysteine-rich structural protein of the mitochondrial capsules. In their experiment they injected {^{75}Se} selenite or {^{35}S} cysteine into rat testicles. Dodecyl sulfate–polyacrylamide gel electrophoresis of purified, double-labeled mitochondrial capsules revealed only a single ^{75}Se-labeled component, whose molecular weight was 17,000. Because most of the ^{35}S label and the major zone of stained protein on the gels coincided with the position of ^{75}Se, this suggests that selenium is associated with a cysteine-rich structural protein in its content of vitamin E and selenium.

The cumulation pattern of ^{75}Se in the testes and epididymis has been observed in rats fed a low-selenium Torula yeast diet (Brown and Burk, 1973). Nearly 40% of whole-body ^{75}Se in the Se-deficient rat was found in the testis three weeks after a single {^{75}Se} selenite injection. After three weeks, the concentration of ^{75}Se declined in the testis but increased in the epididymis, suggesting that selenium was incorporated into the spermatazoa in the testis. Autoradiographs showed the ^{75}Se concentrated in the midpiece of the sperm. This suggests that there may be a specific need for selenium in the mitochondria, which are exclusively located in the sperm midpiece. Subcellular fractionation showed that the mitochondria of the testes contained more.

Smith *et al.* (1979) has demonstrated that in the bull, ^{75}Se retention in the epididymis was highly correlated ($r = 0.92$) with spermatozoal concentration. They also demonstrated that when ejaculated bovine spermatozoa were subjected to repeated freezing and thawing in distilled water, the selenium remained with the spermatozoa. This suggests that selenium was tightly bound to the structural components of the cell. Calvin (1978) has demonstrated that the flagellum of rat spermatozoa contained a specific selenopolypeptide which he named "selenoflagellin." Calvin (1978) suggested that the protein is important in the formation and function of the flagellum. Bartle *et al.* (1980) have injected intramuscularly four groups of three dairy bulls with 5, 10, 20, and 40 mg of selenium as sodium selenite per 90 kg of body weight. Selenium injections significantly increased blood selenium, blood glutathione peroxidase, semen selenium, and seminal plasma glutathione peroxidase. The blood selenium and glutathione peroxidase did not increase until after the 10-mg injection. Seminal plasma glutathione peroxidase appeared highly sensitive in selenium status because enzyme levels tripled within 48 hr following the 5-mg injection. Niemi *et al.* (1981) have investigated the association of selenium with ejaculated bovine spermatozoa. After incubation with dithiothreitol, over 75% of the radioactive selenium was released after 30 min of incubation. Of the selenium-75 released by dithiothreitol, 85% was associated with spermatozoal protein. The selenium-75 containing protein was found predominantly in a single band after polyacrylamide gel electrophoresis. The molecular weight of the selenoprotein was found to approximately 21,500 daltons. The protein was similar in size to the 17,000-dalton selenoprotein isolated from rat sperm (Calvin, 1978).

Trinder *et al.* (1969) have cited the beneficial effects of selenium with vitamin E on the incidence of retained placentas in dairy cows on low-selenium diets.

2.8 Myopathy of the Gizzard

Jungherr and Pappenheimer (1937) have originally reported gizzard myopathy in turkey poults to be the most characteristic, if not the only, vitamin E deficiency symptom in turkeys. However, other investigators have observed that under appropriate conditions vitamin E-deficient poults also display nutritional muscular dystrophy (myopathy of the skeletal muscle), severe myopathies of the smooth muscle (gizzard) and the cardiac muscle. Walter and Jensen (1963) have noted that the addition of sulfur-containing amino acids to the diet had no beneficial effect on these myopathies, whereas selenium as sodium selenite or vitamin E at 20 IU/

kg of diet was entirely effective in preventing all symptoms. Scott *et al.* (1967) have used a selenium-low practical turkey starting diet and a semipurified basal diet to produce a selenium deficiency in young turkey poults. Myopathies of the heart and of the gizzard as well as poor growth and mortality were observed in young poults receiving a practical diet containing all nutrients known to be required, except for supplemental vitamin E and methionine. When vitamin E and methionine were added to the diet improved growth was observed, but gizzard myopathy was not prevented. These results suggest that the lesion of gizzard myopathy and that of poor growth may result from different biochemical processes. Maximal growth was not achieved until the diet also was supplemented with at least 0.1 ppm of selenium as sodium selenite. Under the conditions of these experiments, the selenium requirement in the practical-type diet depended to some extent on the amount of vitamin E or methionine supplementation. When vitamin E was supplemented, the selenium requirement ranged from approximately 0.18 ppm to about 0.28 ppm of selenium as sodium selenite when vitamin E was absent.

The order of prominance of the "selenium-responsive" diseases of the young poult appears to be first, myopathy of the smooth muscle (gizzard); second, myopathy of the myocardium; and third, myopathy of the skeletal muscle. The primary nutritional factor required appears to be selenium with vitamin E of lesser importance, and sulfur amino acids are completely ineffective in the prevention of these myopathies in poults.

2.9 Growth

Several investigators have maintained rats on selenium-deficient diets supplemented with adequate vitamin E for prolonged periods. Even though the animals were fed the ration from the time they were weaned, growth impairment was the only consistent selenium-responsive abnormality reported in these animals (Siami, 1972).

In the offspring of selenium-deficient mothers, additional signs of selenium deficiency have been observed. The rats tended to grow poorly and had sparse or absent hair (Hurt *et al.*, 1971). In addition, cataracts have been reported in the second-generation selenium-deficient rats (Whanger and Weswig, 1975). Selenium administration leads to weight gain as well as hair growth.

Several adult squirrel monkeys were fed a low-selenium semipurified diet with adequate vitamin E and *Candidia utilis* as the protein source (Muth *et al.*, 1971). Body weight loss, alopecia, and listlessness developed in the monkeys after they were on the diet for nine months. When one monkey died, three of the other monkeys were injected with 40 μg of selenium as selenite at two-week intervals. Those monkeys given the selenium salts recovered. However, three were not given selenium and became critically ill and died. Numerous lesions were found in the dead monkeys including hepatic necrosis, skeletal muscle degeneration and nephrosis, and capillary loss. In monkeys the tactile hairs are not affected by the loss of capillaries. The tactile hairs are supplied by cavernous blood sinuses rather than capillaries.

2.10 Selenium—Responsive Unthriftiness of Sheep and Cattle

Selenium-responsive unthriftiness is probably the most widespread and economically significant of all the selenium-responsive diseases of New Zealand livestock (Andrews et al., 1968). Selenium-responsive unthriftiness is characterized by an inability to maintain optimum growth rates. For several months, lambs may appear normal and then show markedly reduced weight gains; others stop eating, stop growing, lose weight, become dejected, and die. In sheep the fleece becomes harsh and dry. There is no increase in SGOT levels. Diarrhea is not a constant feature, but it may occur if the condition is associated with endoparasitism. The only postmortem findings are nonspecific advanced emaciation and osteoporosis. No characteristic microscopic lesions are apparent. In dairy and beef cattle the symptoms vary from a subclinical growth depression to a syndrome characterized by a sudden and rapidly progressive loss of condition, profuse diarrhea, and sometimes high mortality. Occasionally, severe outbreaks occur in adult cattle.

The autumn and winter months are the most critical times for this disease. Selenium has been demonstrated by several investigators to prevent the unthrifty condition in both sheep and cattle. Treated lambs produce up to 30% more wool than do those affected with unthriftiness and not treated.

The effects of a vitamin E- and selenium-deficient diet supplemented with 6.8% cod liver oil on experimental swine dysentery have been studied by Teige et al. (1977). Sixteen growing pigs were fed a vitamin E- and selenium-deficient diet. At that time half of the animals were given a daily supply of vitamin E and selenium. After having been fed these diets for 53 days, the pigs were infected orally with minced colonic material from cases with typical swine dysentery. The incubation times were, however, much shorter and the clinical symptoms were much more pronounced in the vitamin E- and selenium-deficient diet than the group given a daily supply of vitamin E and selenium. The authors conclude that the treatment with vitamin E and selenium in the supplemented group greatly increased resistance to swine dysentery.

2.11 Peridontal Diseases of Ewes

What appears to be the same disease was first described in New Zealand by Salisbury *et al.* (1953), and subsequently by Hart and Mackinnon (1958).

Not uncommonly, in both the North and South islands the disease occurs in areas or on properties where other selenium-responsive diseases have been diagnosed. Peridontal disease is seen in ewes three to five years of age, and causes a loss of physical condition resulting from difficulty in mastication. Peridontal disease is characterized by loosening and shedding of permanent premolars and molars, and sometimes also of incisors. The loosening of teeth is also associated with gingival hyperplasia, gingival infection, resorption, replacement fibrosis of alveolar bone, alveolar infection, and bony exotoses on the adjacent part of the mandible or maxilla.

Other selenium-responsive diseases are also associated with peridontal diseases and the virtual disappearance in affected areas after selenium dosing suggested that selenium might be implicated in the etiology. Two long-term controlled selenium trials have been conducted. One trial was done on a North Island coastal sand area and another was done in the Southland. Both studies have shown that selenium administration will greatly reduce the incidence (Andrews *et al.*, 1968). However, since treatment did not prevent peridontal disease completely, it seems that other factors may also be involved.

2.12 Encephalomalacia

Encephalomalacia is an ataxia resulting from hemorrhages and edema within and/or granular layers of the cerebellum with pyknosis. Eventually the Purkinje cells disappear and the molecular and granular layers of the cerebellar folia separate (Scott and Stoewsand, 1961).

Dam and Granados (1945) have demonstrated that in certain of the body fats of vitamin E-deficient chicks, formation of lipoperoxides can occur *in vivo* and can be prevented by supplementing the diet with vitamin E. Michaelis and Wallman (1950) have shown that vitamin E can be readily changed to a free semiquinone radical. It is believed that the protection by antioxidants against peroxidation occurs through the neutralization by the antioxidant of the free radicals which are produced in an early stage in the auto-oxidation of fat. One important function of vitamin E appears to be the inhibition of the auto-oxidation of fats, not only in the diet of animals but also within the animal body. Several investigators have presented evidence that in the prevention of encephalomalacia in chicks, vitamin E functions as a biological antioxidant closely related to linoleic acid metabolism (Century and Horwitt, 1959; Machlin *et al.*, 1959).

Sugai *et al.* (1960) have postulated that vitamin E protects chicks against encephalomalacia by preventing the breakdown of linoleic acid to 12-oxo-*cis*-9-octadecenoic (keto) acid. Their suggestion of the mechanism is based on their observations that α-tocopherol gave protection against encephalomalacia in the presence of 1.5% methyl linoleate, but not in presence of 0.25% of the ketoacid.

Many workers have presented evidence showing that antioxidants other than tocopherol are capable of preventing encephalomalacia and many other signs of vitamin E deficiency in animals. Dam (1957), Singsen *et al.* (1955), and Bunnell *et al.* (1955) demonstrated that the antioxidant diphenyl-*p*-phenylenediamine (DPPD) prevented encephalomalacia when given by mouth or subcutaneously to chicks with values of vitamin E in blood too low to be detected. It is possible that antioxidants might prevent certain E deficiency diseases by completely replacing vitamin E in the metabolism of the animal. On the other hand, the artificial antioxidants may be effective because they prevent destruction of traces of vitamin E in the diet or animal body. In addition to DPPD, ethoxyquin and methylene blue have been shown to be as effective in preventing encephalomalacia in chicks with very low intakes of vitamin E over prolonged periods (Machlin and Gordon, 1960; Scott and Stoewsand, 1961).

Scott and Stoewsand (1961) have found the DPPD was effective in preventing encephalomalacia even when it was given daily to chicks which received a purified basal diet free from antioxidants and vitamin E. In addition, the diet was allowed to develop complete oxidative rancidity during the course of the experiment. Three quarters of the chicks died of encephalomalacia on the same basal diet, but without DPPD.

Low concentrations of such antioxidants as DPPD and ethoxyquin are capable of preventing encephalomalacia in chicks, but fail to prevent exudative diathesis or muscular dystrophy in the same chicks. This suggests that vitamin E acts as an antioxidant in preventing encephalomalacia. Because several antioxidants are capable of preventing encephalomalacia, this suggests that such action is nonspecific and can be performed by antioxidants which have chemical and biological properties so similar to those of the vitamin that they can enter into metabolic schemes where vitamin E acts as an antioxidant. Phytyl-ubichromenol was also found to prevent encephalomalacia, which suggested that the presence of this compound in brewer's and Torula yeasts may account for the effectiveness of yeasts in the prevention of encephalomalacia (Sondergaard *et al.*, 1962).

Machlin *et al.* (1959) have reported the prevention of exudative diathesis and muscular dystrophy by ethoxyquin but the amount required was so high that it bordered on a toxic level. Since vitamin E at very low concentrations can easily prevent these diseases, its effect in preventing exudative diathesis and muscular dystrophy may be due to its participation in metabolic processes which are more specific than prevention of peroxidation within the animal body.

Selenium has been reported to be ineffective in preventing encephalomalacia (Dam *et al.*, 1957). Century and Horwitt (1964) have reported that addition of 0.13 ppm of selenium as sodium selenite to a semisynthetic diet containing 4% corn oil stripped of tocopherol reduced the incidence of encephalomalacia. When 8% corn oil was fed on the other hand, selenium in levels up to 1.56 ppm had no effect on the percentage of chicks with encephalomalacia. Growth was not affected by the addition of selenium.

From these results the protective action of selenium against nutritional encephalomalacia was shown to be small but significant when the dietary lipid stresses were borderline. The failure of other investigators (Dam *et al.*, 1957) to observe a protective effect by selenium against encephalomalacia may have been due to their use of larger overwhelming levels of the stressor lipid, thereby masking whatever possible protection the added selenium might afford. When Century and Horwitt (1964) fed the chicks 8% corn oil, the protective effect of selenium appeared to be overwhelmed. Even higher levels of selenium in the diet, up to 1.56 ppm, did not overcome the effect of feeding the corn oil.

References

Andrews, E. D., Hartley, W. J., and Grant, A. B., 1968. Selenium-responsive diseases of animals in New Zealand, *New Z. Vet. J.* 16:3–17.
Bartle, J. L., Senger, P. L., and Hillers, J. K., 1980. Influence of injected selenium in dairy bulls on blood and semen selenium, glutathione peroxidase and seminal quality, *Biol. Reprod.* 23:1007–1013.
Bertrand, G., 1912. *Eighth Int. Cong. Appl. Chem. N.Y.* 23:30. Cited by W. Mertz, 1981. The essential trace elements, *Science* 213:1332–1338.

Brown, D. G., and Burk, R. F., 1973. Selenium retention in tissues and sperm of rats fed a torula yeast diet, *J. Nutr.* 103:102–108.

Bunk, M. J., and Combs, Jr., G. F., 1981. Relationship of selenium-dependent glutathione peroxidase activity and nutritional pancreatic atrophy in selenium-deficient chicks, *J. Nutr.* 111:1611–1620.

Burk, R. F., 1976. The significance of selenium levels in blood, in Proceedings of the Symposium on Selenium–Tellurium in the Environment, Industrial Health Foundation, Pittsburgh, pp. 194–203.

Burk, R. J., 1978. Selenium in nutrition, *Wld. Rev. Nutr. Diet.* 30:88–106.

Calvin, H. I., 1978. Selective incorporation of selenium-75 into a polypeptide of the rat sperm tail, *J. Exp. Zool.* 204:445–452.

Calvin, H. I., Cooper, G. W., and Wallace, E., 1981. Evidence that selenium in rat sperm is associated with a cysteine-rich structural protein of the mitochondrial capsules. *Gamete Res.* 4:139–149.

Cantor, A. H., and Scott, M. L., 1974. The effect of selenium in the hen's diet on egg production, hatchability, performance of progeny and selenium concentration in eggs, *Poult. Sci.* 53:1870–1880.

Century, B., and Horwitt, M. K., 1959. Effect of fatty acids on encephalomalacia, *Proc. Soc. Exp. Biol. Med.* 102:375–377.

Century, B., and Horwitt, M. K., 1964. Effect of dietary selenium on incidence of nutritional encephalomalacia in chicks, *Proc. Soc. Exp. Biol. Med.* 117:320–322.

Cregar, C. R., Mitchell, R. H., Atkinson, R. L., Ferguson, T. M., Reid, B. L., and Couch, J. R., 1960. Vitamin E activity of selenium in turkey hatchability, *Poult. Sci.* 39:59–63.

Daft, F. S., Sebrell, W. H., and Lillie, R. D., 1942. Prevention by cystine or methionine of hemorrhage and necrosis of the liver in rats, *Proc. Soc. Exp. Biol. Med.* 50:1–5.

Dam, H., 1957. Influence of antioxidants and redox substances on signs of vitamin E deficiency, *Pharmacol. Rev.* 9:1–16.

Dam, H., and Glavind, J., 1938. Alimentary-exudative diathesis, *Nature* 142:1077–1078.

Dam, H., and Glavind, J., 1940. Vitamin E and kapillarpermeabilitat, *Naturwissenschaften* 28:207.

Dam, H., and Granados, H., 1945. Peroxidation of body fat in vitamin E deficiency, *Acta Physiol. Scand.* 10:162–171.

Dam, H., Hartman, S., Jacobsen, J. E., and Sondergaard, 1957. The albumin globulin ratio in plasma and exudate of chicks suffering from exudative diathesis, *Acta Physiol. Scand.* 41:149–157.

Hart, K. E., and Mackinnon, M. M., 1958. Enzootic paradontal disease in the Bulls–Santoft area, *N.Z. Vet. J.* 6:118–123.

Hartley, W. J., and Grant, A. B., 1961. A review of selenium responsive diseases of New Zealand livestock, *Fed. Proc. Fed. Am. Soc. Exp. Biol.* 20:679–688.

Hartley, W. J., 1963. Selenium and ewe fertility, *Proc. N.Z. Soc. Anim. Prod.* 23:20–27.

Handreck, K. A., and Godwin, K. O., 1970. Distribution in the sheep of selenium derived from ^{75}Se labelled ruminal pellets, *Aust. J. Agr. Res.* 21:71–84.

Hidiroglou, M., Hoffman, I., Jenkins, K. J., and Mackay, R. R., 1971. Control of nutritional muscular dystrophy in lambs by selenium implantation, *Anim. Prod.* 13:315–321.

Hidiroglou, M., Jenkins, K. J., and Corner, A. H., 1972. Control of nutritional muscular dystrophy in lambs by vitamin E implantations, *Can. J. Anim. Sci.* 52:511–516.

Hidiroglou, M., Jenkins, K. J., Wauthy, J. M., and Proulx, J. E., 1972. A note on the prevention of nutritional muscular dystrophy by winter silage feeding of the cow or selenium implantation of the calf, *Anim. Prod.* 14:115–118.

Hurt, H. D., Cary, E. E., and Visek, W. J., 1971. Growth, reproduction and tissue con-centration of selenium in the selenium-depleted rat, *J. Nutr.* 101:761–766.

Jenkins, K. J., and Hidiroglou, M., 1972. A review of selenium vitamin E responsive problems in livestock. A case for selenium as a feed additive in Canada, *Can. J. Anim. Sci.* 52:591–620.

Jenkins, J. J., Hidiroglou, M., Wauthy, J. M., and Proulx, J. E., 1974. Prevention of nutritional muscular dystrophy in calves and lambs by selenium and vitamin E additions to the maternal mineral supplement, *Can. J. Anim. Sci.* 54:49–60.

Jensen, L. S., 1968. Selenium deficiency and impaired reproduction in Japanese quail, *Proc. Soc. Exp. Biol. Med.* 128:970–972.

Jungherr, E., and Pappenheimer, A. M., 1973. Nutritional myopathy of gizzard in turkeys, *Proc. Soc. Exp. Biol. Med.* 37:520–526.

Kuchel, R. E., and Buckley, R. A., 1969. The provision of selenium to sheep by means of heavy pellets, *Aust. J. Agr. Res.* 20:1099–1107.

Kuchel, R. E., and Godwin, K. O., 1976. The prevention and cure of white muscle disease in lambs by means of selenium pellets, *Proc. Aust. Soc. Anim. Prod.* 11:389–392.

Lannek, N., and Lindberg, P., 1975. Vitamin E and selenium deficienices of domestic animals, *Adv. Vet. Sci. Comp. Med.* 19:127–164.

Lannek, N., Lindberg, P., and Tollerz, G., 1962. Lowered resistance to iron in vitamin E deficient piglets and mice, *Nature* 195:1006–1014.

Latshaw, J. D., and Osman, M., 1975. Distribution of selenium in egg white and yolk after feeding natural and synthetic selenium compounds, *Poult. Sci.* 54:1244–1252.

Lindberg, P., and Siren, M., 1965. Fluorometric selenium determinations in the liver of normal pigs and in pigs affected with nutritional muscle dystrophy and liver dystrophy, *Acta Vet. Scand* 6:59–64.

Machlin, L. J., Gordon, R. S., and Meisky, K. H., 1959. The effect of antioxidants on vitamin E deficiency symptoms and production of liver peroxide in the chicken, *J. Nutr.* 67:333–343.

Mathias, M. M., and Hogue, D. E., 1971. Effect of selenium, synthetic antioxidants, and vitamin E on the incidence of exudative diathesis, *J. Nutr.* 101:1399–1402.

Mertz, W., 1981. The essential trace elements, *Science* 213:1332–1338.

Michaelis, L., and Wollman, S. H., 1950. Free radicals derived from tocopherol and related substances, *Biochim. Biophys Acta* 4:156–159.

Michel, R. L., Whitehair, C. K., and Kealey, K. K., 1969. Dietary hepatic necrosis asso-ciated with selenium-vitamin E deficiencies in swine, *J. Am. Vet. Med. Assoc.* 155:50–59.

Muth, O. H., 1970. Selenium-responsive disease of sheep, *J. Am. Vet. Assoc.* 157:1507–1511.

Muth, O. H., and Allaway, W. H., 1963. The relationship of white muscle disease to the distribution of naturally occurring selenium, *Am. Vet. Med. Assoc.* 142:1379–1384.

Muth, O. H., Weswig, P. H., Whanger, P. O., and Oldfield, J. E., 1971. Effect of feeding selenium deficient ration to the subhuman primate (*Saimiri sciureus*), *Am. J. Vet. Res.* 32:1603–1605.

Niemi, S. M., Kuzan, F. B., and Senger, P. L., 1981. Selenium in bovine spermatazoa, *J. Dairy Sci.* 64:853–856.

Obel, A. L., 1953. Studies on the morphology and etiology of so-called toxic liver dystrophy (hepatosis dietetica) in swine, *Acta Path. Microbiol. Scand. Suppl.* 94.

Piper, R. C., Froseth, J. A., McDowell, L. R., Kroening, G. H., and Dyer, I. A., 1975. Selenium-vitamin E deficiency in swine fed peas (Pisum satirum), *Am. J. Vet. Res.* 36:273–281, 1975.

Porta, E. A., de la Iglesia, F. A., and Hartroft, W. S., 1968. Studies on dietary hepatic necrosis, *Lab. Invest.* 18:283–297.

Salisbury, R. M., Armstrong, M. C., and Gray, K. G., 1953. Ulcero-membranous gingivitis in the sheep, *N.Z. Vet. J.* 1:51–52.

Schwarz, K., 1944. Tocopherol Als Leberschutchstoss, *Z. Physiol. Chem.* 281:109–116.

Schwarz, K., 1948. Uber die Lebertranschadigung der Ratte und ihre Verhutung durch Tocopherol, *Z. Physiol. Chem.* 283:106–112.

Schwarz, K., 1951. Production of dietary necrotic degeneration using American torula yeast, *Proc. Soc. Exp. Biol. Med.* 77:818–823.

Schwarz, K., 1965. The role of vitamin E, selenium and related factors in experimental liver disease, *Fed. Proc. Fed. Am. Soc. Exp. Biol.* 24:58–67.

Schwarz, K., 1975. Essentiality and metabolic functions of selenium, *Med. Clin. No. Am.* 60:745–758.

Schwarz, K., and Corwin, L. M., 1960. Prevention of decline of α-ketoglutarate and succinate oxidation in vitamin-E-deficient rat liver homogenates, *J. Biol. Chem.* 235:3387–3392.

Schwarz, K., and Foltz, C. M., 1957. Selenium as an integral part of factor 3 against dietary necrotic liver degeneration, *J. Am. Chem. Soc.* 79:3292–3293.

Schwarz, K., and Pathak, K. D., 1975. The biological essentiality of selenium, and the development of biologically active organoselenium compounds of minimum toxicity, *Chem. Scr.* 8A:85–95.

Schwarz, K., Porter, L. A., and Fredga, A., 1972. Some regularities in the structure–function relationship of organoselenium compounds effective against dietary liver necrosis, *Ann. N.Y. Acad. Sci.* 192:200–214.

Scott, M. L., 1970. Nutritional and metabolic interrelationships involving vitamin E, selenium and cystine in the chicken, *Int. Z. Vitam. Forsch.* 40:334–343.

Scott, M. L., and Stoewsand, G. S., 1961. Ataxias of vitamin A and vitamin E deficiencies, *Poult. Sci.* 40:1517–1523.

Scott, M. L., Hill, F. W., Norris, L. C., Dobson, D. C., and Nelson, T. S., 1955. Studies on vitamin E in poultry nutrition, *J. Nutr.* 56:387–402.

Scott, M. L., Olson, G., Krook, L., and Brown, W. R., 1967. Selenium-responsive myopathies of myocardium and of smooth muscle in the young poult, *J. Nutr.* 91:573–583.

Segerson, E. C., and Ganapathy, S. N., 1980. Fertilization of ova in selenium/vitamin E-treated ewes maintained on two planes of nutrition, *J. Anim. Sci.* 51:386–394.

Segerson, E. C., Riviere, G., Bullock, T. R., Thimaya, S., and Ganapathy, S. N., 1981. Uterine contractions and electrical activity in ewes treated with selenium and vitamin E, *Biol. Repr.* 23:1020–1028.

Siami, G., Schulert, A. R., and Neal, R. A., 1972. A possible role for the mixed function oxidases in the requirement for selenium in the rat, *J. Nutr.* 102:857–862.

Smith, D. G., Senger, P. L., McCutchan, J. F., and Landa, C. A., 1979. Selenium and glutathione peroxidase distribution in bovine semen and selenium-75 retention by the tissues of the reproductive in the bull, *Biol. Reprod.* 20:377–383.

Sondergaard, E., Scott, M. L., and Dam, H., 1962. Effects of ubiquinones and phytylubichromenol upon encephalomalacia and muscular dystrophy in the chick, *J. Nutr.* 78:15–20.

Sprinker, L. H., Harr, J. R., Newberne, P. M., Whanger, P. D., and Weswig, P. H., 1971. Selenium deficiency lesions in rats fed vitamin E-supplemented rations, *Nutr. Rep. Int.* 4:335–340.

Sugai, M., Inone, M., Tsuchiyama, H., and Kummerow, F. A., 1960. The interrelationship of vitamin E, linoleic and long chain keto acids, *Fed. Proc. Fed. Am. Soc. Exp. Biol.* 19:421.

Swahn, O., and Thafvelin, B., 1962. Vitamin E and some metabolic disease of pigs, *Vitam. Horm. (N.Y.)* 20:645–657.

Teige, J., Nordstoga, K., and Aursjo, J., 1977. Influence of a diet on experimental swine dysentery, I. Effects of a vitamin E and selenium deficient diet supplemented with 6.8% cod liver oil, *Acta Vet. Scand.* 18:384–396.

Thafvelin, B., 1960. Role of cereal fats in the production of nutritional disease in pigs, *Nature (London)* 188:1169–1172.

Thompson, J. N., and Scott, M. L., 1969. Role of selenium in the nutrition of the chick, *J. Nutr.* 97:335–342.

Thompson, R. H., 1976. The levels of selenium and glutathione peroxidase activity in blood of sheep, cows and pigs, *Res. Vet. Sci.* 20:229–231.

Thompson, J. N., and Scott, M. L., 1970. Impaired lipid and vitamin E absorption related to atrophy of the pancreas in selenium deficient chicks, *J. Nutr.* 100:797–809.

Tollerz, G., and Lannek, N., 1964. Protection against iron toxicity in vitamin E-deficient piglets and mice by vitamin E synthetic antioxidants, *Nature (London)* 201:846–847.

Trapp, A. L., Keahey, K. K., Whitenack, D. L., and Whitehair, C. K., 1970. Vitamin E-selenium deficiency in swine: differential diagnosis and nature of the field problem, *J. Am. Vet. Med. Assoc.* 157:289–300.

Trinder, N., Woodhouse, C. D., and Renton, C. P., 1969. The effect of vitamin E and selenium on the incidence of retained placentae in dairy cows, *Vet. Rec.* 85:550–553.

Van Vleet, J. F., Carlton, W., and Olander, H. J., 1970. Hepatosis dietetica and mulberry heart disease associated with selenium deficiency in Indiana, swine, *J. Am. Vet. Med. Assoc.* 157:1208–1219.

Van Vleet, J. F., Meyer, K. B., and Olander, H. J., 1973. Control of selenium–vitamin E deficiency in growing swine by parenteral administration of selenium–vitamin E preparations to baby pigs or to pregnant sows and their baby pigs, *J. Am. Vet. Med. Assoc.* 163:452–456, 1973.

Van Vleet, J. F., Ruth, G., and Ferrans, V. J., 1976. Ultrastructural alterations in skeletal muscles of pigs with selenium–vitamin E deficiency, *Am. J. Vet. Res.* 37:911–922.

Walter, E. D., and Jensen, L. J., 1963. Effectiveness of selenium and noneffectiveness of sulfur amino acids in preventing muscular dystrophy in the turkey poult, *J. Nutr.* 80:327–331.

Walter, E. D., and Jensen, L. L., 1964. Serum glutamic–oxalacetic transaminase levels, muscular dystrophy, and certain hematological measurements in chicks and poults as influenced by vitamin E, selenium and methionine, *Poult. Sci.* 43:919–926.

Weichselbaum, T. E., 1935. Cystine deficiency in the albino rat, *Q. J. Exp. Physiol.* 25:363–367.

Whanger, P. D., and Weswig, P. H., 1975. Effects of selenium, chromium and antioxidants on growth, eye cataracts, plasma cholesterol and blood glucose in selenium deficient vitamin E supplemented rats, *Nutr. Rep. Int.* 12:345–358.

Whanger, P. D., Weswig, P. H., Oldfield, J. E., Cheeke, P. R., and Schmitz, J. A., 1976. Selenium and white muscle disease in lambs: effects of vitamin E and ethoxyquin, *Nutr. Rep. Int.* 13:159–173.

Wu, S. H., Oldfield, J. E., Whanger, P. D., and Weswig, P. H., 1973. Effect of selenium, vitamin E and antioxidants on testicular function in rats, *Biol. Reprod.* 8:625–629.

Wu, A. S. H., Oldfield, J. E., Shull, L. R., and Cheeke, P. R., 1979. Specific effect of selenium deficiency on rat sperm, *Biol. Reprod.* 20:793–798.

Metabolism of Selenium

3.1 Absorption

Selenium is absorbed from the gastrointestinal tract to various extents, and the distribution within the body varies with the species and with the chemical form and the amount of the element ingested.

Selenium is rapidly and efficiently absorbed from naturally toxic or near-toxic seleniferous diets and from soluble salts of the element added to normal diets. Rats, consuming a seleniferous wheat diet containing 18 ppm Se, retained 63% of the ingested selenium during the first week as well as a similar proportion of sodium selenite fed at the same levels (Anderson and Moxon, 1941; Moxon, 1937). Studies designed to show toxicity in rats suggested a somewhat greater absorption from seleniferous grains than from selenites and selenates, but there was a very low absorption from selenides and elemental selenium (Franke and Painter, 1938). Certain organic compounds, including selenodiacetic and selenopropionic acids, are considerably less toxic to rats, per unit of selenium, then is selenite. This observation may be a consequence of lower absorption (Moxon *et al.*, 1941). Different chemical forms of selenium have greatly varied effects in their ability to prevent liver necrosis in rats and exudative disthesis (ED) in chicks. The variations probably reflect differences in absorbability as well as the ability of the body to metabolize the compound.

Cantor *et al.* (1975) have studied the biological availability of selenium in feedstuffs and selenium compounds for the prevention of exudative diathesis in chicks. A basal diet consisting of casein-soy protein, and torula yeast deficient in both vitamin and selenium was used. Graded levels of selenium as supplied by sodium selenite which was standard, or by test ingredients, were fed for periods of 12–21 days. Selenium in most of the feedstuffs of plant origin was highly available, ranging from 60% to 90%. These included wheat (70.7%), brewer's yeast (88.6%), brewer's grain (79.8%), corn (86.3%), soybean meal (59.8%), cottonseed meal (86.4%), dehydrated alfalfa meal (210.0%), and distiller's dried grains plus solubles (65.4%). The finding that selenium in wheat was fairly highly available is similar to the results of Bragg and Seier (1972), who found that determinations of the selenium content in wheat obtained by bioassays were in good agreement with those from chemical assays.

Selenium was less than 25% available from feedstuffs of animal origin which included tuna meal (22.4%), poultry by-product meal (18.4%), menhaden meal (15.6%), fish solubles (8.5%), herring meal (24.9%), and meat and bone meal. In addition, high availability values were obtained for sodium selenate and selenocystine, while low values were observed for selenomethionine, sodium selenide, selenomethionine, and selenopurine. The gray form of elemental selenium was almost completely unavailable. Plasma glutathione peroxidase activity in chicks fed sodium selenite or selenomethionine was highly correlated with protection against (ED). The results of Cantor et al. (1975) suggest that biological availability is determined by the ability of the chick to utilize the various forms of selenium for enzyme activity. Douglass et al. (1981) have reported that the availability of selenium was only 54%–58% as great from tuna as from selenite for the induction of glutathione peroxidase in liver and in red cells. Low availability of selenium from fish or other substances should be considered when assessing selenium status of human beings from dietary intake.

At physiological levels studies with ^{75}Se indicate that the duodenum is the main site of absorption of selenium and there is no absorption from the rumen or abomasum of sheep, or the stomach of pigs (Wright and Bell, 1966). When rations containing 0.35 and 0.50 ppm Se, respectively, were consumed, about 35% of the ingested ^{75}Se isotope was absorbed in sheep and 85% in pigs. Thomson and Stewart (1973) have estimated that intestinal absorption after oral administration was estimated to be 91%–93% for selenite and 95%–97% for selenomethionine. In three young women, the intestinal absorption of ^{75}Se selenite by the three subjects was 70%, 64%, and 44% of the dose (Thomson and Stewart). Other investigations have also observed that monogastric animals have a higher intestinal absorption of selenium then ruminants (Butler and Peterson, 1961; Cousins and Cairney, 1961). One possible explanation is that fecal selenium is mostly present in insoluble forms, and selenite is apparently reduced to insoluble compounds in the rumen (Butler and Peterson, 1961; Cousins and Cairney, 1961). Ehlig et al. (1967) have observed that selenium retention in lambs is greater for selenomethionine than for selenite. This could have been explained by the fact that there was a higher rate of urinary excretion from the selenite than from the selenomethionine treatments instead of differences in apparent rates of absorption from the two selenium sources. Hidiroglou *et al.* (1968) have observed that sheep rumen bacteria are capable of metabolizing inorganic ^{75}Se and incorporating the element into the microbial protein in the form of ^{75}Se-selenomethionine. It is likely that selenomethionine would be more available to protein than selenite. After the incorporation of ^{75}Se-selenomethionine, it is likely extensively broken down and that selenium then enters other metabolic pathways which are influenced by the selenium status of the animal (Burk *et al.*, 1972).

If animals were selenium deficient, radioselenium was more efficiently retained than animals on selenium-supplemented diets. This general pattern has been observed in chicks (Jensen *et al.*, 1963), rats (Burk *et al.*, 1968), and sheep (Lopez *et al.*, 1968; Wright and Bell, 1964). This increased retention probably reflects greater tissue demands for selenium, but it is likely the pattern of excretion is also affected by the level of dietary selenium. Lopez *et al.* (1968) have measured the whole body of retention of selenium, urine, and fecal losses in several tissues following the administration of ^{75}Se to lambs fed graded levels of dietary selenium. Whole-body loss of ^{75}Se 48–336 hr after administration of the isotope was inversely proportional to the dietary level of selenium. In addition, the concentration of ^{75}Se in various tissues was also inversely related to the dietary selenium level. A similar inverse experiment was demonstrated by Hopkins *et al.* (1966), who also observed a similar inverse relationship in whole-carcass retention of injected ^{75}Se in rats which had previously been fed diets low or high in selenium.

Medinsky *et al.* (1981) have studied the organ distribution and retention of selenium in rats after inhalation of selenious acid and elemental selenium aerosols. Although the rate of absorption of these two compounds into blood was different, once absorbed, both chemical forms behaved identically. The results indicated that both compounds joined the same metabolic pool.

After selenium is absorbed, it is first carried mainly in the plasma (Buescher *et al.*, 1960), where selenium is apparently associated with the plasma proteins (McConnell and Levy, 1962). From the plasma proteins, selenium enters all of the tissues, including the bones, the hair, and the red blood cells and leukocytes (Buescher *et al.*, 1960; Cousins and Cairney, 1961).

In dogs, time-distribution studies of dog serum proteins have demonstrated that albumin is the immediate acceptor of injected ^{75}Se. After some time, the selenium is released and bound by other serum proteins, mainly α-2- and β-1-globulins (Schwarz and Foltz, 1957). Jenkins *et al.* (1969) have studied the distribution of ^{75}Se among various electrophoretic serum protein fractions at different time intervals after the administration of the isotope to chicks. Within the first 2 hr after crop intubation with $H_2{}^{75}SeO_3$, about 40% and 22% of the serum protein radioactivity was located in the α-2- and α-3-globulins, respectively. The percentage of serum protein with radioactivity bound to the γ-globulin increased rapidly during the first 24 hr. During the subsequent 24–173-hr interval, 52%–70% of the ^{75}Se was carried by the α-2- and the γ-globulin fractions.

When ^{75}Se, as selenite, is added to human blood, it is rapidly taken up by the cells (50%–70% within 1–2 min) and then is released into the plasma so that most of the radioactivity is in the plasma by 15–20 min (Lee *et al.*, 1969). Dickson and Tomlinson (1967) have analyzed the subfractions of plasma and cells from 254 normal individuals and found the highest selenium concentration to be in the plasma α- and β-globulins. Hirooka and Galambos (1966) found that after the administration of $^{75}SeO_4^{2-}$, there was a greater association of ^{75}Se with the α-2-globulins than with other serum protein fractions in rats and humans. The lipid fraction of serum lipoproteins was also firmly labeled with ^{75}Se both in rats and in humans. In patients with liver disease, which was associated with alcoholism, there was an increased binding of ^{75}Se by serum lipoproteins. McConnell and Levy (1962) have found that selenium-75 is also incorporated into the α- and β-lipoproteins of rat and dog serum, with the greatest in the α-lipoprotein. The binding of ^{75}Se-selenite onto human plasma proteins has also been studied by Sandholm (1975). The most heavily labeled proteins were the β-lipoprotein and an unidentified fraction located electrophoretically between the α-1- and α-2-globulin fractions. Burton *et al.* (1977) have demonstrated that radioactive selenite reacts with purified human and goat immunoglobulins at acidic and neutral pH. The antigenic properties of the immunoglobulins were retained during the selenium labeling as shown by immunoelectrophoresis and autoradiography.

Enhanced *in vitro* uptake of ^{75}Se by the red blood cells has been observed in selenium-deficient sheep (Lopez *et al.*, 1968) as well as in children suffering from kwashiorkor (Burk *et al.*, 1967). The rate of disappearance of ^{75}Se from whole blood, plasma, and red cells in dogs has been studied by McConnell and Roth (1962). After subtoxic amounts of ^{75}Se were administered, the isotope could be detected in various blood proteins for as long as 310 days. Disappearance of radioactivity from the red cells was the greatest at 100–120 days after the initial injection. The data suggest that once selenium enters these cells, it remains there throughout their life span. Incorporation of selenium into myoglobin, cytochrome c, the muscle enzymes, myosin, aldolase, and nucleoproteins has also been observed (McConnell, 1963).

The intracellular distribution of radioselenium varies with the tissue and with the level of selenium. Wright and Bell (1964) have observed that ^{75}Se was distributed rather evenly among the particulates and soluble fraction in liver. In contrast, nearly 75% of the activity in kidney cortex was found in the nuclear fraction. The intracellular distribution of ^{75}Se was altered depending on the dietary level of selenium (Wright and Bell, 1964). Among the subcellular fractions, the microsomes appear to be the initial site for incorporating selenium into protein. Shortly after injection in rats, ^{75}Se became associated with a large number of proteins of the liver and kidney, as shown by the distribution obtained after chromatography on Sephadex G-200 (Millar, 1972). However, after longer periods, the ^{75}Se became concentrated in only one or two peaks, and in this form was resistant to attack by several sulphydryl reagents and ascorbic acid. Millar (1972) suggested that the α-globulins are mainly involved in ^{75}Se binding.

Selenium occurs in all the cells and tissues of the body at levels that vary with the tissue as well as the level of selenium in the diet. The kidney, and especially the kidney cortex, has by far the greatest selenium concentration, followed by the glandular tissues, especially the pancreas, and the pituitary, and the liver. Muscles, bones, and blood are relatively low and adipose tissue very low in selenium. Cardiac muscle is consistently higher in selenium than skeletal muscle. The kidney and liver are the most sensitive indicators of the selenium status of the animal, and the selenium concentrations in these organs can provide valuable diagnostic information. Andrews *et al.* (1968) have suggested that selenium levels of less than 0.25 ppm (fresh basis) in the kidney cortex, and 0.02 ppm in the liver, indicate a marked selenium deficiency in sheep. Normal concentrations are greater than 1.0 ppm of selenium in the kidney cortex and 0.1 ppm in the liver. About one half of this quantity indicates a marginal selenium deficiency.

Dickson and Tominson (1967) have examined autopsy specimens of the liver, skin, and muscles of ten adults with the following results: liver, range 0.18–0.66, average 0.44; skin, range 0.12–0.62, mean 0.27; muscle, range 0.26–0.57, mean 0.37 μg selenium/g of whole tissue.

At toxic intakes of selenium, i.e., 10–100 times or more greater than those normally ingested, tissue selenium concentrations rise steadily until levels as high as 5–7 ppm in liver and kidneys and 1–2 ppm in the muscles are reached. Beyond these tissue levels, excretion begins to keep pace with absorption (Cousins and Cairney, 1961). Selenium was not, therefore, continuously cumulative in the tissues.

The selenium that is deposited in the tissues is highly labile. Once animals are transferred from seleniferous to nonseleniferous diets (Anderson and Moxon, 1941), or following injections of stable or radioactive selenium (Blincoe, 1960; Lopez *et al.*, 1968; Yousef *et al.*, 1968), selenium is removed from the body through excretion into the feces, the urine, and the expired air. The amounts and proportions that appear in the feces, urine, or air depend on the level and form of the intake, the nature of the rest of the diet, and the species. At high dietary intakes of the element, exhalation of selenium is an important route of excretion (Ganther *et al.*, 1966; McConnell, 1948), but exhalation of selenium is much less so at low intakes (Ganther *et al.*, 1966). The pulmonary excretion of injected ^{75}Se was increased when the protein and methionine contents of the diet were increased.

Fecal excretion of ingested selenium is generally greater than urinary excretion in ruminants (Butler and Peterson, 1961; Cousins and Cairney, 1961; Paulson *et al.*, 1966), but not in monogastric species (McConnell, 1948).

When selenium was injected into sheep or other nonruminant species, the urine was the major pathway of excretion (Lopez *et al.*, 1968). Lopez *et al.* (1968) have studied the effect of the route of administration and of increasing dietary levels of selenium upon the pattern of ^{75}Se excretion in lambs. Increasing dietary selenium levels did not substantially affect volatile, urine, or fecal excretion of ^{75}Se when it was administered orally. On the other hand, increasing dietary selenium levels markedly increased the amounts of ^{75}Se in the urine or feces. The selenium in the feces mostly consisted of selenium which had not been absorbed from the diet, together with small amounts excreted into the bowel with the biliary, pancreatic and intestinal secretions (Levander and Baumann, 1966).

Greger and Marcus (1981) have conducted two 51-day human metabolic studies in eight adult males fed a combination of high- or low-protein or phosphorous diets and have studied the effect on selenium metabolism. Subjects lost significantly less selenium in their feces, but significantly more selenium in their urine, when they were fed high-protein, low-phosphorous and high-protein, high-phosphorous diets rather than low-protein, low-phosphorous and low-protein, high-phosphorous diets. When the diets were additionally supplemented with methionine and cystine, subjects lost even more selenium in the urine when they were fed low-protein, low-phosphorous and low-protein, high-phosphorous diets rather than high-protein, low-phosphorous diets.

Selenium metabolism and excretion can be affected by interactions with other elements (see Chapter on Interactions). In addition, selenium metabolism and excretion can be altered appreciably by the presence of sulfate. The urinary excretion of selenium following a parenteral dose of sodium selenate was increased nearly threefold in rats given sulfate parenterally and in the diet (Ganther and Baumann, 1962). When 2% sodium sulfate was added to a diet containing 5 ppm as selenate, the urinary excretion of selenium over a 16-day period increased to 68% of that ingested, compared with 51% in the absence of added sulfate (Halverson *et al.*, 1962). Sulfate, however, has only a slight effect on the urinary excretion of selenium administered in the form of selenite (Ganther and Baumann, 1962). In other experiments, sulfate was also very much more effective against selenate than against selenite or seleniferous grain (Halverson *et al.*, 1962).

3.2 Placental Transfer

Selenium is transmissible through the placenta to the fetus, whether supplied in inorganic or organic forms. This transmission has been demonstrated in the mouse (Hansson and Jacobson, 1966), rat (Westfall *et al.*, 1938), dog (McConnell and Roth, 1964), and sheep (Lopez *et al.*, 1968; Wright and Bell, 1964). Further evidence can be derived from reports that selenium administration to the mother during pregnancy prevents WMD (white muscle disease) in lambs and calves. The placenta apparently presents something of a barrier to the transfer of selenium in inorganic forms as several workers have demonstrated that the concentration of ^{75}Se in the blood and most organs of the fetus, following injection of the ewe with ^{75}Se sodium selenite, is lower than in the mother (Jacobson and Oksanen, 1966; Wright and Bell, 1964). Jacobson and Oksanen (1966) have also observed that when ewes are injected with ^{75}Se-selonomethionine or ^{75}Se-selenocystine, the ^{75}Se concentration in the lambs is higher than when selenite is injected, and is nearly as high as in the mother. Hansson and Jacobson (1966) have also obtained similar results with ^{75}Se-selenomethionine in mice, indicating that the selenoamino acid forms of selenium more readily pass through the placental barrier than do inorganic forms.

3.3 Mechanism of the Antioxidant Action of Selenium

Until the discovery of glutathione peroxidase, hypotheses concerning the biochemical bases of the nutritional interrelationships of selenium and vitamin E have suggested either that both nutrients function as nonspecific antioxidants or that each nutrient has a different and specific function in metabolism. After it was observed that several vitamin E deficiency diseases were prevented by synthetic antioxidants and also that there was increased lipid peroxidation in tissues of deficient animals, the biological antioxidant hypothesis was developed (Tappel, 1962). Recent evidence has demonstrated that vitamin E and selenium have several specific metabolic functions, the most important of which is the protection of the biological membranes from lipid peroxidation (Figure 3-1). The main opposition to the biological antioxidant hypothesis has resulted from difficulties in detecting lipid peroxides in vitamin E-deficient animals (Green, 1972). However, the existence of peroxides *in vivo* in adipose tissue has been reported (Chvapil *et al.*, 1974), and numerous *in vitro* studies have demonstrated the protection by vitamin E and selenium against peroxidation of unsaturated membrane lipids under oxidizing conditions (Combs *et al.*, 1974). Therefore, several investigators believe that controlled lipid peroxidation may be a continuous metabolic process in all tissues.

Erythrocyte plasma membranes are quite labile to lipid peroxidation because of their high content of polyunsaturated fatty acids and to their direct exposure to molecular oxygen. Hemolysis of the cells results from peroxidation of the erythrocyte membrane lipids. Horn *et al.* (1974) have observed that the vitamin E-deficient rabbit, which is an animal very resistant to hemolysis, can be made sensitive by feeding increased levels of unsaturated fats, thus increasing the unsaturated fatty acid content of the erythrocyte, thereby making it more sensitive to oxidative hemolysis. Increasing erythrocyte fragility has been associated with the anemias of vitamin E-deficient chicks, rats, monkeys, fish, and humans with protein-calorie malnutrition and premature infants. The action of vitamin E in protection of erythrocytes from oxidative hemolysis appears to be that of a lipid–soluble antioxidant with biological specificity.

In addition, dietary selenium has also been observed to prevent oxidative hemolysis in the vitamin E-deficient rat. Rotruck *et al.* (1971, 1972) demonstrated a selenium-dependent factor present in erythrocytes that protected the cells from hemolysis in the presence of glucose. This work led to the discovery of the first well-documented biochemical function of selenium in animals when they observed that selenium is an essential constituent of the enzyme glutathione peroxidase (glutathione: H_2O_2 oxidoreductase, E.C. 1.11.1.9). This enzyme is recognized to be important in the metabolic destruction of peroxides, utilizing reducing equivalents from glutathione to reduce hydrogen peroxide as well as fatty acid hydroperoxides (Flohe, 1971). Therefore, it functions in cellular protection from pro-oxidant stressors.

Many cellulor oxidations have been shown to generate the reactive superoxide anion (O_2^-) by the univalent reduction of oxygen (Fridovich, 1974). If levels of superoxide are unchecked, lipid peroxidation has been increased (Kellogg and Fridovich, 1975). Superoxide dismutase reduce

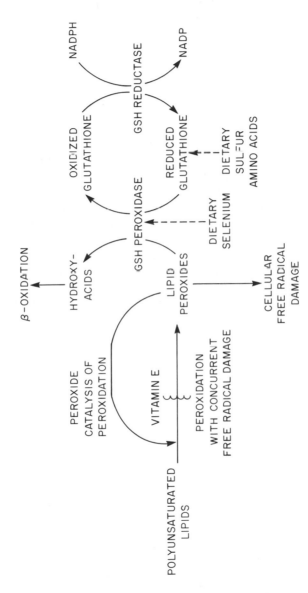

Figure 3-1. Function of selenium in the GSH peroxidation system in inhibition of lipid peroxidation damage.

superoxide levels to hydrogen peroxide (Fridovich, 1974), which in turn is reduced to water by catalase and peroxidases, including glutathione peroxidase.

Hydrogen peroxide in the absence of adequate glutathione peroxidase may react further with excess superoxide to produce the powerful oxidant, hydroxyl radical which is capable of initiating severe oxidative degradation of cellular components (King *et al.*, 1975). McCay *et al.* (1976) have demonstrated that partially purified glutathione peroxidase inhibits lipid peroxidations in hepatic mitochondria and microsomes. Glutathione peroxidase seems to have a critical function in the protection of the cell from normal oxidative stress. The activity of glutathione peroxidase is dependent upon the amount of biologically available selenium in the diet.

Membranes of subcellular organelles contain relatively high concentrations of polyunsaturated lipids as well as hemoproteins which are strong catalysts of lipid peroxidation (Tappel, 1973). The membranes of subcellular organelles, especially those of the mitrochondria and microsomes of the liver, are labile to damage from lipid peroxidation. Lipid peroxidation in these membranes can result in structural damage which interferes with cell function. In mitochondria peroxidation is associated with inhibition of ascorbic acid synthesis in vitamin E-deficient rats (Chatterjee and McKee, 1965). Severe disruption of normal cell function by the release of hydrolytic enzymes can result from peroxidative damage to lysosomal membranes. Lysosomal membranes, however, do not contain hemoprotein pro-oxidants and tend to be less labile to lipid peroxidation than the membranes from mitochondria and microsomes (Tappel, 1973).

Vitamin E and selenium have a cooperative function in protecting subcellular membranes in preparations of chick liver from lipid peroxidation (Combs and Scott, 1974). In the hepatic microsomes of vitamin E- and selenium-deficient chicks, lipid peroxidation was found to be very high. At the time of hatching, when the lipid content of the liver was high, these membranes had very little innate protection against peroxidation. By six and nine days of age, peroxidation had declined sharply. If the chicks received an adequate level of vitamin E (100 IU/kg), there was an almost complete inhibition of peroxidation at nine days, but this quickly reversed, and by 26 days peroxidation in these tissues was almost equal to that of microsomes from chicks receiving the basal diet. If the basal diet was supplemented with sodium selenite, there was no inhibition of lipid peroxidation. However, lipid peroxidation was completely prevented when both vitamin E and selenium were fed. The hepatic mitochondria was also influenced in the same manner by vitamin E and selenium.

Membrane protection has been used as a parameter for the dietary requirements for both vitamin E and selenium (Combs and Scott, 1974). If chicks were fed diets containing an adequate constant amount of selenium (0.15 ppm), incremental levels of vitamin E showed progressive inhibition of hepatic microsomal lipid peroxidation according to the concentration of vitamin E activity in the plasma (Combs *et al.*, 1975). The data implied that a level of vitamin E of about 1 mg/100 ml of plasma is optimal for membrane stability. In the presence of selenium, the amount of vitamin E required in the diet to produce this level of vitamin E was 30 IU/kg diet. With adequate vitamin E, the amount of selenium required for the chick was 0.6 ppm.

Using such parameters as growth and efficiency of feed utilization, the amounts of vitamin E necessary for membrane protection were similar to the chicks' nutritional requirements. Selenium-adequate chicks had optimal growth and efficiency of feed utilization when fed at least 30 IU vitamin E per kilogram (Combs *et al.*, 1975).

The cooperative effect of vitamin E and selenium in protecting biological membranes appears to be due to (a) the action of vitamin E as a membrane-bound antioxidant and (b) the prevention by glutathione peroxidase of hydroxyl radical formation and attack on unsaturated membrane lipids, and also possibly the destruction of any lipid hydroperoxides that may form.

The interaction of vitamin E and selenium with other dietary components such as cysteine remain to be elucidated. Chow and Tappel (1974) have suggested that sulfur-containing amino acids may spare requirements for vitamin E. They hypothesized that by providing cysteine for the synthesis of the tripeptide glutathione (L-glutamyl-L-cysteinylglycine). The hypothesis was not supported by the studies of Hull and Scott (1976), who reported that glutathione peroxidase activities in normal and in dystrophic muscle were equivalent and that reduced glutathione levels were elevated significantly in the dystrophic tissue.

3.4 Effect of Paraquat

Burk *et al.* (1980) have studied the effect of the herbicides, paraquat, and diquat, administration on liver necrosis, and lipid peroxidation in the rat. Paraquat and diquat facilitate formation of superoxide anion formation in biological systems, and lipid peroxidation has been postulated to be their mechanism of toxicity. In general, paraquat and diquat were both much more toxic to selenium-deficient rats than to control rats. Diquat was more toxic than paraquat, but both caused rapid and massive liver and kidney necrosis and very high ethane production rates in selenium-deficient rats. These results suggest that paraquat and diquat exert their acute toxicity largely through lipid peroxidation in selenium-deficient rats.

Selenium deficiency had no effect on superoxide dismutase activity in erythrocytes or in the 105,000-g supernate of liver or kidney. Glutathione peroxidase, which is a well-documented selenium-dependent enzyme in animals, is probably not an important factor. Selenium injection protected against diquat poisoning and also the concomitant lipid peroxidation and mortality. However, this treatment caused no increase in glutathione peroxidase activity of liver, kidney, lung, or plasma after ten hours. These results suggest that either a selenium-dependent factor in addition to glutathione peroxidase exists that protects against lipid peroxidation or some other unknown process may take place.

3.5 Effect on Cytochrome P-450

Burk and Masters (1975) have studied the effects of selenium deficiency on the hepatic microsomal cytochrome P-450 system in the rat. Cytochrome P-450 and b_5 contents, NADPH-cytochrome c reductase, ethylmorphine demethylase, 2- and 4-biphenyl hydroxylase activities, and pentobarbital sleeping time were measured in male rats fed a control diet and a selenium-deficient Torula yeast diet and the effect of phenobarbital pretreatment was determined on these parameters. Phenobarbital treatment in selenium-deficient rats produced an increase in cytochrome P-450 content of only 70% as compared to 150% in similarly treated controls. The induction of ethylmorphine demethylase activity by phenobarbital was also impaired in selenium-deficient rats. Cytochrome b_5 content, NADPH-cytochrome c reductase activity, biphenyl hydroxylase activity, or pentobarbital sleeping time were not affected. In addition, both the cytochrome P-450 ethyl isocyanide-binding spectra and the percentage of the carbon monoxide-reactive cytochrome P-450 which bound metyrapone was the same in the phenobarbital-treated controls and the phenobarbital-treated selenium-deficient rats. Burk and Masters data (1975) indicate a specific effect of selenium in the hepatic microsomal cytochrome P-450 system. The biochemical function of selenium giving rise to this effect is unknown.

Eaton *et al.* (1980) have examined the dose–response of selenium and several other metal ions on rat liver metallothionein, glutathione, heme oxygenase, and cytochrome P-450. Adult male rats received intraperitoneal injections of selenium at the maximum tolerable dose (MTD) of 25 μmol/kg/day and at least three serial dilutions ($\frac{1}{2}$, $\frac{1}{4}$, $\frac{1}{8}$, etc.) of selenium. Selenium significantly increased heme oxygenase at all levels of the trace element. Selenium had no effect on cytochrome P-450 and metallothionein, but increased hepatic GSH at the two highest doses.

3.6 Selenium and Hepatic Heme Metabolism

Maines and Kappas (1976) have studied the effect of selenium on the regulation of hepatic heme metabolism and the induction of δ-aminolevulinate synthase and heme oxygenase. Selenium was found to be a novel regulator of both the mitochondrial enzyme δ-aminolevulinate synthase [succinyl-CoA: glycine C-succinyltransferase (decarboxylating); E.C. 2.3.1.37] and the microsomal enzyme heme oxygenase [heme, hydrogen donor: oxygen oxidoreductase (α-methene-oxidizing, hydroxylating); E.C. 1.14.99.3] in liver. The effect of selenium on inducing these enzyme activities was prompt, reaching a maximum within 2 hr after a single injection. The cellular content of heme was significantly increased 30 min after injection. However, this value decreased within 2 hr, coinciding with the period of rapid induction of heme oxygenase. Mainis and Kappas (1976) postulate that selenium may only indirectly increase heme oxygenase, but may mediate an increase in heme oxygenase through increased production and cellular availability of free heme, which results from an increased heme synthetic activity of hepatocytes.

Correia and Burk (1976) have examined hepatic heme metabolism in selenium-deficient and selenium-adequate control rats. Administration of phenobarbital stimulated heme synthesis in the liver in both the selenium-deficient and selenium-adequate rats. Microsomal heme oxygenase activity was increased six- to eightfold by phenobarbital in selenium-deficient but not control rats. Correia and Burk (1976) suggest that the previously reported abnormalities of cytochrome P-450 induction in selenium-deficient rats may be related to increased degradation of hepatic heme. In contrast, Mackinnon and Simon (1976) have observed that hepatic heme content and synthesis in the selenium-deficient rats did not appreciably differ from the levels in selenium-replete animals. However, in the selenium-deficient animals, phenobarbital treatment did not increase hepatic heme synthesis to the extent documented in selenium-replete rats. Total hepatic heme content increased 68% of the control value in the selenium-replete phenobarbital-treated rats, but only a 10% increase was observed in the selenium-deficient group. Changes in δ-amino [^{14}C] levulinic acid incorporation were similar to those of the hepatic heme content.

Further experiments by Burk and Correia (1981) have indicated that the defect observed in their experiments in their defect in heme utilization was not restricted to the phenobarbital-stimulated state. Other agents such as tryptophan and methemalbumin, which also raise heme content, also caused a greater stimulation of microsomal heme oxygenase activity in selenium-deficient livers than in controls.

Selenium accelerated the reduction of methemoglobin in rat erythrocytes (Iwata *et al.*, 1977). The mode of action was suggested as a catalysis of the methemoglobin reduction by glutathione. In other species of animals, the enhancement by selenite of methemoglobin in erythrocytes was about the same in humans and guinea-pigs (Iwata *et al.*, 1978). In dogs, there was smaller enhancement and in rabbits there was greater enhancement by selenite of methemoglobin reduction in erythrocytes.

Masukawa and Iwata (1979) have also observed that selenite decreased nitrite-induced mortality in male dd strain mice in a dose-dependent manner. Its effect seemed to be due to its action in reducing methemoglobin formed by nitrite. Selenite was also found to reduce methemoglobinemia induced by aniline and phenylhydrazine in rats (Iwata *et al.*, 1979).

References

Anderson, H. D., and Moxon, A. L., 1941. The excretion of selenium by rats on a seleniferous wheat ration, *J. Nutr.* 22:103–108.

Andrews, E. D., Hartley, W. J., and Grant, A. B., 1968. Selenium-responsive disease of animals in New Zealand, *N.Z. Vet. J.* 16:3–17.

Blincoe, C., 1960. Whole body turnover of selenium in the root, *Nature* 186:398.

Bragg, D. B., and Seier, L. C., 1972. The biological activity of selenium in wheat for broiler checks, *Poult. Sci.* 51:1786.

Buescher, R. G., Bell, M. C., and Berry, R. K., 1960. Effect of excessive calcium on selenium[75] in swine, *J. Anim. Sci.* 20:368–372.

Burk, R. F., and Correia, M. A., 1981. Selenium and hepatic heme metabolism, in *Selenium in Biology and Medicine*, J. E. Spallholz, J. L. Martin, and H. E. Ganther (eds.), AVI Publishing Co., Westport, Connecticut, pp. 86–97.

Burk, R. F., and Master, B. S. S., 1975. Some effects of selenium deficiency on the hepatic microsomal cytochrome P-450 system in the rats, *Arch. Biochem. Biophys* 170:124–131.

Burk, R. F., Pearson, W. N., Wood, R. F., and Viteri, F., 1967. Blood-selenium levels and in vitro red blood cell uptake of ^{75}Se in kwashiorkor, *Am. J. Clin. Nutr.* 20:723–733.

Burk, R. F., Whitney, R., Frank, H., and Pearson, N., 1968. Tissue selenium levels during the development of dietary liver necrosis in rats fed Torula yeast diets, *J. Nutr.* 95:420–428.

Burk, R. F., Brown, D. G., Seely, R. J., and Scaief, C. C., 1972. Influence of dietary and injected selenium on whole-body retention, route of excretion, and tissue retention of $^{75}SeO_3^{2-}$ in the rat, *J. Nutr.* 102:1049–1056.

Burk, R. F., Lawrence, R. A., and Lane, J. M., 1980. Liver necrosis and lipid peroxidation in the rat as the result of paraquat and diquat administration. Effect of selenium deficiency, *J. Clin. Invest.* 65:1024–1031.

Burton, R. M., Higgins, P. J., and McConnell, K. P., 1977. Reaction of selenium with immunoglobulin molecules, *Biochim. Biophys. Acta* 493:323–331.

Cantor, A. H., Scott, M. L., and Noguchi, T., 1975. Biological availability of selenium in feedstuffs and selenium compounds for prevention and exudative diathesis in chicks, *J. Nutr.* 105:96–105.

Chatterjee, I. B., and McKee, R. W., 1965. Lipid peroxidation and biosynthesis of L-ascorbic acid in rat liver microsomes, *Arch. Biochem. Biophys.* 110:254–164.

Chow, C. K., and Tappel, A. L. 1974. Response of glutathione peroxidase to dietary selenium in rats, *J. Nutr.* 104:444–451.

Chvapil, M., Peng, Y. M., Aronson, A. L., and Zukoski, C., 1974. Effect of zinc on lipid peroxidation and metal content in some tissues of rats, *J. Nutr.* 104:434–443.

Combs, G. F., and Scott, M. L., 1974. Dietary requirements for vitamin E and selenium measured at the cellular level in the chick, *J. Nutr.* 104:1292–1296.

Combs, G. F., Noguchi, T., and Scott, M. L., 1975. Mechanisms of action of selenium and vitamin E in protection of biological membranes, *Fed. Proc. Fed. Am. Soc. Exp. Biol.* 34:2090–2095.

Correia, M. A., and Burk, R. F., 1976. Hepatic heme metabolism in selenium-deficient rats: Effect of phenobarbital, *Arch. Biochem. Biophys* 177:642–644.

Cousins, F. B., and Cairney, I. M., 1961. Some aspects of selenium metabolism in sheep, *Flust. J. Agr. Res.* 12:927–942.

Dickson, R. C., and Tomlinson, R. H., 1967. Selenium in blood and human tissues, *Clin. Chim. Acta* 16:311–321.

Douglas, J. S., Morris, V. C., Soares, J. H., and Levander, O. A., 1981. Nutritional availability to rats of selenium in tuna, beef kidney and wheat, *J. Nutr.* 111:2180–2187.

Eaton, D. L., Stacey, N. H., Wong, K. L., and Klaasen, C. D., 1980. Dose–response effects of various metal ions on rat liver metallothionein, glutathione, heme oxygenase, and cytochrome P-450, *Toxicol. Appl. Pharmacol.* 55:393–402.

Ehlig, C. F., Hogue, D. E., Allaway, W. H., and Hamm, D. J., 1967. Fate of selenium from selenite or selenomethionine, with or without vitamin E in lambs, *J. Nutr.* 92:121–126.

Flohe, L., 1971. Die glutathioneperoxidase: epidemiologie und biologische aspekte, *Klin. Wochenschr.* 49:669–683.

Franke, K. W., and Painter, E. P., 1938. A study of the toxicity and selenium content of seleniferous diets: with statistical consideration, *Cereal Chem.* 15:1–24.

Fridovich, I., 1974. Superoxide dismutases, *Annu. Rev. Biochem.* 44:147–159.

Ganther, H. E., and Baumann, C. A., 1962a. Selenium metabolism. I. Effects of diet, arsenic and cadmium, *J. Nutr.* 77:210–216.

Ganther, H. E., Levander, O. A., and Baumann, C. A., 1966. Dietary control of selenium volatilization in the rat, *J. Nutr.* 88:55–60.

Green, J., 1972. Vitamin E and the biological antioxidants theory, *Ann. N.Y. Acad. Sci.* 203:29–44.

Greger, J. L., and Marcus, R. E., 1981. Effect of dietary protein, phosphorous and sulfur amino acids on selenium metabolism of adult males, *Ann. Nutr. Metab.* 25:97–108.

Halverson, A. W., Guss, P. L., and Olson, O. E., 1962. Effect of sulfur salts on selenium poisoning in the rat, *J. Nutr.* 77:459–464.

Hansson, E., and Jacobson, S. O., 1966. Uptake of (^{75}Se) selenomethionine in the tissues of the mouse studied by whole-body autoradiography, *Biochim. Biophys. Acta* 115:285–293.

Hidiroglou, M., Heaney, D. P., and Jenkins, K. J., 1968. Metabolism of inorganic selenium in rumen bacteria, *Can. J. Physiol. Pharmacol.* 46:229–232.

Hirooka, T., and Galambos, J. T., 1966. Selenium metabolism. III. Serum proteins, lipo-proteins and liver injury, *Biochim. Biophys. Acta* 130:321–328.

Horn, L. R., Barker, M. O., Reed, G., and Brin, M., 1974. Studies on peroxidative hemolysis and erythrocyte fatty acids in the rabbit. Effect of dietary PUFA and vitamin E, *J. Nutr.* 104:192–201.

Hull, S. J., and Scott, M. L., 1976. Studies on the changes in reduced glutathione in chick tissues during onset and regression of nutritional muscular dystrophy, *J. Nutr.* 106:181–190.

Iwata, H., Masukawa, T., Kasamatsu, S., Inoue, K., and Okamoto, H., 1977. Acceleration of methemoglobin reduction in erythrocytes by selenium, *Experientia* 15:678–680.

Iwata, H., Masukawa, T., Kasamatsu, S., and Komemushi, S., 1978. Stimulation of meth-emoglobin reduction by selenium: A comparative study with erythrocytes of various animals, *Experientia* 34:534–535.

Iwata, H., Masukawa, T., Nakaya, S., 1979. Effect of selenite on drug-induced methem-aglobinemia in rats, *Biochem. Pharmacol.* 28:2209–2211.

Jacobson, S. O., and Oksanen, H. E., 1966. The placental transmission of selenium in sheep, *Acta Vet. Scand.* 7:66–76.

Jenkins, K. J., Hidiroglou, M., and Ryan, J. F., 1969. Intravascular transport of selenium by chick serum proteins, *Can. J. Physiol. Pharmacol.* 47:459–467.

Jensen, L. S., Walter, E. D., and Dunlap, J. S., 1963. Influence of dietary vitamin E and selenium distribution of ^{75}Se in the chick, *Proc. Soc. Biol. Med.* 112:899–901.

Kellog, E. W., and Fridovich, I., 1975. Superoxide, hydrogen peroxide, and singlet oxygen in lipid peroxidation by a xanthine oxidase system, *J. Biol. Chem.* 250:8812–8817.

King, M. M., Lai, E. K., and McCay, P. B., 1975. Singlet oxygen production associated with enzyme-catalyzed lipid peroxidation in liver microsomes, *J. Biol. Chem.* 250:6496–6502.

Lee, M., Dong, A., and Yano, J., 1969. Metabolism of ^{75}Se-selenite by human whole blood *in vitro*, *Can. J. Biochem.* 47:791–797.

Levander, O. H., and Baumann, C. A., 1966a. Selenium metabolism. V. Studies on the distribution of selenium in rats given arsenic, *Toxicol. Appl. Pharmacol.* 9:98–105.

Lopez, P. L., Preston, R. L., and Pfander, W. H., 1968a. *In vitro* uptake of selenium-75 by red blood cells from the immature ovine during varying selenium intakes, *J. Nutr.* 94:219–226.

Mackinnon, A. M., and Simon, F. R., 1976. Impaired hepatic heme synthesis in the phen-obarbital-stimulated selenium-deficient rat, *Proc. Soc. Exp. Biol. Med.* 152:568–572.

Maines, M. D., and Kappas, A., 1976. Selenium regulation of hepatic heme metabolism: Induction of delta-aminolevulinate synthase and heme oxygenase, *Proc. Natl. Acad. Sci.* 73:4428–4431.

Masukawa, T., and Iwata, H., 1979. Protective effect of selenite on nitrite toxicity, *Exper-ientia* 35:1360.

McCay, P. B., Gibson, D. D., Fong, K. L., and Hornbrook, K. R., 1976. The effect of glutathione peroxidase activity in lipid peroxidation in biological membranes, *Biochem. Biophys. Acta* 431:459–468.

McConnell, K. P., 1948. Passage of selenium through the mammary glands of the white rat and the distribution of selenium in the milk proteins after subcutaneous injection of sodium selenate, *J. Biol. Chem.* 173:653–657.

McConnell, K. P., and Levy, R. S., 1962. Presence of selenium-75 in liproteins, *Nature* (*London*) 195:774–776.

McConnell, K. P., and Roth, D. M., 1962. [75]Se in rat intracellular liver fractions, *Biochim Biophys Acta* 62:503–508.

McConnell, K. P., 1963. Metabolism of selenium in the mammalian organism, *J. Agr. Food Chem.* 11:385–388.

McConnell, K. P., and Roth, D. M., 1964. Passage of selenium across the placenta and also into the milk of the dog, *J. Nutr.* 84:340–344.

Medinsky, M. A., Cuddihy, R. G., Griffith, W. C., and McClellan, R. O., 1981. A stimulation model describing the metabolism of inhaled and ingested selenium compounds, *Toxicol. Appl. Pharmacol.* 59:54–63.

Millar, K. R., 1972. Distribution of Se^{75} in liver, kidney, and blood proteins of rats after intravenous injection of sodium selenite, *N.Z.J. Agr. Res.* 15:547–564.

Moxon, A. L., Dubois, K. P., and Potter, R. L., 1941. The toxicity of optically inactive *d*- and *l*-selenium-cystine, *J. Pharmacol. Exp. Ther.* 72:184–195.

Rotruck, J. T., Hoekstra, W. G., and Pope, A. L., 1971. Glucose-dependent protection by dietary selenium against hemolysis of rat erythrocytes *in vitro, Nature (London)* 231:223–224.

Rotruck, J. T., Pope, A. L., Ganther, H. E., and Hoekstra, W. G., 1972. Prevention of oxidative damage to rat erythrocytes by dietary selenium, *J. Nutr.* 102:689–696.

Sandholm, M., 1975. Function of erythrocytes in attaching selenite-Se onto specific plasma proteins, *Acta Pharmacol et Toxicol.* 36:321–327.

Schwarz, K., and Foltz, C. M., 1957. Selenium as an integral part of factor 3 against dietary necrotic liver degeneration, *J. Am. Chem. Soc.* 79:3292–3293.

Tappel, A. L., 1962. Vitamin E as the biological lipid antioxidant, *Vitam. Horm. (N.Y.)* 20:493–510.

Tappel, A. L., 1973. Lipid peroxidation damage to cell components, *Fed. Proc. Fed. Soc. Am. Exp. Biol.* 32:1870–1874.

Thomson, C. D., and Stewart, R. D. H., 1973. Metabolic studies of (^{75}Se) selenomethionine and (^{75}Se) selenite in the rat, *Brit. J. Nutr.* 30:139–147.

Thomson, C. D., and Stewart, R. D. H., 1974. The metabolism of (^{75}Se) in young women, *Br. J. Nutr.* 32:47–57.

Westfall, B. B., Stohlman, E. F., and Smith, M. I., 1938. The placental transmission of selenium, *J. Pharmac. Exp. Ther.* 64:55–57.

Wright, E., 1965. The distribution and excretion of radioselenium in sheep, *N.Z.J. Agr. Res.* 8:284–291.

Wright, P. L., and Bell, M. C., 1964. Selenium-75 metabolism in the gestating ewe and fetal lamb, *J. Nutr.* 84:49–57.

Wright, P. L., and Bell, M. C., 1966. Comparative metabolism of selenium and tellurium in sheep and swine, *Am. J. Physiol.* 211:6–10.

Yousef, M. K., Coffman, W. J., and Johnson, H. D., 1968. Total rate of body turnover of selenium-75 in rats, *Nature* 219:1173–1174.

Comparative Metabolism and Biochemistry of Selenium and Sulfur

4.1 Introduction

The chemical and physical characteristics of selenium and sulfur are similar (Table 4-1). Selenium and sulfur have similar configurations of electrons in their outermost valence shells, even though the third shell of selenium is completely filled. The bond energies, ionization potentials, the sizes of the atoms whether they are in the covalent or ionic state, the electronegativities, and the polarizabilities are essentially identical.

Our previous experience with living systems tells us that these two elements, which are similar chemically and physically, cannot always substitute for one another *in vivo*. One major clue to these differences may be observed in the following equation:

$$H_2SeO_3 + 2H_2SO_3 \rightarrow Se + 2H_2SO_4 + H_2O \qquad (4\text{-}1)$$

This equation states that the quadrivalent selenium in selenite tends to undergo reduction, whereas the quadrivalent sulfur in sulfite tends to undergo oxidation. The chemical difference in the ease of reduction of selenite versus the ease of oxidation of sulfite is also reflected in mammalian metabolism of these compounds. In mammals selenium compounds also generally tend to be reduced, whereas sulfur compounds have a general tendency to be oxidized.

Another chemical difference between selenium and sulfur can be seen in the relative acidic strength of H_2Se and H_2S. Even though the analogous oxyacids of selenium and sulfur are of similar strength (Table 4-1), H_2Se is a much stronger acid than H_2S. The hydrides of selenium and sulfur also have a different acidic strength. This difference is seen in the dissociation behavior of the selenohydryl group of selenocysteine (pk 5.24) compared with that of the sulfhydryl group of cysteine (pk 8

77

Table 4-1. Selected Chemical and Physical Properties of the Group VIA
Elements[a]

Property	O	S	Se	Te
Electron configuration	$2s^22p^4$	$3s^23p^4$	$4s^23d^{10}4p^4$	$5s^24d^{10}5p^4$
Covalent radius (Å)	0.66	1.04	1.17	1.37
Ionic radius (M^{2-}) (Å)	1.45	1.90	2.02	2.22
Ionic radius (M^{6+} in MO_4^{2-}) (Å)	—	0.34	0.40	—
Bond energy (M–M) (kcal/mol)	33	63	44	33
Bond energy (M–H) (kcal/mol)	111	88	67	57
Ionization potential (eV)	13.61	10.36	9.75	9.01
Electron affinity (eV)	−7.28	−3.44	−4.21	—
Electronegativity (Pauling)	3.5	2.5	2.4	2.1
Polarizability (M^{2-}) ($cm^3 \times 10^{-25}$)	39	102	105	140
pK_a				
$\quad$ MO(OH)$_2$ aqueous	—	1.9	2.6	2.7
$\quad$ MO$_2$(OH)$_2$ aqueous	—	−3	−3	—
$\quad$ (H$_2$M), aqueous	16	7.0	3.8	2.6
$\quad$ (HM−), aqueous	—	12.9	11.0	11.0

[a] Permission granted by O. A. Levander, Metabolism of selenium and sulfur, trace elements in Human
Health and Disease, Vol. 11, *Essential and Toxic Elements,* A. S. Prasad and D. Oberleas (eds.),
Academic Press, New York, 1976.

.25). This difference is biologically important because at physiological
pH, the sulfhydryl group in cysteine (or other thiols) exists mainly in the
protonated form; the selenohydryl group in selenocysteine (or other selenols) exists largely in the dissociated form.

Even though several superficial similarities in the metabolism of selenium and sulfur do exist, the physiologic role of selenium may result
from the small chemical and physical differences between these two elements. The purpose of this chapter will be to focus on some unique
metabolic and chemical properties of selenium that will increase our understanding of the biological importance of this element.

4.2 Comparative Metabolism of Selenium and Sulfur

4.2.1 Microorganisms

Microbial systems offer opportunities to study the metabolism of
selenium and sulfur. The interactions include competition between sulfur
and selenium; relationships between methionine and cystine and their
corresponding seleno-amino acids; reductive vs. oxidative pathways; and
dimethyl selenide formation by microorganisms.

4.2.1.1 Competitive Sulfur–Selenium Phenomena

Although there are certain similarities in the metabolism of selenium and sulfur by microorganisms, there are also differences in the individuality of these elements. Shrift (1954) showed that sulfate interfered with selenate uptake by *Chlorella vulgaris*. The uptake of sulfate by *Penicillium chrysogenum* was blocked by selenate (Yamomoto and Segal, 1966). They postulated that this inhibition was mediated through the sulfate permease. Selenate toxicity in yeast could also be decreased by sulfate (Fels and Cheldelin, 1949a).

No one has demonstrated that selenium could totally replace sulfur in microorganisms or other living organisms. The most extensive replacement of sulfur with selenium has been obtained using strains of *Escherichia coli,* which were adapted to high levels of 0.01 M of inorganic selenate (Shrift and Kelley, 1962). Huber *et al.* (1967) have observed that cells utilize selenium only if sulfur is present as a contaminant and that selenium can replace only 30%–40% of the normal sulfur requirement. Linear growth rather than exponential growth on the selenate medium was observed and with the sulfate medium the lag period was longer. Total growth on the selenate medium was markedly reduced and the efficiency of growth (grams of cells produced per gram of glucose utilized) on selenate was less than on sulfate.

The competitive antagonism between selenate and sulfate is also known for a variety of other microorganisms (Kylin, 1967). Sulfur compounds such as methionine, cysteine, and even thiamin also antagonized selenate toxicity but not competitively (Widstrom, 1961).

Even though sulfur compounds are almost always antagonistic to selenium toxicity, enhanced selenite toxicity has been observed in several strains of *E. coli* (Scala and Williams, 1963). The experimental conditions as well as the choice of organism probably had much to do with these observations.

Some bacteria have the ability to adapt to selenium compounds that are added in high concentrations. *Bacillus coli-communis* and *Streptococcus pyogenes-aureus* were grown on media containing high concentrations of selenite, selenate, and selenocyanate. No growth in high concentrations was observed at 24 hr, but growth was later observed at 72 hr (Levine, 1925). In an experiment with *E. coli* which were grown on a medium containing 0.025 M of selenite, there was a long phase followed by renewed growth (Leifson, 1936). In another experiment with *E. coli*, *Proteus vulgaris,* and *Salmonella Thompson*, viable bacterial counts decreased and then increased later (Weiss *et al.* 1965). *E. coli* were trained to growth with $4 \times 10^2\ M\ K_2SeO_4$ and $2 \times 10^{-4}\ M\ K_2SO_4$. The adaptation was permanent (Shrift and Kelly, 1962). *Candida albicans* colonies from selenite plates subcultured with 10^{-2} M selenite showed permanent adaption (Falcone and Nickerson, 1960). *Anacystis nidulans* were trained to grow in 20 mg/100 ml of selenate. However, the resistance was lost after subculture without selenate (Kumar, 1964).

4.2.1.2 Selenomethionine and Selenocystine

Pure β-galactosidase has been isolated from *E. coli* grown on media containing sulfate and media also containing 0.01 M inorganic selenite. Huber and Criddle (1967a) found that 20 of the 27 mol of methionine residues found per 135,000 g of protein had been substituted by selenomethionine. On the other hand, the cysteic acid content of the selenium enzyme was almost the same as the sulfur enzyme. Although the purified enzymes had similar specific activities, the content of enzyme per milligram of cells grown on selenate was only 10% of that for cells grown on sulfate. *E. coli* has been grown on a low-sulfate medium. The resulting enzyme appeared to be more sensitive to sulfur replacement by selenium having a specific activity less than half that of the normal enzyme even though only 8%–10% of the sulfur was replaced by selenium.

A number of studies have compared methionine and selenomethione as substrates for enzymes. Before transmethylation reactions involving methionine can occur, methionine should be activated in a reaction with ATP to form 5-adenosyl methionine. The yeast transferase which catalyzes this reaction is more active with selenomethionine than with methionine (Mudd and Cantoni, 1957). Se-adenosylselenomethionine was also demonstrated to transfer readily its methyl group to guanidoacetic acid in the presence of a methylpherase isolated from pig liver. Bremer and Natori (1960) reported S-adenosylselenomethionine to be as good a substrate as its sulfur analog in rat liver microsome catalyzed choline synthesis and could also transfer its methyl group to sulfhydryl or selenol compounds, including methane–selenol and hydrogen selenide. Appreciable nonenzymic methylation occurred with Se-adenosylselenomethionine in the presence of boiled microsomes. In contrast to the earlier results (Pan and Tarrer, 1967), the maximum velocity of the selenium analogs, selenomethionine, and selenoethionine as substrates was lower in comparison with the sulfur substrates, and the maximum velocity was always lower for the selenium compounds, even though they had a lower Km (Michaelis Constant). Pan *et al.* (1964) have studied the incorporation of the alkyl group of selenomethionine and selenomethionine into various lower constituents and found that the incorporation pattern was similar to that obtained with the sulfur analogs. The ethyl group incorporation was less than that of the methyl, but incorporation continued for a longer time. Wu and Wachsman (1971) have found that selenomethionine was as efficient a methyl donor as methionine for nucleic acid methylation in both *Bacillus megaterium* and *E. coli.*

Seleno-amino acids appear to be incorporated into proteins by the same enzymic reactions as their corresponding amino acids. Early studies with *E. coli* showed that both methionine and selenomethionine were incorporated into protein (Nisman and Hirsch, 1958). Selenomethionine could completely replace methionine for the normal exponential growth of a methionine-requiring mutant of *E. coli* (Cowie and Cohen, 1957), whereas selenomethionine only partially satisfied the methionine requirement of methionineless strains of *E. coli* or *B. megaterium* (Wu and Wachsman, 1970). A methionine-requiring mutant of *E. coli* was subcultured for more than a hundred generations with selenomethionine in place of methionine, and the sulfur needs of these cells were supplied with sulfate (Cowie and Cohen, 1957). Growth was somewhat slower than with methionine, but was exponential. The enzymes necessary for growth and division were apparently functioning despite their alterations. Strain differences appeared to influence the extent to which methionine can be replaced. Similar replacement of methionine by selenomethionine was found in another strain of *E. coli* (Coch and Greene, 1971), but in another strain selenomethionine supported growth for only about five generations (Wu and Wachsman, 1970). In *Chlorella vulgaris* selenomethionine inhibited division at first, but growth of the cell continued to a point where division resumed. During the period of cell enlargement, methionine whose sulfur came from sulfate no longer was incorporated into cell proteins. The resistance correlated with decreased permeability to methionine and to a selenium analog, but the adaptation was permanent. Apparently, selenomethionine which replaced the normal sulfur metabolite gave rise to altered proteins, some of which were unable to function in division (Shrift, 1954; Shrift and Sproul, 1963).

In *Chlamydomonas reinhardi* and *Chlorella pyrenoidosa* experimental growth curves indicated an adaptive response with selenomethionine (Shrift, 1960).

Replacement without loss of function was also seen with the precursor of coenzyme A, pantethine (Mautner and Gunther, 1959). This sulfur-containing cofactor is required for the growth of *Lactobacillus helveticus,* but the selenium analog can replace it, mole for mole.

As measured by ATP–pyrophosphate exchange and by hydroxamic acid formation, the purified methionyl–tRNA synthetase from *Sarcina lutea* was shown to have similar activity with both methionine and selenomethionine (Hahn and Brown, 1967). Identical values for maximum velocity and for the Michaelis constant were observed for both substrates.

E. coli methionyl–tRNA synthetase cannot distinguish between methionine and selenomethionine as this enzyme aminoacylates methionine–tRNA with either compound, the Km being 1.1×10^{-5} M for selenomethionine and 0.7×10^{-5} M for methionine (Hoffman *et al.*, 1969).

Several workers have demonstrated that microorganisms can biosynthesize selenomethionine from inorganic selenium salts (Huber *et al.*, 1967). Paulson *et al.* (1968) have investigated the incorporation of inorganic selenium into the protein of mixed rumen bacteria cultures and found that most of the selenium was removed after dialysis of the acid-insoluble fraction. Thiols also released the selenium in this fraction indicating little or none of the selenium was present as selenomethionine. These results indicate that nonamino forms of selenium may exist in proteins. Weiss *et al.* (1965) have reported that selenocystine is present in microbial proteins. However, because of the extreme chemical instability of this seleno-amino acid, reports such as these should be regarded with caution.

4.2.1.3 Reductive vs. Oxidative Pathways

Several examples of selenium reduction have been reported, but few well-documented cases of selenium oxidation. This illustrates the principle that selenium tends to be reduced rather than oxidized in living systems. *Micrococcus lactilyticus* has been shown to use molecular hydrogen to reduce selenite, but not selenate to selenide (Woolfolk and Whiteley, 1962). Hydrogen uptake was measured manometrically for comparison with elemental selenium and selenide formation. When selenite was used an initial phase of hydrogen uptake occurred accompanied by the formation of elemental selenium. Two moles of hydrogen being consumed per mole of selenite reduced to the elemental state. Selenium was subsequently reduced to the selenide level. This process consumed an additional mole of H_2. After colloidal selenium was incubated in this experimental system, it was stoichiometrically reduced to selenide similar to the second phase of selenite reduction. Although selenate was not reduced in this system, tellurite, tellurate, and a number of other substances were reduced. Whiteley and Woolfolk (1962) have postulated that this is a low-potential system which mediates the flow of electrons from hydrogen to a variety of reducible substances in this organism. This low-potential system may involve ferredoxin.

Cell-free preparations of yeast also are able to reduce selenite to elemental selenium, but a flavin moiety appears to be required for adequate electron flow (Nickerson and Falcone, 1963). The optimal pH for selenite reduction was about 7 in the cell-free system. Dialysis caused almost complete loss of activity from the heat labile, soluble, nonparticulate fraction. This activity could be restored by addition of the dialyzable fraction of heated extracts. Extensive reduction of selenite by the dialyzed enzyme source was observed with the addition of glucose-6-phosphate, nicotinamideadenine-dinucleotide phosphate, and oxidized glutathione menadione. Glucose-6-phosphate apparently served as the primary source of reducing equivalents for the NADP-linked enzymes. Because extraction with hexane abolished the activity of the boiled cell-free extract that was used as a coenzyme source, menadione was also included. The addition of menadione or the hexane-extracted material partially restored activity. The reconstituted system was somewhat sensitive to arsenite. A concentration of 10^{-3} M produced about 50% inhibition in the presence or absence of 10^{-3} glutathione. Arsenite sensitivity was interpreted by Nickerson and Falcone (1963) as evidence for involvement of an essential dithiolenzyme. They suggested that selenite was bound to vicinal thiol groups of the protein. The glutathione requirement may be necessary to maintain the thiol groups of the protein in the reduced state. Selenite at 10^{-2} M was slowly reduced to elemental selenium by 2×10^{-6} M glutathione, but this reaction was too sluggish compared with the reaction of selenite with reduced menadione. The results of the experiment suggest that the electron donor for selenite reduction was a quinone linked to nicotinamide-adenine-dinucleotide phosphate and glucose-6-phosphate by a specific dehydrogenase. Nickerson and Falcone (1963) did not test other thiols to see if they were as active as glutathione. This would be the case if the role of glutathione was to keep protein thiols in a reduced state. Glutathione is known to be specific for the reduction of selenite in mammalian systems (Ganther, 1966). On the other hand, glutathione and glutathione reductase may be mediators of selenite reduction in yeast extracts. Selenite may react nonenzymatically with glutathione to form selenodiglutathione, which undergoes a NADPH-dependent reduction to elemental selenium in the presence of yeast crystalline glutathione reductase (Ganther, 1968). This enzyme is inhibited by arsenite and by other heavy metals and may possess a dithiol prosthetic group (Massey and Williams, 1965).

Another flavin requirement for optimal reduction of selenite to elemental selenium was reported for cell-free extracts of *Streptococcus faecalis* and *S. faecium* (Tilton *et al.*, 1967) in relation to the use of this criterion in differentiating between the species. The difference in ability to reduce selenite was greatest for whole cells incubated with high concentrations of selenite, but became smaller when reduction was studied in cell-free extracts or when low concentrations of selenite were used. The cell-free extract selenite-reducing activity was inhibited by mercuric

chloride and iodoacetate and was very sensitive to oxygen. Dialysis resulted in complete inactivation which was partially overcome by thiols.
Evidence for the involvement of a flavoprotein was obtained. Selenite is
reduced to the insoluble elemental selenium by *Salmonella heidelberg*
and this may be the basis for the tolerance of *Salmonella* to selenite
(McCready *et al.*, 1966). In some instances, adaptation of microorganisms
to selenium toxicity may be due to increased levels of "selenoreductase"
in the selenium-tolerant strains (Letunova, 1970).

Levine (1925) was the first to show the inhibitory effect of selenium
on the growth of certain microorganisms and cited earlier work showing
that bacteria could reduce certain selenium compounds. Levine (1925)
confirmed the ability to reduce selenite could be used for differentiating
anaerobes. In addition, he also used it in a medium-selective (or typhoid)
bacillus. Levine (1925) has found a relationship between the ability of an
organism to reduce selenite and to grow in its presence. McCready *et al.*
(1966) showed that selenite toxicity is primarily associated with prolon-
gation of the lag phase of growth, and that the tolerance of salmonellae
to the selenite ion involves an active process whereby selenite is converted to insoluble selenium, the medium being detoxified in this way.
Lapage and Bascomb (1968) have investigated the ability of 548 strains
of gram-negative rods to reduce selenite. This characteristic was studied
in conjunction with the ability of the gram-negative rods to produce H_2S
and reduce nitrate. The reduction of selenite was shown to be a reasonably
good test compared with other tests used in classifying these organisms.
The ability to reduce selenite often paralleled the ability to produce H_2S,
but it did not parallel the ability to reduce nitrate.

Oxidation by microbes of the reduced form of selenium has received
limited investigation. *Aspergillus niger*, a fungus, has been reported to
have oxidized selenite to selenate (Bird *et al.*, 1948). In addition, two
bacterial species have converted elemental selenium to selenate (Sapozhnikov, 1937).

4.2.1.4 Dimethyl Selenide Formation

Challenger (1955) has reviewed the formation of the volatile dimethyl
selenide by certain molds from inorganic selenium salts. Challenger and
North (1934) have shown that *Scopulariopsis brevicaule* could convert
selenite or selenate to dimethyl selenide, but experiments to produce
dimethyl sulfide from sulfite, thiosulfate, thiourea, elemental sulfur, or
thiodiglycolic acid were negative. When dimethyl sulfoxide, but not the
sulfone, was added to cultures of the mold, dimethyl sulfide could be
produced. A species difference was observed with *Schizoophyllum com-*

mune. This bacterium was able to convert inorganic sulfate to a variety of volatile sulfur compounds which included hydrogen sulfide, methyl mercaptan, dimethyl sulfide, and dimethyl disulfide (Challenger and Charlton, 1947). *S. brevicaule,* in addition to synthesizing dimethyl selenide from inorganic selenium salts, could also produce trimethylarsine and dimethyltelluride from inorganic arsenic and tellurium salts, respectively. Challenger (1951) has postulated a mechanism which consisted of alternate methylation and reduction steps to account for the biosynthesis of dimethyl selenide from selenite:

$$H_2SeO_3 \rightarrow HSeO_3^- \xrightarrow{CH_3^+} CH_3SeO_3H \xrightarrow{reduction}$$

$$CH_3SeO_2^- \xrightarrow{CH_3^+} (CH_3)_2SeO_2 \xrightarrow{reduction} (Ch_3)_2Se \quad (4\text{-}2)$$

Eleven microorganisms have been isolated from the soil that are capable of producing $(CH_3)_2Se$ (Doran and Alexander, 1977). A strain of *Penicillium* that produced $(CH_3)_2Se$ from inorganic compounds was isolated from raw sewage (Flemming and Alexander, 1972). $(CH_3)_2Se$ was also evolved from soils containing added inorganic selenium and glucose in the presence of air (Francis *et al.*, 1974).

Reamer and Zoller (1980) have observed that inorganic selenium compounds are converted to volatile methylated species (dimethyl selenide, dimethyl diselenide, and dimethyl selenone) by microorganisms in sewage sludge and soil. When no selenium was added, no volatile selenium compounds were detected. All samples were evaluated without the addition of nutrients and in the presence of nitrogen or air. The methylation process may be an important step in the detoxification process for microorganisms exposed to high selenium concentrations. These as well as the studies of others have shown that microbial methylation of selenium often occurs in various media under different conditions. Thus, biomethylation may be a pathway for the mobilization of selenium to the atmosphere. The data presented by Reamer and Zoller (1980) cannot verify or dispute Challenger's (1951) mechanism of methylation. His mechanism included no steps for the formation of $(CH_3)_2Se_2$ which was observed. Reamer and Zoller (1980) modified the mechanism to include a concentration-dependent branch to produce $(CH_3)_2Se_2$ (Figure 4-1). $(CH_3)_2Se_2$ production could occur at the CH_3SeO_2 intermediate, where reduction could form either CH_3SeOH or CH_3SeH, which would rapidly produce $(CH_3)_2Se_2$. This type of mechanism is similar to that of the corresponding sulfur species. From the data of Reamer and Zoller (1980) it appears that the right-hand pathway in Figure 4-1 may be just as active as a detoxification mechanism for microorganisms present in the sludge. They have two possible explanations for the usefulness of this right hand pathway:

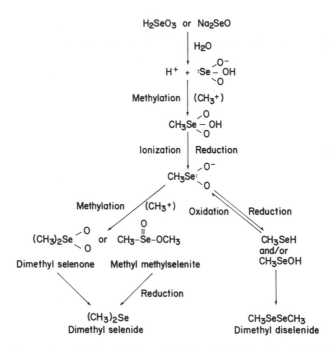

Figure 4-1. Proposed mechanism for the methylation of selenium. Permission Donald Reamer, *Science* 208:500–502, 1981, by the American Association for the Advancement of Science.

When the selenium concentration is sufficiently high, the first mechanism may proceed rapidly to reduce the $(CH_3)_2Se_2$ and the left-hand mechanism is abbreviated at the $(CH_3)_3SeO_2$ step where this compound is expelled. It may be energetically easier for the organism to expel the $(CH_3)_2SeO_2$ than to reduce it to $(CH_3)_2Se$; a second and a more plausible theory is that different microorganisms are responsible for the formation of each species and that the microorganisms responsible for the production of $(CH_3)_2Se$ are hindered at high selenium concentrations, whereas the microorganisms producing $(CH_3)_2Se_2$ are more selenium tolerant.

4.2.2 Plants

4.2.2.1 Sulfur–Selenium Antagonism

Because of the resemblance in chemical and physical properties between the S and Se atoms, most studies that involve the effects of substances on selenium uptake and toxicity have been concerned with antagonism by sulfur compounds, particularly sulfate. In some cases other sulfur compounds also counteract selenate toxicity. The ability of other sulfur metabolites to counteract selenate toxicity in a noncompetitive manner could be interpreted to mean that the selenate is assimilated into a series of analogs, each of which may contribute to the overall toxic effect. The sulfur compounds provided may externally counteract the effects of the internally produced selenium analogs.

Unlike the competition between sulfate and selenate, competitive antagonism between selenite and any other sulfur compounds has never been demonstrated. Certain sulfur compounds, especially sulfate, counteract the effects of low selenite concentrations, but they are not effective at higher concentrations (Trelease and Trelease, 1938). Because selenite can have a nonmetabolic chemical association with sulfhydryl compounds (Ganther and Corcoran, 1969), effects should be carefully studied to distinguish between a metabolic competition and a nonmetabolic chemical association with selenite.

From these observations of sulfur–selenium competition, one might predict that selenium uptake by crops and forage crops and forage plants should be reduced by the application of sulfur compounds to seleniferous soils. With soils under greenhouse conditions antagonism between selenium and sulfur have been observed (Hurd-Karrer, 1937; Hurd-Karrer, 1938), but when gypsum ($CaSO_4$) or elemental sulfur was added to field plots, negligible effects were observed (Beath et al., 1937; Moxon, 1937; Franke and Painter, 1937). The difference in the effects by the forms of sulfur may have been that the selenium in the soil was physiologically and structurally unrelated to the form of sulfur added.

Nonsulfur compounds such as phosphate (Bonhorst and Palmer, 1957; Mahl and Whitehead, 1961), arsenite (Mahl and Whitehead, 1961), arsenate (Bonhorst and Palmer, 1957), (Mahl and Whitehead, 1961), and histidine (Opienska-Blauth and Iwanowski, 1952) counteract selenite and selenate toxicity. The way these substances act is unknown. On the other hand, phosphate has also been reported to enhance the toxicity of selenite.

4.2.2.2 Assimilation of Sulfur and Selenium

Both Hurd-Karrer (1934, 1937a, 1938) in extensive studies with wheat and Trelease and Trelease (1938) have observed that selenium accumulation in A. racemosus depended upon both selenium and sulfur concentrations in the growth medium. It is also known that selenate uptake by A. crotalariae roots is an active transport process (Ulrich and Shrift, 1968; Shrift and Ulrich, 1969; Leggett and Epstein, 1956). In barley root selenate also competes with sulphate (Leggett and Epstein, 1956). The resistance of a transport site for which two molecules compete is derived from experiments in which the ratio between the concentrations of the two molecules determines the degree of which each is absorbed. If the ratio of sulfur and selenium is kept the same despite changes in the total amounts, the growth response or the amount of selenate or sulfur absorbed will remain constant. A change in this ratio will vary the growth response or absorption.

The Michaelis–Menten analysis, which is usually applied to the study of enzyme action, has also been used successfully to assess ion transport by plant roots (Leggett and Epstein, 1956). The interaction between the two ions was found to be competitive. These data indicate that the two ions were bound with the same affinity to the cell membrane at a common binding site. The common binding site for sulfate and selenate, which has also been termed *carrier* or *permease*, has been found in several fungi (Tweedie and Segal, 1970; Yamamoto and Segal, 1966).

The internal ratio of sulfur to selenium should approximate the external ratio, if binding by the carrier were the only factor for transport into the cell. Additional factors must be instrumental in the uptake of these two ions by the intact plant because earlier plant studies by Hurd-Karrer (1938) had demonstrated that the internal S/Se ratio was much greater than the external ratio.

Active transport process requires that a substance accumulate unchanged. Excised roots of *Astragalus* species have been found to accumulate selenate unchanged (Ulrich and Shrift, 1968). After 1 hr a gram of root tissue from one species of *Astragalus* had accumulated 0.4 μmol of selenium from a solution that contained 0.01 μmol/ml, which is about a 40-fold concentration. Selenate uptake was inhibited by sulfate and by the respiratory inhibitors azide and dinitrophenol, and by zero degree temperature. Most selenium was extractable in the selenate form (Shrift and Ulrich, 1969). Comparable studies with selenomethionine also show a similar active transport in alga *C. vulgaris* (Shrift, 1967) and in hamster intestine cells (McConnell and Cho, 1967).

Selenite transportation is more complicated. When selenite is supplied to various organisms, selenium accumulates, but uptake can be prevented by respiratory inhibitors (Ulrich and Shrift, 1968; Shrift, 1967). In excised *Astragalus* roots a high proportion of the extractable selenium was no longer selenite, but had been converted to other forms (Shrift and Ulrich, 1969; Rosenfeld, 1962). Whether or not selenite enters excised *Astragalus* roots by a continuous diffusion process or selenate enters excised *Astragalus* by a carrier mechanism comparable to that for the active transport of many other substances is not known. Whichever the way of entry, the selenite must be converted to different forms, either metabolically, by energy-dependent reactions, or in part nonmetabolically.

Many gaps exist in our knowledge of the intermediate steps by which selenate and sulfate are assimilated. In many cases reliance is placed on comparative biochemistry. One could assume that metabolic steps that occur with sulfate and selenate are similar. Often such assumptions prove reliable, but at times other pathways are discovered and the assumptions of biochemical units are not confirmed.

In both plants and microorganisms the first step in sulfate assimilation is the reaction of the ion with ATP to yield adenosine-5′-phosphosulfate:

$$\text{ATP} + \text{SO}_4 \xrightleftharpoons{\text{ATP-sulfurylase}} \text{PAS} + \text{PP} \qquad (4\text{-}3)$$

ATP sulfurylase has been characterized in both plants and microorganisms (Ellis, 1969; Adams and Rinne, 1969) and also converts selenate to APSe (Akagi and Campbell, 1962). Although selenate prevents the formation of APS by the enzyme preparation from several higher plants (Ellis, 1969), APSe has not been directly described from these sources. However, because the activation of sulfate by ATP-sulfurylase is so well established, its existence in plants is likely. Perhaps APSe is less stable than APS in the presence of the sulfurylase. This might be due to the tendency of enzyme-bound APSe to transfer its selenate moiety to a group on the enzyme, or to water, more rapidly than it could be liberated as APSe. Its existence in plants is likely. Otherwise another mechanism for the assimilation of selenate to amino acids would have to be elucidated.

The second step in sulfate assimilation is generally recognized to be a reaction between ATP and another molecule of APS catalyzed by the enzyme APS-kinase to give 3′-phosphoadenosine-5′phosphosulfate:

$$\text{APS} + \text{ATP} \xrightleftharpoons{\text{APS-kinase}} \text{PAPS} + \text{ADP} \qquad (4\text{-}4)$$

In microbial cells the occurrence of this enzyme is well established, but its existence in higher plants has been questioned and is unconfirmed (Ellis, 1969). Efforts to elucidate PAPSe with a yeast APS-kinase have not been successful (Wilson and Bandurski, 1958). Perhaps the enzyme could not convert APSe to PAPSe or PAPSe was too unstable under the experimental conditions used. Nissen and Benson (1964) have suggested that selenate could be metabolized to amino acids through activation as adenosine phosphoselenate, rather than as 3′-phosphoadenosine-5′-phosphoselenate. Steps for the assimilation of selenite or selenate are also uncertain. Chen et al. (1970) have used radioactive precursors and found that serine serves as an intermediate in the synthesis of both cysteine and selenocysteine in *Astragalus bisulcatus,* and that methylation of Se-methylselenocysteine and S-methylcysteine occurs from the methyl group of methionine, but the mechanisms of the Se atom insertion is virtually unknown.

Differences in the biochemistry of sulfur and selenium have been demonstrated. No selenium analogs have been found for glutathione in *Astragalus* (Virupaksha and Shrift, 1963), plant sulfolipid in *Chlorella* and *Euglena* (Nissen and Benson, 1964), or choline sulfate in *Aspergillus* (Nissen and Benson, 1964). There are also reactions involving selenium for which sulfur analogs have not been found (Shrift, 1969). For example, selenocystathionine has been found in the accumulators *Stanleya pinnata* (Virupaksha and Shrift, 1963) and in *Neptunia amplexicaulis* (Peterson and Butler, 1962). Cystathionine, which is the sulfur analog, has not been found to any extent in higher plants, even though it is a major metabolite of sulfur in animals and microorganisms.

Although there was little incorporation of selenium into the sulfur amino acids in plant proteins of the selenium accumulator *Neptunia amplexicaulis* (Peterson and Butler, 1962), extensive incorporation of seleno amino acids occurred in ryegrass, wheat and clover (Peterson and Butler, 1962). Huber and Criddle (1967b) have compared the chemical properties of selenocysteine and selenocystine with their sulfur analogs. They concluded that the chemical differences between these analogs are sufficiently great that it is unlikely that the selenium analog could replace cysteine or cystine in a protein molecule without marked alteration of the protein structure and function. This should be especially true for the "sulfhydryl proteins" whose activity seems directly dependent on the presence of a free cysteine side chain. Near pH 7 the cysteine side chain exists predominantly in the sulfhydryl form, whereas while selenocysteine is found almost exclusively as the selenide ion. The pH dependence of a selenocysteine-containing "sulfhydryl protein" should therefore be greatly changed if not lost completely (Huber and Criddle, 1967b). Peterson and Butler (1962) have suggested that the selenium accumulator species have evolved a detoxification mechanism whereby selenium may be excluded from protein incorporation. Nonaccumulator species do not have this mechanism. If selenium were incorporated into proteins, alteration of the protein structure could result; this could cause inactivation of the protein and eventual poisoning of the plant.

The principle seleno-organic compound in the leaves of *A. bisulcatus* has been identified as Se-methylselenocysteine (Nigam and McConnell, 1969). The leaves of this plant also contains S-methylcysteine, which is not found in the nonaccumulator species of *Astragalus* (Nigam and McConnell, 1969.; Chow *et al.* (1971) have determined the effects of selenium and sulfur concentrations in the growth medium upon the bio-synthesis of these two amino acids. Increasing concentrations of selenate at a constant sulfate concentration results in increasing quantities of Se-methylselenocysteine and decreasing quantities of S-methylcysteine. Similarly, increasing concentrations of sulfate at a constant selenate concentration results in increasing quantities of Se-methylselenocysteine. Chow *et al.* (1971) suggest that a common enzyme system is involved in the synthesis of both S-methylcysteine and Se-methylselenocysteine.

Plants grown in a medium containing 129 ppm of Se and no S developed red rings at stem nodes and a pink coloration of the roots. This phenomenon may be due to the deposition of elemental selenium and was not observed in the presence of sulfate even with higher selenate concentrations. This latter observation may represent reduced selenate uptake in the presence of sulfate. Pink coloration in the roots of selenite-injured wheat has been observed by Hurd-Karrer (1937), who believed the coloration was due to the deposition of elemental selenium. In the study of Chow *et al.* (1971) the plants were not visibly harmed by the excessive amounts of selenium.

4.2.2.3 Volatile Forms

California investigators became interested in the assumption that selenium accumulator plants gave off volatile selenium in some form (Lewis *et al.*, 1966). This assumption was based on the characteristic "garlicky" odor of *Astragalus* species growing on seleniferous soils. Biological production of volatile selenium has been known for decades mainly as a result of experiments with microorganisms and plants. Additionally, losses of selenium have been observed during storage or drying of accumulator and nonaccumulator plant tissues containing selenium (Asher *et al.*, 1967; Gissel-Nielsen, 1970). Collection of volatile selenium was first made from growing plants, including the nonaccumulator alfalfa (*Medicago sativa L.*) (Lewis *et al.*, 1966). Later, dimethyl diselenide $(CH_3)_2Se_2$, was identified in volatile material collected from the accumulator *Astragalus racemosus*. However, dimethyl selenide was not found in the volatiles from this species (Evans *et al.*, 1968).

Studies utilizing nonaccumulator plant species indicated that the sunflower (Helianthus anuus), spinach (Spinacia olerarea), cabbage (Brassica oleracea var. capitata), and four other cruciferous species, gave off volatile selenium during growth on nutrient media which contained up to 3 mg/liter selenium as selenite or selenate. Volatile selenium production varied from less than 0.05% of leaf selenium per hour to 0.40% per hour. The proportion of selenium has little correlation with the concentration of selenium in the tissues of the plant or with selenium concentration in the growth medium. The intensity of the sunlight appeared to be the primary influence on the production of volatile selenium. However, this effect was later shown to be due to increased temperatures of the leaf tissues, rather than to an effect of light *per se* (Lewis *et al.*, 1974). Experiments with fresh leaf homogenates showed that tissues from plants cultured in selenite-containing solutions released 10–16 times more volatile selenium as leaves from plants whose culture medium contained selenate. Fresh leaf homogenates (Lewis *et al.*, 1974) produced the most volatile selenium when cultured at a pH of 7.8 and a temperature of 40°C.

Using cold trap techniques and gas chromatography, dimethyl selenide was identified as the single selenium volatile compound produced by cabbage leaves. Dimethyl sulfide, the sulfur analog, was also released by these leaves (Lewis *et al.*, 1974). A crude enzyme fraction was isolated from fresh cabbage leaves. This enzyme fraction was shown to cleave Se-methylselenomethionine selenonium salt to dimethyl selenide and homoserine. No additional cofactors were apparently required for the cleavage reaction. The enzyme fraction from cabbage also cleaved S-methyl methionine sulfonium salt to dimethyl sulfide and homoserine. Similar cleavage of the sulfonium compound has been observed with enzyme fractions separated from soil bacteria (Mazelis *et al.*, 1965). The precursor of dimethyl selenide, Se-methylselenomethionine selenonium compound, was isolated from cabbage plants cultivated in nutrient solutions containing up to 3 mg/liter selenium as selenite or selenate (Lewis, 1974). This selenonium compound has been found in other species of nonaccumulators, but has not been found in species of accumulator plants. Shrift (1969) has suggested its presence or absence as a biochemical method for distinguishing selenium nonaccumulators from accumulators.

It is interesting that no dimethyl disulfide was detected in the volatile material collected from freshly harvested cabbage leaves in the experimental system of Lewis *et al.* (1971). The characteristic odor of crucifers has been attributed in part to the production of dimethyl disulfide. Bailey *et al.* (1961) have also reported this compound to occur in sulfur-volatile compounds from cabbage. In the above-described experiments, dimethyl disulfide was found only in volatile material collected from cabbage leaves that had been harvested the previous day. It is possible that this compound is a decomposition product that is not produced by growing cabbage plants or freshly harvested leaves from these plants. This possibility remains to be confirmed.

The release of volatile selenium compounds which occur in apparently healthy plants under physiological conditions does not require that high or toxic levels of selenium be present in the growth medium. The release of dimethyl selenide by nonaccumulator species of higher plants seems to be simply a by-product of normal, biochemical reactions within the plant, where selenium is metabolized along the pathways of sulfur. This process is probably different from the animal exhalation of dimethyl selenide. This latter process is likely a detoxification mechanism.

Dimethyl diselenide production by accumulator species of higher plants is twice as efficient for removal of selenium as the reaction in the nonaccumulator species. Two atoms of selenium are removed per two methyl groups in accumulator species, compared to one atom of selenium removed per two methyl groups in nonaccumulator species. The release of dimethyl diselenide by accumulator plants is more likely the result of reactions in which selenium and sulfur behave analogously rather than a detoxification mechanism. Volatilization of selenium appears to play only a small role, if any, in the relative tolerance of accumulator species for seleniferous growth media.

The release of volatile selenium compounds by intact plants and the possibility of subsequent assimilation of the volatile compound(s) by adjacent plants are factors which should be carefully required when planning controlled experiments on selenium requirements. The transfer of selenium between plants may be a significant factor as the transfer of ^{75}Se between separate alfalfa plants in a greenhouse has been reported (Asher *et al.*, 1967).

4.2.2.4 Seleno-Amino Acids

Purified cysteine syntheses from Se-accumulator and -nonaccumulator plants both catalyze the incorporation of Se^{-2} into selenocysteine. In this case trace amounts of Se^{-2} can be readily incorporated, since the enzyme from plants has a greater affinity for Se^{-2} than S^{-2} (Ng and Anderson, 1978a). Chloroplasts also incorporate Se^{-2} into selenocysteine. They contain most of the cysteine synthase activity associated with leaf tissue (Ng and Anderson, 1978b). These data suggest that chloroplasts have the potential to synthesize selenocysteine provided that they possess a mechanism for the formation of Se^{-2}. The fact that plants can readily incorporate SeO_3^{-2} and SeO_4^{-2} into various amino acids (Peterson and Butler, 1962; Nigam and McConnell, 1953) suggests that mechanisms may exist for the reduction of the more oxidized forms of Se to oxidation state -2. The chloroplast seems the most likely site for these reduction processes, because this organelle is the primary site of production of reducing substances to be used in the reduction of SO_4^{-2} and NO_2. Any Se^{-2} produced could be readily assimilated via the cysteine synthase activity associated with chloroplasts.

The forms of inorganic Se most likely to be available for plants is not known and is probably dependent on many factors (e.g., soil type) (Schrift, 1973). SeO_3^{-2}, on the other hand, is known to be more readily assimilated than SeO_4^- or elemental Se (Shrift and Ulrich, 1969) . SeO_3^{-2} could be reduced either by the bound assimilatory SO_4^{-2} reduction pathway (Schiff and Hodson, 1973) by exchange with carrier–S–SO_3^-, or, by free (unbound) SO_3^{-2} reductase. These possibilities require that Se is metabolized by processes analogous to those for inorganic S. However, an alternate independent pathway could be responsible for the reduction of SeO_3^{-2}. The possibility that SeO_3^{-2} is reduced by the bound pathway has not been investigated. It is known, however, that SeO_3^{-2} is reduced by an NADPH-specific SO_3^{-2} reductase (E.C. 1.8.1.2) from E. coli (Kemp et al., 1963). The maximum rate of SeO_3^{-2} reduction catalyzed by the E. coli enzyme was approximately tenfold faster than for SO_3^{-2}. The very high Km value for SeO_3^{-2} (90 mm) suggests that this mechanism is unlikely to be important in vivo. In plants, SO_3^{-2} reductase activity is known to be associated with chloroplasts (Asada et al., 1969), but the reduction of SeO_3^{-2} by this reductase has not been reported.

An alternative pathway for SeO_3^{-2} reduction in yeast has been described by Hseih and Ganther (1975). They have reported that yeast GSH reductase (E.C. 1.6.4.2), in the presence of GSH and NADPH, catalyzes the reduction of SeO_3^{-2} to Se^{-2}. Because at least some of the GSH reductase activity of plants is associated with chloroplasts (Jablonski and Anderson, 1978) and reduction of oxidized GSH (GSSG) is coupled to photosynthetic electron transport (Jablonski and Anderson, 1978), it is possible that SeO_3^{-2} could be reduced to Se^{-2} in chloroplasts in a light-dependent reaction involving GSH reductase. Ng and Anderson (1979) have illuminated intact pea chloroplasts in the presence of O-acetylserine (OAS) and incorporated SeO_3^{-2} and SO_3^{-2} into selenocysteine and cysteine, respectively. Sonicated chloroplasts catalyzed SeO_3^{-2} and SO_2^{-2} incorporation at ~3.9% and 32%, respectively, of the rates of intact chloroplasts. Addition of GSH and NADPH increased rates to ~91% and 98% of the intact rates, but SeO_3^{-2} incorporation under these conditions was essentially light independent. Ng and Anderson (1979) concluded that SeO_3^{-2} and SO_3^{-2} are reduced in chloroplasts by independent light-requiring mechanisms. They proposed that SeO_3^{-2} is reduced by light-coupled GSH reductase and that the Se^{-2} produced is incorporated into selenocysteine by cysteine synthase.

Ng and Anderson (1979) have proposed the mechanism shown in Figure 4-2 which provides one explanation for the light-dependent incorporation of SeO_3^{-2} into selenocysteine in chloroplasts. Although OAS has not yet been reported in chloroplasts, all the other conditions required for this pathway occur in this organelle. They include light-coupled re-

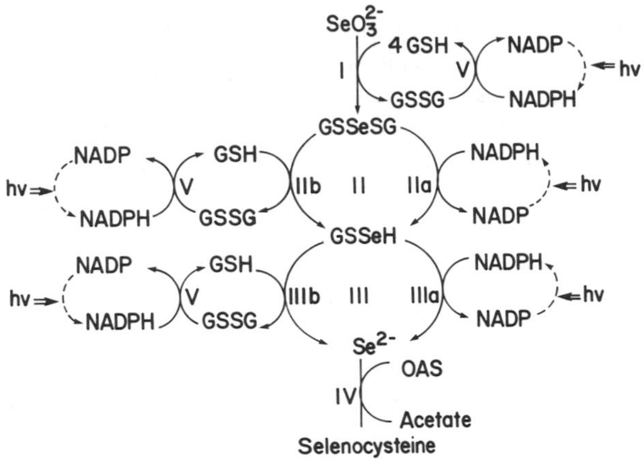

SeO_3^{2-}

4 GSH — NADP ⇐ hv

I — GSSG — V — NADPH

GSSeSG

NADP — GSH — NADPH ⇐ hv

hv ⇒ V — IIb — II — IIa

NADPH — GSSG — NADP

GSSeH

NADP — GSH — NADPH ⇐ hv

hv ⇒ V — IIIb — III — IIIa

NADPH — GSSG — NADP

Se^{2-}

OAS

IV

Acetate

Selenocysteine

Figure 4-2. Proposed pathway for the incorporation of SeO_3^{2-} into selenocysteine by il-luminated chloroplasts. Permission *Phytochemistry* 18:573–580, B. H. Ng, and J. W. Anderson. Light-dependent incorporation of selenite and sulphite into selenocysteine and cysteine by isolated pea chloroplasts, 1979, Pergamon Press, Ltd.

duction of NADP, an internal pool of GSH (estimated at 3.5 mM) (Foyer and Halliwell, 1976), cysteine synthase (Frankhauser *et al.*, 1976; Ng and Anderson, 1978a), and GSH reductase (Jablonski and Anderson, 1978). This pathway requires that the internal pool of GSSG produced in reaction I (Figure 4-2) be reduced to GSH. The light-dependent GSH-reductase (Jablonski and Anderson, 1978) could accomplish this. The latter reaction, together with the requirement for NADPH for reaction II and III, might explain the requirement for light. However, if reactions I–III proceed nonenzymically using GSH as reductant, light would be necessary for the reduction of GSSG by GSH reductase (Jablonski and Anderson, 1978). Ng and Anderson (1979) have also found that none of the component reactions, including the light reaction, are inhibited by KCN, but GSH reductase is inhibited by $Zn(Cl)_2$. These inhibition studies are consistent with the proposed scheme in Figure 4-2.

4.2.3 Animals

4.2.3.1 Competitive Sulfur Selenium Phenomena

Only a few studies have been undertaken on the transport of selenium compounds in mammalian systems. Selenomethionine has been shown by McConnell and Cho (1965) to be actively transported by everted hamster intestinal sacs, but selenite and selenocystine were not. Methionine blocked the transport of selenomethionine, but sulfite and cystine did not block the respective transport of selenite or selenocysteine. High-protein diets have the ability to decrease selenium toxicity. This might be related to such a methionine–selenomethionine transport antagonism at the intestinal level (McConnell and Cho, 1967). Since sulfate increases urinary excretion rather than decreases intestinal absorption of selenium, a similar mechanism apparently does seem not operate in the protection of sulfate against selenate toxicity (Ganther and Baumann, 1962a). Results of the above studies show that analogs of sulfur compounds can compete effectively for sulfur transport mechanisms in animal systems.

   White and Somers (1977) have shown that a reduction in dietary concentration from 2 to 0.7 g/kg resulted in increased wool and plasma selenium concentrations. Likewise, Pope *et al.* (1979) have also demonstrated that decreased dietary sulfate resulted in decreased urinary excretion and increased apparent retention of ^{75}Se in pair fed sheep given radioactive selenate. In the experiment of White and Somers (1977) feed intake was not significantly influenced by sulfur treatment. White (1980) has studied the effect of a sulfur deficiency on the metabolism of selenium and sulfur. The sheep were fed high-sulfur (2 g S/kg) or low-sulfur (0.5 g S/kg) diets for two periods of 35 days each, and received selenium as selenomethionine at dietary concentrations 0.02, 0.06, 0.09, and 0.67 mg Se/kg. Sheep fed the low-sulfur diet had reduced feed intake, reduced nitrogen, sulfur, and selenium balance, but elevated plasma and wool selenium concentrations. The selenium concentrations in organs and tissues of the animals when slaughtered paralleled the selenium intake of the animal. The renal cortex contained the highest concentration and bone the lowest. The effect of the 0.5 g S/kg diet on feed intake is in contrast with the results from the experiment of White and Somers (1977). This difference in feed intake may have been responsible for many of the effects on selenium metabolism observed in this experiment. However, once the feed intake effects are accounted for, the implications for sulfur–selenium interactions remain as before, i.e., more selenium is incorporated into wool and plasma protein when dietary sulfur is limiting than when it is not. In contrast Acuff and Smith (1981) have found that in rats fed 15% of casein-based diets containing varying levels (0.0002%, 0.02%, and 0.42%) of inorganic sulfate and given a pulse dose of $Na^{75}SeO_4$ that there was a 42% reduction in carcass ^{75}Se in those rats fed the diet containing 0.0002% inorganic sulfate compared with rats fed the diet containing 0.02% of inorganic sulfate. There was a further 22% decrease in the carcass ^{75}Se in those rats fed the diet containing the high level of inorganic sulfate (0.42%) compared with those fed the 0.02% level of inorganic sulfate.

Further evidence for the biological interaction between these two elements is supported by work with several species of animals, which indicates that sulfur does alter the metabolism of selenium. Muth *et al.* (1961) have reported that the addition of sulfur as Na_2SO_4 to the diet decreased the effectiveness of dietary selenium as Na_2SeO_3. Schubert *et al.* (1961) observed an increase in the incidence of white muscle disease (WMD) in sheep grazed on alfalfa after the field was treated with gypsum. Analysis showed an increase in the total sulfur of the plants, largely as sulfate. Whanger *et al.* (1969) have conducted two trials with sheep to study the effect of sulfur on the incidence of WMD. Neither methionine nor sulfate increased the incidence of WMD in either trial, but sulfate significantly increased the number of lambs with degenerative lesions of the heart. However, in another experiment, sulfate had a slight effect on the metabolic fate of ^{75}Se-selenate in the gestating or lactating ewe (Paulson *et al.*, 1966).

4.2.3.2 Reductive vs. Oxidative Pathways

Diplock *et al.* (1973) and Caygill *et al.* (1971) administered radioselenite, usually orally, to rats under different nutritional conditions. After the liver subcellular organelle fractions were prepared, the oxidation state of the radioselenium present was characterized by the following criteria: acid-volatile selenium, thought to be selenide; Zn + HCl-reducible selenium, assumed to be selenite; and the residual nonvolatile selenium, called selenate. These designations may not be entirely correct as these investigators leave the reader with the impression that selenate had indeed been formed *in vivo* from selenite. Because this process is unlikely thermodynamically, the nonvolatile fraction of selenium more likely represents organic selenium derivatives that were not easily cleaved by the Zn + HCl treatment. Rhead *et al.* (1974) reported that selenate, as well as selenite, can be reduced to H_2Se by treatment with zinc and HCl. About 36% and 43% of the radioselenium present in the liver microsomes and mitochondria, respectively, was in the form of selenide (Diplock *et al.*, 1973; Caygill *et al.*, 1971). The percentage of selenide present was dependent upon the vitamin E status of the animals. Microsomes and mitochondria isolated from vitamin E-deficient rats contained only 30% and 26% of the selenium, as selenide. If antioxidants were omitted from the homogenization medium (*d*-tocopherol and mercaptoethanol), additional decreases were observed in the percentage of selenide present. Under these conditions microsomes and mitochondria from vitamin E-deficient rats contained only 17% and 11% selenide (Diplock and Lucy, 1973).

Diplock and Lucy (1973) have interpreted these results to be consistent with their hypothesis that selenide may be the biologically active form of selenium in the active site of certain nonheme iron proteins. These workers first showed that a portion of the selenium in the liver organelles of rats given small doses of the element behaved as acid-labile protein-bound selenide. The proportion of selenium present as selenide was greater when vitamin E was included in the diet than when it was not included. Zonal centrifugation techniques indicated that the acid-labile selenide was mainly associated with the mitochondria and the smooth endoplasmic reticulum (Caygill *et al.*, 1971). In vitamin E-deficient rats this association was absent, but could be restored by refeeding the vitamin to deficient rats for five days. In addition, Caygill *et al.* (1973) showed that the large increase in selenide observed in the smooth reticulum of rats treated with phenobarbital was not seen in animals deficient in vitamin E (Caygill *et al.*, 1973). The initial rate and extent of microsomal aminopyrine demethylation was very depressed in rats deficient in vitamin E and selenium, but full restoration of activity occurred only if both nutrients were available together (Giasuddin *et al.*, 1975). The authors concluded that there may be a specific vitamin E-dependent role for selenium as selenide in the smooth reticulum. The selenide may form part of the active site of a nonheme iron-containing protein that functions between the cytochrome P450 and the flavoprotein in the NADPH-dependent electron transfer chain of rat liver microsomal fractions. The work of Levander *et al.* (1973) lends credibility to this concept. They have demonstrated that selenium was an effective catalyst for the reduction of cytochrome C by thiols.

In the Diplock–Lucy hypothesis the *in vivo* antioxidant effects of vitamin E would be mainly directed toward oxygen-labile selenide-containing proteins rather than toward polyunsaturated lipids. Experiments by Caygill and Diplock (1973) have suggested the existence of oxidant-labile nonheme iron in rat liver microsomes that is dependent on the presence of dietary vitamin E and selenium. However, the nutritional relationship between vitamin E and polyunsaturated lipids might be explained by the formation of specific tocophenol–phospholipid complexes that would be necessary for normal membrane stability and permeability. The ultimate proof of his idea must await the discovery of a selenoheme iron protein.

However, investigations on sulfur metabolism, which may be similar to the metabolism of selenium, indicate that the reduction of sulfite to sulfide does not occur in animal tissues. Reduced sulfur is generally acquired in the form of the thiol groups of amino acid, and their subsequent metabolic fate is oxidative. There could be a specific, unique mechanism which exists in animal tissues for the reduction of selenite administered orally to the selenide that had been detected in liver subcellular fractions. Therefore, we have a distinct difference in the manner by which selenium and sulfur are metabolized. Diplock *et al.* (1973) have essentially ruled out the possibility that the intestinal microorganisms were reponsible for the reduction of selenite to selenide. Work with different trapping agents for the volatile selenium liberated from tissues by acid treatment established that this material was not dimethyl selenide, but was probably hydrogen selenide. The selenide found in rat liver subcelluar organelles might have been an artifact formed *in vitro* as a decomposition product of selenotrisulfide generated by reaction between selenite and the mercaptoethanol which was added to the homogenization medium. This possibility seems to be eliminated by studies that showed a similar proportion of the selenium was found to be present in the selenide whether or not the thiol was added to the solution used for homogenization (Diplock *et al.*, 1973). The rat livers, however, were first chilled in a solution that contained mercaptoethanol even if later homogenizations were carried out in a medium devoid of the thiol. It is possible that some thiol diffused into the liver during the chilling process, which means that an artificial production of selenide cannot be discounted completely.

4.2.3.3 Amino Acid Metabolism

4.2.3.3.1 Biosynthesis. An early investigation concerned with the presence of selenium in the tissues of animals poisoned with the element established that most of the selenium was associated with protein, but the precise form of selenium in these animal proteins was not established (Smith *et al.*, 1938). Later, McConnell and Wabnitz (1957) used a combination of radioactive tracer and chromatographic techniques to try to characterize the selenium present in tissue proteins. These investigators found radioactivity in those areas that would correspond to cystine or selenocystine and methionine or selenomethionine. The methodology employed by McConnell to prepare tissue protein hydrolysates has been questioned because both selenocystine (Huber and Criddle, 1967b) and selenomethionine (Shepherd and Huber, 1969) are known to be destroyed by the techniques utilized. Another interpretation of McConnell's results has been offered by Schwarz and Sweeny (1964). They found that the selenite could bind to a variety of sulfur compounds *in vitro* to give products with chromatographic properties similar to the parent sulfur compound. Therefore, one could question the use of chromatographic criteria as the sole means of characterizing selenium compounds. Whether or not selenocystine or selenomethionine is synthesized *in vivo* from sodium selenite in mammals has been studied by Cummins and Martin (1967). These investigators feed high levels of selenium as radioselenite to a rabbit for 35 days. In addition, the day before sacrifice, the animal received orally and by injection large doses of radioactive selenite. The liver was homogenized and after the nuclear fraction was removed, the supernate was dialyzed at pH 11 for a period of two weeks. After this drastic treatment, the purified hepatic protein was then subjected to a mild enzymatic hydrolysis. After ion exchange chromatography, no radioactivity was found in those fractions that would correspond to the elution volumes of the selenium amino acids. Similar ion exchange chromatography was also used to characterize the urinary selenium metabolites of a rabbit which had been injected 24 hr previously with radioselenite. Similar radioactive peaks could have been obtained merely by adding radioselenite to a chemically defined mixture of sulfur compounds or to a normal urine sample *in vitro*. Because the presence of selenocystine and selenomethionine could have been inferred by the appearance of two radioselenium peaks that eluted at the appropriate volumes, the results of this experiment demonstrate the danger of relying exclusively on chromatographic criteria for identification of selenium compounds. Cummins and Martin (1967) concluded from their experiments that there was no biosynthetic pathway by which selenium as selenite could replace sulfur in cystine or methionine.

Godwin and Fuss (1972) have claimed that a small proportion of a dose of radioselenite given orally to rabbits was possibly converted to a selenocystine. Fuss and Godwin (1976) have also administered $Na_2{}^{75}SeO_3$ to lambs. Again, small but significant amounts of selenium were incorporated as seleno-amino acids into the proteins of liver, kidney, and pancreas, as well as into the proteins of milk and plasma. In a similar experiment with ewes, Fuss and Godwin (1976) identified both selenomethionine and selenocystine chromatographically in enzyme digests of defatted liver and kidney. It was not possible to say whether this conversion was of any physiologic significance. One might expect to find various selenide derivatives in the tissues of animals shortly after they had been given relatively large doses of selenite as a result of the nonspecific reaction of selenite with sulfhydryl groups. Millar (1972) has reported a transient appearance of radioselenium in several rat liver proteins. The rats had been given submicrogram quantities of the element, but after two weeks only one or two different proteins contained radioactivity.

4.2.3.3.2 Substitution. A close relationship between the metabolism of selenomethionine and methionine has been observed in a variety of mammalian systems. Selenomethionine could inhibit the transport of methionine by hamster intestine and vice versa (McConnell and Cho, 1965). Selenomethionine was also incorporated into polypeptides in an *in vitro* rat liver protein-synthesizing system (McConnell and Hoffman, 1972). Although this may not be representative of the normal physiologic state, McConnell *et al.* (1974) found that there are certain parallels in the metabolism of selenium and sulfur amino acids in the skin of zinc-deficient rats.

Owing to this apparent metabolic similarity between methionine and selenomethionine, Yousef and Johnson (1970) postulated that the total body turnover rate of radioactive selenomethionine might be a convenient index for protein metabolism during growth and aging in rats. Said and Hegsted (1970) reported that this technique was of no value in assessing overall changes in protein metabolism in animals fed different amounts of protein or methionine. They found changes in the body radioactivity were not affected by changes in these dietary variables but rather were greatly affected by the selenium content of the diet. Additional evidence for the lack of metabolic substitution of selenomethionine and methionine derives from the work of Millar and Shepphard (1973), who injected rats with a mixture of (^{35}S) methionine and (^{75}Se) selenomethionine and found that the distribution of ^{35}S and ^{75}Se in the liver proteins and kidney supernatant fractions was different after gel filtration. Differences were especially noticed in the proteins of intermediate molecular weight. Millar *et al.* (1973) also compared the metabolism of intravenously injected sodium selenite, sodium selenate, and selenomethionine in rats. The ^{75}Se distribution patterns were so similar that Millar *et al.* (1973) concluded that some breakdown of the selenomethionine takes place, and that selenium is released in a form which is metabolized in a way that cannot be distinguished from that of selenite or selenate.

Thus, the hypothesis that selenomethionine necessarily follows the same course as methionine in protein metabolism appears to be false, but needs further study. The observations of Schwarz (1961) may still hold. He believed that selenium when given in large amounts may travel with sulfur in its metabolic pathways, but when supplied at small physiologic levels available evidence suggests that it follow pathways of its own. The metabolic differences between methionine and selenomethionine also apply for cystine versus selenocystine. Huber and Criddle (1967b) noted that the chemical differences between the latter two amino acids are so great that if one substituted for the other, there would be vast alterations in the properties of any proteins which might contain them. Olson and Palmer (1976) have also studied the incorporation of $^{75}SeO_3^{2-}$ into rat liver and kidney tissue. Pronase digests of acetone powders of rat liver and kidney administered $^{75}SeO_3^{2-}$ were subjected to fractionation by cation exchange chromatography. Very little, if any, selenocystine was found in the digests. However, good evidence was obtained for the occurrence of 2,7-diamino-4-thia-5-selenaoctanedioic acid. Olson and Palmer (1976) suggested that the selenocysteine portion of this compound was formed by the reduction of the selenite to selenide with its subsequent incorporation into the amino acid by the action of serine hydrolase (E.C. 4.2.1.22). These data may explain why some have not found selenium present in tissues as selenocystine. No selenomethionine was found under the conditions of this study.

   4.2.3.3.3 Forms of Selenium in Tissue Proteins. If one assumed that (a) the endogenous biosynthesis of selenomethionine or selenocystine from selenite or other inorganic selenium forms by monogastric mammals is nonexistent or only a minor pathway of selenium metabolism and (b) the selenomethionine supplied exogenously obtained in the diet from wheat (Olson *et al.,* 1970) consists largely of catabolism to the selenite, what are some of the possibilities for the forms of selenium in tissue proteins? One possibility was postulated by Painter (1941):

$$4RSH + H_2SeO_3 \rightarrow RSSeSR + RSSR + 3H_2O \qquad (4\text{-}5)$$

Ganther (1968) has characterized several compounds of the RSSeSR family (selenotrisulfides), after reacting selenious acid nonenzymatically with cysteine, 2-mercaptoethanol, or coenzyme A. The possibility of incorporating selenium into proteins through such a reaction was first investigated by Holker and Speakman (1958). They generated cysteine residues in wool by using thioglycolic acid to reduce about 65% of the disulfide bridges. After the washed wool was treated with selenious acid, they observed a loss of sulfhydryl groups, an increase in dry weight, and recovery of resistance to stretching in the treated wool. The restrengthening seems to result from the formation of cross-linkages containing selenium, rather than from an oxidation of cysteine residues to the disulfides, because no reduction of selenious acid to elemental selenium occurred during the treatment. Liberation of elemental selenium during acid hydrolysis in an amount nearly equal to the increase in dry weight of the treated wool (2.11%) indicated the presence of large amounts of selenium. This increase in dry weight corresponded to an uptake of 26.7 mM of Se per 100 g of wool. The loss of sulfhydryl amounted to only 46 mM/100 g, or 1.72 SH per Se. Holker and Speakman (1958) concluded that each atom of selenium combined with two cysteine residues. The data for the weight gain favored S–Se–S linkages over S–Se–S. However, the stoichiometry of selenious acid reduction to S–Se–S requires not two but four thiols to be oxidized per selenium. The investigators based their conclusions on data that do not account for the necessary reducing equivalents. The reducing equivalents may have been obtained by the oxidation of some other protein groups in addition to the cysteine residues. This discrepancy may reflect uncertainties in the experimental data, such as incomplete titration of sterically hindered SH groups in the protein that might have been still accessible to selenious acid.

The selenotrisulfide derivative of glutathione (GSSeSG) reacted further to form the glutathione selenopersulfide (Ganther, 1971) under physiologic conditions of pH and reactant concentrations:

$$GSSeSG + GSH \rightarrow GSSeH + GSSG \qquad (4\text{-}6)$$

Selenopersulfides may play a side role in the biological function of selenium (see Section 4.3.1 as an electron transfer catalyst). Ganther and Corcoran (1969) have prepared selenopersulfides by cross-linking reduced pancreatic ribonuclease with selenite *in vitro* to form an intramolecular R–S–Se–S–R linkage in place of a disulfide bridge. Jenkins and Hidiroglou (1971) have produced circumstantial evidence for the *in vivo* incorporation of selenium into proteins by selenotrisulfide. They found a good correlation between the cysteine content of several proteins and their ability for selenium uptake.

Even though selenotrisulfide formation provides a reasonable explanation for the initial binding of selenite by tissue proteins, the nature of the binding of selenite changes with time in a manner that is not understood (Millar, 1972). About 70% of the serum protein radioactivity taken from chicks 4 hr after dosing with radioactive selenite could be released either by alkali treatment, or by reduction with thiols or sulfitolysis. However, four days after radioactive selenite administration, much less ^{75}Se was removed by reduction or sulfitolysis, whereas the alkali treatment was equally effective. The average half-life of thick serum proteins is too long to allow for this relatively rapid change in ^{75}Se. Therefore, seleno-amino acid formation was not considered a reasonable explanation for this occurrence. An influence on the strength of selenium binding of the amino acid residues adjacent to the selenium binding site was thought to be a reasonable explanation. The selenium from the more labile binding sites would be quickly lost at first, and then a greater proportion of the more resistant selenium-binding complexes would remain.

Another possibility concerning the nature of selenium in tissue proteins is that selenium might exist in an unsuspected or unique form. Ganther *et al.* (1974) have worked with glutathione peroxidase and have suggested that the selenium present at the active site may be present as a seleninic or selenenic acid derivative. This report seems to indicate that the metabolism of physiological amounts of selenium is unrelated to the metabolism of sulfur.

4.2.3.4 Methylated Metabolites

4.2.3.4.1 Volatile Compounds. Hofmeister (1894) first suggested that animals given inorganic selenium salts can produce methylated selenium compounds which are eliminated from the lungs. The formation of these volatile metabolites seems important only when the animals are given subacute doses of the element (Olson et al., 1963). The volatile selenium has been characterized as dimethyl selenide (McConnell and Portman, 1952). Work using gas chromatographic techniques suggests that some dimethyl diselenide may also be formed (Vlasakova et al., 1972). Since dimethyl selenide is about 500 times less toxic than selenite (McConnell and Portman, 1952) (the LD_{50} of dimethyl selenide in the rat is 1.6 g of Se/kg (Frank and Moxon, 1936), the methylation of selenium is usually regarded as a highly effective mechanism for detoxifying the element. However, dimethyl selenide despite its lack of toxicity does enhance the toxicity of mercury of mercury (Parizek et al., 1971) and arsenic (Obermeyer et al., 1971) through a synergitic mechanism. McConnell and Roth (1966) have found that the amount of volatile selenium formed depends on the form of selenium given and can be increased by certain dietary variables such as high protein level, supplemental methionine, previous exposure to selenium, unknown "volatization factors" present in some crude diets, and the genetic strain of the animal (Ganther et al., 1966). These unidentified factors in crude diets fed to rats prior to an injection of labeled selenite cause a threefold increase in the formation of volatile selenium (Ganther and Baumann, 1962b; Ganther et al., 1966). The form of selenium is important since neither sodium selenate nor selenomethionine is converted to volatile selenium as readily as sodium selenite (Hirooka and Galambos, 1966; McConnell and Roth, 1966). Arsenic, mercury, cadmium and thallium can block the formation of volatile selenium, but lead had no such inhibitory effect (Levander and Argnett, 1969).

Rosenfeld and Brath (1948) first suggested that the biosynthesis of volatile selenium might be an enzymatic process. They showed that liver minces that had been autoclaved could not convert selenium into volatile selenium, but that fresh bovine liver minces could. Ganther (1966) first studied in detail the enzymatic synthesis of volatile selenium from sodium selenite in mouse liver extracts. He observed that the 9000-g liver fraction was more active than either the 165,000-g fraction from the supernatant or the washed microsomes alone. The probable methyl donor was thought to be 5-adenosyl-L-methionine. Other important cofactors needed for optional activity included reduced nicotinamide adenine dinucleotide phosphate (NADPH), coenzyme A, adenosine 5'-triphosphate and magnesium. Arsenite and cadmium, which are dithiol blockers, were effective inhibitors of volatile selenium formation, but N-ethylmaleimide and p-mercuribenzoate, which are monothiol reagents, were much less effective.

Volatile selenium biosynthesis had a specific requirement for glutathione. Other thiols or dithiothreitol could not satisfy the glutathione requirement. With this information and the knowledge that glutathione can react with selenious acid to form the selenotrisulfide derivative of glutathione (Ganther, 1968),

$$4GSH + H_2SeO_3 \rightarrow GSSeSG + GSSG + 3H_2O \qquad (4\text{-}7)$$

Ganther (1971) postulated a reaction pathway whereby the selenotrisulfide would be reduced further by either NADPH via glutathione reductase or by an excess of glutathione to yield glutathione selenopersulfide:

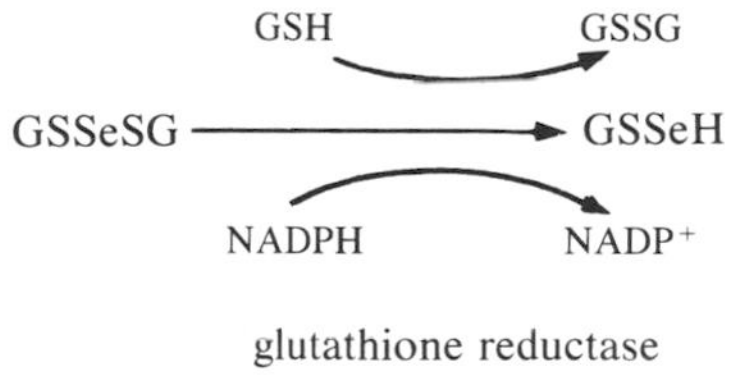

$$(4\text{-}8)$$

glutathione reductase

The selenopersulfide gives rise to hydrogen selenide, which is methylated by S-adenosylmethionine to dimethyl selenide. This pathway for the biosynthesis of dimethyl selenide is quite different from that proposed by Challenger (1951) (discussed in Section 4.2.1.4). In the mechanism proposed by Ganther (1971) no methylation occurs until the selenium is first reduced to the $2-$ oxidation state, but in the mechanism of Challenger (1951) consists of alternate methylation and reduction steps. Since certain compounds that contain selenium in the $2-$ valence state are readily methylated by microsomal systems from rat liver (Bremer and Natori, 1960). In addition, the first step of Challenger's (1951) pathway of dimethyl selenide formation (methylation of the biselenite to yield methaneselenonic acid) an oxidation of the $4+$ to $6+$ state seems to be required. In mammalian systems this appears to be a highly unlikely process. There is the possibility, however, that the selenopersulfide could act as the methyl acceptor in Ganther's (1971) proposed pathway rather than the highly toxic hydrogen selenide (Diplock et al., 1973). In mammalian systems the overall concept of the reduction of selenium to the selenide level followed by methylation seems to be the most realistic explanation of the mechanism. Certainly, other mechanisms such as that of Challenger (1951) might apply to plants and microorganisms.

    4.2.3.4.2 Urinary Methylated Compounds. Identification of trimethylselenonium ion $(CH_3)_3Se^+$, a major urinary metabolite of selenite, has been reported, (Byard, 1969). The compound was isolated by absorption on Dowex 50, followed by elution and precipitation as the Reineckate derivative. Identification was based on the comparison with synthetic trimethyl selenonium ion using paper and ion exchange chromatography, nuclear magnetic resonance, infrared and mass spectrometry, and cocrystalization. This report constituted the first chemical characterization of any selenium metabolite in urine. When a variety of selenium sources were tested (Palmer *et al.*, 1970), trimethylselenonium ion was found to be a general excretory product of selenium ion metabolism in the rat. Byard and Baumann (1967) noted that the amount of trimethyl selenium ion excreted relative to a second unidentified metabolite increased with increasing selenite dosage. Palmer *et al.* (1969) found that trimethyl selenonium comprised about 40% of the urinary selenium (10% of the dose) when rats were injected with either subtoxic doses (0.8 mg Se/kg) of selenite or with microgram quantities. This percentage was increased by prefeeding the animals with nonradioactive selenium before the injection of labeled selenite. Palmer *et al.* (1969) found that mixing normal rat urine with selenite did not result in the formation of trimethyl selenonium ion, and Byard (1969) found that trimethyl selenonium ion could be detected in bladder urine indicating that trimethyl selenonium ion is not an artifact, and is a major metabolite of selenite. Trimethylselenonium ion may be a detoxification product by selenium just as in the case of dimethyl selenide. Because trimethylselenonium ion has little or no ability to prevent the liver necrosis due to selenium deficiency (Tsay *et al.*, 1970), this compound is also inactive biologically in a nutritional sense. However, like dimethyl selenide, trimethylselenonium ion exhibits a synergistic toxicity with arsenic (Obermeyer *et al.*, 1971). A product–precursor relationship seems to exist between dimethyl selenide and trimethylselenonium ion; however, this possibility cannot be stated with certainty. An appreciable amount of radioactivity was found in the urine of rats given small doses of dimethyl (^{75}Se) selenide (Parizek *et al.*, 1971). On the other hand, only very small amounts of radioactive selenium were excreted in the urine of rats given large doses of this compound (Obermeyer *et al.*, 1971).

The excretion of (^{75}Se) trimethylselenonium ion in rat urine injected with radioselenomethionine seems to be another difference in the metabolism of selenium and sulfur, as no trimethylsulfonium ion was found in the urine of rats injected with methionine (Palmer, 1973).

4.3 Comparative Biochemistry of Selenium and Sulfur

4.3.1 Selenopersulfide as an Electron Transfer Catalyst

Schwarz (1961a) had shown that weaning male rats lacking both selenium and vitamin E developed a massive hepatic necrosis within 3–4 weeks. Chernick *et al.* (1955) had previously noted that livers from selenium and vitamin E-deficient rats demonstrated a respiratory decline. During this time the prevailing hypothesis of the mode of action of vitamin E was that tocopherol acted on a fat-soluble antioxidant (Tappel, 1972; Green, 1972). Selenium was thought to exert its biological effect through its antioxidant properties (Tappel and Caldwell, 1967).

Hunter *et al.* (1963) have demonstrated that vitamin E could protect against the mitochondrial swelling induced by various chemical agents that initiated lipoperoxidation. If selenium acted as an antioxidant *in vivo,* the same type of mitochondrial protection should be observed. Levander *et al.* (1973a) also found that dietary vitamin E protected against the swelling of rat liver mitochondria induced by ferrous ion, ascorbate, and thiols, but dietary selenium had no such protective effect. Levander *et al.* (1973a; 1974b) found that selenium seemed to have a marked accelerating effect on the mitochondrial caused by sulfide or by certain thiols, such as cysteine. In addition, when liver mitochondria were prepared from rats deficient in selenium (selenium-deficient mitochondria) *in vitro* added selenite was found to enhance cysteine-induced swelling. Other selenium compounds could also accelerate the swelling of selenium-deficient mitochondria caused by either cysteine or glutathione (GSH). The most active form of selenium was selenite, followed by selenocystine and then selenate or selenomethionine. Selenium compounds were specific because a wide variety of other oxyanions or cations could not promote GSH-induced *in vitro* mitochondrial swelling. Dithiol agents such as Hg^{2+}, Cd^{2+}, arsenite, iodoacetate, and *N*-ethylmaleimide inhibited the swelling of selenium-deficient mitochondria caused by the addition of GSH plus selenite to the incubation medium. Dinitrophenol and dicoumarol, which are uncoupling agents of oxidative phosphorylation, had little or no effect on swelling induced by GSH plus selenite. Amytal and antimycin A, which are respiratory inhibitors, only partially blocked the swelling caused by GSH and selenite. Cyanide completely blocked such swelling, which indicates that the swelling might be at least partially mediated at the cytochrome level. Selenite has also been shown to be an effective catalyst for the reduction of cytochrome C by thiols in model chemical systems (Levander *et al.,* 1973b). In this latter system other forms of selenium were catalytically active: they included selenocystine, and to a lesser extent selenate, but selenomethionine was inactive. In this model system Cd^{2+} and Hg^{2+} were fully inhibitory, but arsenite had little inhibitory effect. Cyanide also caused a 50% inhibition of the selenite-catalyzed reduction of cytochrome C by GSH. Selenocyanate was found to be a relatively poor catalyst for cytochrome reduction. These results were interpreted by a series of reactions in which the selenite along with GSH formed the selenopersulfide derivative of glutathione (Ganther,

1971). The active species in bringing about reduction of the cytochrome is glutathione selenopersulfide:

$$GSSeSG + GS \rightleftarrows GSSe^- + GSSG \qquad (4\text{-}9)$$

$$2GSSe^- + 2cyt\ C^{3+} \rightleftarrows 2GSSe + 2cyt\ C^{2+} \qquad (4\text{-}10)$$

$$2GSSe \rightleftarrows GSSeSeSB \qquad (4\text{-}11)$$

$$GSSeSeSG + GS \leftrightarrows GSSeSG + GSSe^- \qquad (4\text{-}12)$$

The mechanism of the inhibitory effect of cyanide is thought to be through the destruction of the catalytically active selenotrisulfide and formation of the relatively inactive selenocyanate:

$$GSSeSG + CN^- \rightleftarrows GSSG + SeCN^- \qquad (4\text{-}13)$$

Rhead and Schrauzer have studied kinetically the selenium-catalyzed reduction of methylene blue by a variety of thiols. Under conditions whereby $RS^- > Se$, the selenopersulfide may undergo further reactions with RS to form a diselenide species:

$$RSSe + RSs\text{;-} \rightleftarrows Se^{2-} + RSSR \qquad (4\text{-}14)$$

A one- or two-electron process was involved when the Se^{2-} anion reduced the methylene blue. Rhead and Schrauzer (1974) also found that much like the results of Levander *et al.* (1973b), the selenium-catalyzed reduction of methylene blue could be inhibited by Hg^{2+} or Cd^{2+} and cyanide, but that arsenite was without an observable effect. This lack of effect by arsenite against the selenium-catalyzed reduction of either methylene blue or cytochrome C is sharply contrasted to the pronounced inhibitory effect of arsenite against the mitochondrial swelling caused by selenite and GSH. This difference suggests that the swelling phenomenon induced by selenite and GSH is not a direct process whereby the cytochrome is reduced. A dithiol group may serve instead as an intermediate electron carrier. In addition, Rhead and Schrauzer (1974) also showed selenium could catalyze the reduction of a variety of dyes other than methylene blue by thiols. This reduction rate correlated well with the reduction potential of the dyes. The observations that selenium can act as a catalyst for the reduction of central metal ions in cytochromes by thiols, plus the fact that selenium can also act as a catalyst for the transfer of electrons from thiols to a number of acceptors, suggest that a selenopersulfide may function *in vivo* as an electron transfer catalyst.

Levander and co-workers (1974b) have attempted to characterize the chemical form of selenium in rat liver mitochondria. They gave young rats a vitamin E-supplemented but selenium-deficient diet as well as physiologic amounts of radioselenium (0.1 ppm) in the drinking water for several weeks. After that time, liver mitochondria were prepared and fractionated, and isolation procedures for cytochrome C were applied. In addition to the report by Levander *et al.* (1973b) that selenium was a good catalyst for cytochrome C reduction, Whanger *et al.* (1973) had claimed the existence of a selenium-containing cytochrome. No appreciable quantities of radioselenium were found to be associated with the cytochrome C. However, 60% of the mitochondrial radioselenium was shown by chromatography to be associated with glutathione peroxidase (Chapter 1). It is possible that the selenium in the glutathione peroxidase mitochondria may be directly responsible for the enhanced thiol-induced swelling. Additional work is necessary on characterizing the forms of selenium in mitochondria.

Like cytochrome C selenium can also catalyze the reduction of methemoglobin by glutathione (Masukawa and Iwata, 1977). Selenite, selenate, and selenocystine catalyzed the reduction of methemoglobin by GSH, whereas selenomethione did not. Selenite also catalyzed the reduction of methemoglobin with cysteine or 2-mercaptoethylamine in place of GSH. The effect of selenite was completely inhibited by heavy metals and arsenite. These findings suggest that certain seleno compounds catalyze the reduction of methemoglobin by thiol compounds. Hemoglobin, like cytochrome C, is a hemoprotein and so the mechanism of the reduction of the two compounds by GSH catalyzed by selenium may well be similar.

Most of the selenium in erythrocytes is known to be in the form of glutathione peroxidase (Rotruck *et al.,* 1973), and this enzyme together with GSH protects against the oxidative damage of hemoglobin (Rotruck *et al.,* 1973). However, it is not probable that the peroxidase activity is involved in the reductive process of methemoglobin. In this regard, the findings of Masukawa and Iwata (1977) showed that the catalytic effect of selenite on the reduction of methemoglobin by GSH probably proceeded without the enzyme. Caygill *et al.* (1971) have suggested that selenium, particularly selenide, might have a role in the electron transfer functions associated with mitochondria and smooth endoplasmic reticulum. The fact that certain mitochondrial nonheme iron proteins, which serve an electron carrier function adjacent to the main electron transfer chain, contain sulfide led Caygill *et al.* (1971) to suggest that selenide might form a part of the active center of a class of nonheme iron selenide proteins. More work is needed to elucidate the possible role of selenium in an electron transport system.

4.3.2 Iron–Sulfur Proteins

Several investigators have demonstrated that selenium can be sub-stituted for the labile sulfur of certain nonheme iron proteins such as adrenodoxin (Orme-Johnson *et al.*, 1968), putidaredoxin (Tsibris *et al.*, 1968), and ferredoxin (Fee and Palmer, 1971). These homologs of selenium have physical properties that are similar to those of the native proteins and in some cases have biological activity. The selenium derivative of adrenal iron–sulfur protein was shown to possess biological activity, since the selenoprotein had 60% of the steroid–11β hydroxylase activity and 75% of the NADPH–cytochrome C reductase activity of the native protein (Mukai *et al.*, 1974). In the case of parsely ferredoxin, the selenium de-rivative was prepared by reconstituting the apoprotein with selenious acid and excess dithiothreitol under anaerobic conditions (Fee and Palmer, 1971). The selenious acid probably first reacted with dithiol to produce a compound of the RS–Se–SR type. Excess dithiol then may have reduced the selenotrisulfide to RS^- and Se^{2-} which was incorporated into the protein. The parsley ferredoxin which was selenium substituted had spec-troscopic properties similar to those of the native parsley ferrodoxin. The selenium-substituted protein had about 80% of the activity as the native material in regard to the ferredoxin-mediated reduction of cytochrome C by NADPH in the presence of ferredoxin–NADP reductase. Because biologically active selenium derivatives of some iron–sulfur proteins can be prepared chemically lends credence to the theory of Diplock and Lucy (1973) that a biochemically active form of selenium may be present as a selenide in the active site of certain uncharacterized nonheme iron pro-teins. Tsibris *et al.*, (1968) have substituted ^{80}Se for sulfur in putidaredoxin, an iron–sulfur protein from *Pseudomonas putida,* in order to make elec-tron spin resonance studies of the active center of the protein. After dialysis, the reconstituted selenoprotein contained approximately 2 mol each of iron and selenium per mole of protein, equivalent to the content of iron and sulfur in the native enzyme. The maxima of the visible ab-sorption spectrum of the Se protein was shifted about 20 mμ to longer wavelengths. The biological activity of the selenoprotein was almost equal to that of the sulfur protein. If the ^{80}Se was substituted, there was a marked change in the electron spin resonance spectrum. When the se-lenium was displaced by ^{32}S, the original ultraviolet absorption spectrum was restored. Similar results were found in a subsequent study comparing the hyperfine structure of the ESR spectra obtained when ^{77}Se and ^{80}Se were substituted for sulfur in both putidaredoxin and in the adrenal iron–sulfur proteins (Orme-Johnson *et al.*, 1968). The authors also ob-served that a red (Se^0) was formed if the selenoproteins were exposed to air in the absence of 2-mercaptoethanol, even at temperatures near freezing. Apparently the selenium in these proteins is presumably bound to iron in a manner which is analogous to the sulfur. However, the exact structure of the iron–sulfur complexes has not been established and must be studied further.

4.3.3 Sulfur Salts and Selenium Toxicity in Animals

Results of studies done previously with plants and microorganisms have shown that the toxicity of selenium as selenate is ameliorated by sulfate (Shrift, 1961). Bonhorst and Palmer (1957) have found that the mortality of rats injected with selenate was reduced by injecting magnesium sulfate first before the selenate. Halverson and Monty (1960) have reported that adding sodium sulfate or potassium sulfate to the diet at least in part overcame the growth depression of selenate at a level of 10 μg/g selenium. The effect of sulfate on distribution in tissue and the excretion of selenium was not studied.

Ganther and Baumann (1962a) and Halverson et al. (1962) have studied in greater detail the effect of sulfate on the metabolism of toxic levels of selenate. Ganther and Baumann (1962a) have tested the effect of dietary sulfite or sulfate on selenium toxicity using growth and the metabolism of radioactive selenium. Male weanling rats were fed a purified diet containing 12% casein, 78.7% sucrose, 5% corn oil, and 4% of a low-sulfate mineral mixture. A low-casein diet was used, since a high-protein diet can reduce selenium toxicity, an effect perhaps related to the sulfur-amino acid content. Selenium, at a level of 5 μg/g, was added either as sodium selenite or sodium selenate. The sulfur salts tested included potassium sulfate, sodium sulfate, and sodium sulfite in equimolar proportions. When these salts were added, they replaced an equal weight of sucrose in the diet.

Addition of the selenate reduced the weight gain from 136 to 61 g in a four-week period, while in the case of selenite the gain was reduced to 74 g. Addition of 1% sodium sulfate or 1% of the sodium and potassium sulfate mixture or an equivalent weight of sulfite increased the weight gain of the selenate fed groups to approximately 90 g. The growth of the groups fed selenite was also improved, but to a lesser extent. No effect against selenite was observed by sodium sulfite. Sodium arsenite (5 ppm) also improved growth slightly when added to the diets containing either form of selenium. Ganther and Baumann (1962b) have previously found that the injection of arsenite before the injection of a subtoxic amounts of sodium selenite reduced selenite toxicity by increasing its excretion in the gastrointestinal tract. Arsenite also has been shown to be partially effective against the toxic effects of dietary selenium added as selenite or seleniferous wheat (Dubois *et al.*, 1940).

Those groups fed diets containing selenium showed liver damage, with selenate causing the more severe damage. Arsenite and high sulfate yielded some protection. Spleen enlargement was also observed, but sulfate prevented the enlargement caused by selenate, and arsenic prevented the enlargement caused by either selenate or selenite. No protective action was observed with sulfite.

In the studies with radioactive selenate, feeding radioactive selenate together with sulfate increased the urinary excretion of radioactive selenium from 39% to 48% of the amount given, but did not change the gastrointestinal and fecal excretion, which was about 25% of the injected radioactivity (Ganther and Baumann, 1962a). Sulfate feeding caused small decreases in the selenium in blood, liver, kidney, and carcass. These decreases are consistent with increased urinary excretion of selenium.

After radioactive selenate was injected, the greatest increase in urinary excretion occurred in the groups that were previously fed sulfate and then injected with additional sulfate at the time of selenate injection. Urinary selenium increased under these conditions to up to 46% of the administered radioactivity. However, sulfate had little effect on the excretion of injected selenite. When sulfate was given, only slight increases of urinary selenium were observed, and gastrointestinal excretion decreased. The ineffectiveness of sulfate against injected selenite was consistent with the smaller sulfate effect in counteracting the growth depression caused by selenite. It is possible that the effect of sulfate might result from its diuretic properties, with the increased selenium excretion resulting from increased urine formation. On the other hand, the greater effectiveness of sulfate against selenate as against selenite could indicate a more specific role of sulfate as a selenate antagonist. It is also possible that the greater toxicity of selenite reflects the fact that selenite is a stronger oxidizing agent in addition to being a source of selenate (Shrift, 1961).

Halverson *et al.* (1962) have also studied the effect of dietary sulfur supplements on the toxicity of selenium in rats including the growth, liver storage of selenium, and excretion of selenium. The selenium concentration tested was 10 μg/g. This amount was to be added to the diet as potassium selenate, sodium selenite, and as seleniferous wheat. Sulfur compounds were tested and they included sodium sulfate, sodium sulfite, and sodium thiosulfate. A diet containing 12% casein, 3% corn oil, 2% salts, and 14% and 80.9% wheat or corn were fed to male rats weighing approximately 70–90 g. The sulfur compounds replaced an equal amount of grain. The rats were fed the diets for three weeks.

At a level of 10 ppm selenium as selenate, growth was reduced from 158 to 42 g. The type of grain in the basal diet also affected the toxicity of the selenate. Rats on the corn basal diet showed a greater growth depression and liver selenium accumulation. Feeding 2% sodium sulfate increased the weight gain to about 75 g with the corn diet and to 100 g with the wheat diet. Selenite was somewhat more toxic than the selenate. The most toxic diet included seleniferous wheat. The rats fed this diet gained only 15 g in three weeks. When 2% sodium sulfate was added to the diet, there was no effect on the weight gain. These results support the possibility that the form of selenium in seleniferous grains may be better absorbed. About half or more of the selenium in seleniferous grains is likely to be present in the organic form which should be absorbed more easily.

Among the sulfur compounds tested by Halverson et al. (1962), sodium sulfate was again the most effective against selenate, whereas none of the sulfur compounds were effective against selenite and even appeared to increase its toxicity. These results differ from the results of Ganther and Baumann (1962a), in which sulfate counteracted selenite to some extent, and sulfite did not increase selenite toxicity. At a dietary level of 2% sodium sulfate increased urinary selenium from approximately 50% of the ingested selenium to 69% but did not change the percentage of ingested selenate in the feces. Gastrointestinal or carcass selenium was not measured. All of the sulfur compounds unexpectedly decreased growth when added to the basal diet in the absence of selenite, with sodium sulfite the most toxic (Halverson et al., 1962). These results are in contrast to the study of Ganther and Baumann (1962), who found no growth reduction when 2% sodium sulfite was added to the basal diet. A possible explanation for the different growth effects of sulfite in these two experiments may be the composition of the mineral mixture in the basal diet.

Ganther and Baumann (1962a) used a mineral mixture which omitted alum and sodium fluoride and substituted oxides or chlorides for the corresponding sulfate salts in the original mixture. Halverson et al. (1962) fed salts at 14% and 2% of the diet. These mineral levels supplied by this salt mixture at a level of 2% would have been approximately one half the intake of the mineral mixture of Ganther and Baumann (1962a). Even though some additional minerals were contributed by the dietary wheat and corn, the mineral intake from these diets would have been less than that from the diets of Ganther and Baumann (1962a).

Despite these minor differences, the results of the two studies were very similar. Selenium was more toxic in the form of selenite than selenate. The compound most effective against selenate toxicity was sodium sulfate, but was must less effective against selenite. The results with the seleniferous wheat may have little practicability because of the ineffectiveness of sulfate against selenium provided by seleniferous wheat and because the nature of the selenium compounds in the wheat was not defined.

4.3.4 Other Selenium–Sulfur Interactions

4.3.4.1 Cartilage Interactions

Studies by Ehlig *et al.* (1967) on the incorporation of selenium-75 in various tissues of sheep have shown that cartilage is a site of appreciable selenium incorporation. Selenium in the form of selenite was administered to the animals in both of these studies. Selenium toxicity in animals is sometimes associated with limb malformations and growth disturbances (Wright and Mraz, 1965), which suggests that cartilage is involved secondary to altered chondrotin sulfate metabolism. Indeed, small amounts of selenium depress *in vitro* sulfate incorporation into the chondroitin sulfate or cartilage (Hilz and Lipmann, 1955). Campo *et al.* (1966) have isolated ^{75}Se- and ^{35}S-labeled protein polysaccharides from slides of bovine costal cartilege which had been incubated in media containing ^{75}SeO$_4^{2-}$ or ^{35}SO$_4^{2-}$. Analysis of the protein polysaccharides revealed that ^{75}Se was attached to the protein moiety. The chondroitin sulfate was labeled with ^{35}S but not with ^{75}Se. Only very small amounts of ^{75}Se were fixed to the protein polysaccharides of preboiled slices of cartilage or onto protein polysaccharide itself if these were incubated in the media containing ^{75}SeO$_4^{2-}$. These results suggest that the binding to protein of ^{75}Se derived from ^{75}SeO$_4^{-}$ is dependent upon cellular activity.

In another study, Campo *et al.* (1967) used both *in vitro* and *in vivo*, L-phenylalanine-^{14}C, and L-leucine-^{14}C to study their incorporation into the protein moieties of the protein polysaccharide and collagen of cartilage.Through the use of these radioactive compounds it was possible to study autoradiographically the turnover of the protein portion of the epiphyseal plate of the rat concurrently with a study of the turnover of the ^{35}S-labeled chondroitin sulfate. Because ^{75}S derived from ^{75}SeO$_4^{2-}$ is also fixed to the protein moieties of the protein polysaccharides of cartilage rather than to chondroitin sulfate (Campo *et al.*, 1966), Campo *et al.* (1967) studied autoradiographically the fate of ^{75}Se in rat epiphyseal cartilage and compared its disposition with that of ^{35}SO$_4^{-}$ and the ^{14}C-labeled amino acids. Protein polysaccharides were isolated from cartilage slices and it was found that the ^{75}Se was affixed to the protein moiety. The disposition of ^{75}Se in long bones resembles that seen after the injection of ^{14}C-labeled amino acids. The failure in this experiment to remove significant amounts of radioactivity of ^{75}Se from slices of bovine costal cartilage following digestion with hyaluronidase supported the observation that ^{75}Se is not incorporated into chondroitin sulfate (Campo et al., 1966).

Biological sulfation of chondroitin sulfate is effected by a two-step reaction in which "active sulfate," PAPS, is formed (Robbins and Lipmann, 1956). The first step in the reaction sequence is the formation of APS from ATP and sulfate ion in the presence of the enzyme, sulfate, adenylyltransferase (E.C. 2.7.7.4). PAPS (3'-phosphoadenosine-5-phosphosulfate) is formed from another mole of APS (adenosine-5'-phosphosulfate) and ATP catalyzed by adenylylsulfate kinase (E.C. 2.7.1.25). The sulfotransferase enzyme then transfers the sulfate group to an acceptor molecule. However, when selenate ion is substituted for sulfate ion in an "active sulfate" enzyme system obtained from bakers yeast, no PAP Se (3'-phosphoadenosine-5'-phosphoselenate) is detectable. The APSe (adenosine-5'-phosphoselenate) which is formed in this system hydrolyzes spontaneously, thereby precluding formation of PAP Se (Wilson and Bandurski, 1958). If these results were extended, it would not be expected that the selenium analog of chondroitin sulfate would be formed by the enzymatic mechanism. In one report when ^{75}SeO$_4^{2-}$ was injected into rabbits, ^{75}Se was found to be associated with the sulfomucopolysaccharides of the urine (Galambos and Green, 1964). However, this association may be due to protein also present in this fraction.

Cipera and Hidiroglou (1969) have made in vivo injections of three types of anions (^{75}SeO$_3^{2-}$, ^{75}SeO$_4^{2-}$, ^{35}SO$_4^{2-}$) in order to differentiate between the metabolic behavior of the selenium anions and that of sulfur anion in cartilage. Whereas sulfate was incorporated into glycosaminoglycans, the entry of both selenium ions into this fraction was almost nonexistent. On the other hand, protein and lipid fractions isolated from cartilage had higher levels of activity when selenium ions were injected than when sulfate was injected. In cartilage, the selenate activity disappeared at a faster rate than either sulfate or selenite. In blood sera the rate of activity disappearance was higher with both selenate and sulfate than with selenite.

The inability of cartilage to incorporate ^{75}SeO$_4^{2-}$ into chondroitin sulfate is in agreement with the reported failure to detect active selenate, PAP Se, in an "active sulfate" enzymic system to which selenate ions had been added. This lack of selenate incorporation into chondroitin sulfate is also consistent with the absence of choline selenate in extracts of the conidiospores of *A. niger* when the plant is grown on agar which contained $^{75}SeO_4^{2-}$. Under similar conditions the plant activates and transfers $^{35}SO_4^{2-}$ to form choline sulfate (Nissen and Benson, 1964).

Since urinary sulfated polysaccharides have been found to be bound to protein (DiFerrante and Rich, 1956) in the urine, the finding that ^{75}Se was fixed to the protein moiety of protein-polysaccharide is not surprising. The ^{75}Se found in the urine fraction after injection of $^{75}SeO_4^{2-}$ into rabbits (Galambos and Green, 1964) may also be linked to protein.

The sulfite ion is not directly activated by the "active sulfate" enzymic system. Therefore, it is not expected that the selenite ion would be activated and transferred to chondroitin sulfate. After administration of $^{75}SeO_3^{2-}$, no ^{75}Se was found in the chondroitin sulfate from cocks (Patrick *et al.*, 1964), and very small amounts were found in protein polysaccharides from the condyles of fetal chicks (Halverson *et al.*, 1964). But in view of the report of Campo *et al.* (1966), these small amounts also might have been associated only with the protein portion of the protein polysaccharide.

4.3.4.2 Rhodanese

Cannella *et al.* (1975) have found that rhodanese is able to use selenosulfate in place of thiosulfate producing selenocyanide as the final product. An intermediate seleno-containing enzyme is formed which is the analog of the sulfur-charged enzyme when thiosulfate is the substrate. The typical absorption at 330 nm detected in the active site of the selenoenzyme as a result of the persulfide formation (Finazzi Agro *et al.*, 1972) is shifted to 375 nm, indicating the formation of a perselenosulfide group (Cannella *et al.*, 1975). The selenorhodanese form may be the intermediate for the transfer of selenium from inorganic to organic compounds of biological interest. This interesting approach warrants further investigation.

References

Acuff, R. V., and Smith, J. T., 1981. Selenium absorption and dietary inorganic sulfate status in the rat, *Fed. Proc. Fed. Am. Soc. Exp. Biol.* 40:902.

Adams, C. A., and Rinne, R. W., 1969. Influence of age and sulfur metabolism on ATP sulfurylase activity in the soybean and a survey of selected species, *Plant Physiol.* 44:1241–1246.

Akagi, J. M., and Campbell, L. L., 1962. Studies on thermophilic sulfate-reducing bacteria. III. Adenosine triphosphate-sulfurylase of *Clostridium nigrificans* and *Desulforibrio desulfuricans, J. Bacteriol.* 84:1194–1201.

Asada, K., Tamura, G., and Bandurski, R. S., 1969. Methyl viologen-linked sulfite reductase from spinach leaves, *J. Biol. Chem.* 244:4904–4915.

Asher, C., Evans, C., and Johnson, C. M., 1967. Collection and partial characterization of volatile selenium compounds by *Medicago sativa L.*, *Aust. J. Biol. Sci.* 20:737–748.

Bailey, S., Bazinet, M., Driscoll, J., and McCarthy, A., 1961. The volatile sulfur components of cabbage, *J. Food Sci.* 26:163–170.

Beath, O., and Eppson, H., 1947. The form of selenium in some vegetation, *WHO Agr. Exp. Sta. Bul.,* No. 278.

Bird, M. L., Challenger, F., Charlton, P. T., and Smith, J. O., 1948. Studies on biological methylation. XI. The action of molds on inorganic and organic compounds of arsenic, *Biochem. J.* 43:78–83.

Bonhorst, C. W., and Palmer, I. S., 1957. Metabolic interactions of selenate, sulfate and phosphate, *J. Agr. Food Chem.* 5:931–933.

Bremer, J., and Natori, Y., 1960. Behavior of some selenium compounds in transmethylation, *Biochim. Biophys. Acta* 44:367–370.

Byard, J. L., 1969. Trimethyl selenide, a urinary metabolite of selenite, *Arch. Biochem. Biophys.* 130:556–560.

Byard, J. L., and Baumann, C. A., 1967. Selenium metabolites in the urine of rats given a subacute dose of selenite, *Fed. Proc. Fed. Am. Soc. Exp. Biol.* 26:476.

Campo, R. D., Wengert, Jr., P. A., Tourtellotte, C. D., and Kirsch, M. A., 1966. A comparative study of the fixation of ^{75}Se and ^{35}S onto protein-polysaccharides of bovine costal cartilage, *Biochim. Biophys. Acta* 124:101–108.

Campo, R. D., Tourtellotte, C. D., and Ledrick, J. W., 1967. Selenium-75: An autoradiographic study of its disposition in cartilage and bone, *Proc. Soc. Exp. Bio. Med* 125:512–515.

Cannella, C., Pecci, L., Finazzi Agro, A., Federici, G., Pensa, B., and Cavallini, D., 1975. Effect of sulfur binding on rhodanese fluorescence, *Eur. J. Biochem.* 55:285–289.

Caygill, C. P. J., and Diplock, A. T., 1973. The dependence on dietary selenium and vitamin E of oxidant-labile liver microsomal non-haem iron, *FEBS Lett* 33:172–176.

Caygill, C. P. J., Lucy, J. A., and Diplock, A. T., 1971. The effect of vitamin E on the intracellular distribution of the different oxidation states of selenium in rat liver, *Biochem. J.* 125:407–416.

Caygill, C. P. J., Diplock, A. T., and Jeffery, E. H., 1973. Studies on selenium incorporation into, the electron-transfer function of, liver microsomal fractions from normal and vitamin E-deficient rats given phenobarbitone, *Biochem. J.,* 136:851–858.

Challenger, F., 1955. Biological methylation, *Q. Rev. Chem. Soc.* 9:255–286.

Challenger, F., 1951. Biological methylation, *Adv. Enzymol.* 12:429–491.

Challenger, F., and North, H. E., 1934. The production of organometalloidal compounds by microorganisms. II. Dimethylselenide, *J. Chem. Soc. London,* 68–71.

Challenger, F., and Charlton, P. T., 1947. Studies of biological methylation. X. The fission of the mono- and disulfide links by molds, *J. Chem. Soc. London,* 424–429.

Chen, D. M., Nigam, S. N., and McConnell, W. B., 1970. Biosynthesis of Se-methylselenocysteine and S-methylcysteine in *Astragalus bisulcatus, Can. J. Biochem.* 48:1278–1283.

Chernick, S. S., Moe, J. G., Rodnan, G. P., and Schwarz, K., 1955. A metabolic lesion in dietary necrotic liver degeneration, *J. Biol. Chem.* 217:829–843.

Chow, C. M., Nigam, S. N., and McConnell, W. B., 1971. Biosynthesis of Se-methylselenocysteine and S-methylcysteine in *Astragalus Bisulcatus*: Effect of selenium and sulfur concentrations in the growth medium, *Phytochemistry* 10:2693–2698.

Cipera, J. D., and Hidinoglou, M., 1969. Comparative study of the metabolic fate of selenate, selenite, and sulfate ions in cartilage, *Can. J. Physiol. Pharmacol.* 47:591–595.

Coch, E. H., and Greene, R. C., 1971. The utilization of selenomethionine by *Escherichia coli*, *Biochim. Biophys. Acta* 230:223–236.

Cowie, D. B., and Cohen, G. N., 1957. Biosynthesis by *Escherichia coli* of active altered proteins containing selenium instead of sulfur, *Biochim. Biophys. Acta* 26:252–261.

Cummins, L. M., and Martin, J. L., 1967. Are selenocystine and selenomethionine synthesized *in vivo* from sodium selenite in mammals?, *Biochemistry* 6:3162–3168.

DiFerrante, N., and Rich, C., 1956. The mucopolysaccharide of normal human urine, *Clin. Chim. Acta* 1:519–524.

Diplock, A. T., and Lucy, J. A., 1973. The biochemical modes of action of vitamin E and selenium: A hypothesis, *FEBS Lett.* 29:205–210.

Diplock, A. T., Caygill, C. P. J., Jeffrey, E. H., and Thomas, C., 1973. The nature of the acid-volatile selenium in the liver of the male rat, *Biochem. J.* 134:283–293.

Doran, J. A., and Alexander, M., 1977a. Microbial formation of volatile selenium compounds in soil, *Soil Sci. Soc. Am. J.* 41:70–73.

Dubois, K. P., Moxon, A. L., and Olson, O. E., 1940. Further studies on the effectiveness of arsenic in preventing selenium poisoning, *J. Nutr.* 19:477–482.

Ehlig, C. F., Hogue, D. E., Allaway, W. H., and Hamm, D. J., 1967. Fate of Se from selenite or seleno-methionine with or without vitamin E in lambs, *J. Nutr.* 92:121–126.

Ellis, R. J., 1969. Sulfate activation in higher plants, *Planta* 88:34–42.

Evans, C., Asher, C., and Johnson, C. M., 1968. Isolation of dimethyl diselenide and other volatile selenium compounds from *Astragalus racemosus* (Pursh), *Aust. J. Biol. Sci.* 21:13–20.

Falcone, G., and Nickerson, W. J., 1960. Metabolisms of selenite and mechanism of inhibitory action of selenite on yeasts, *Giorn. Microbiol.* 8:129–150.

Fee, J. A., and Palmer, G., 1971. The properties of parsley ferredoxin and its selenium-containing homolog, *Biochem. Biophys. Acta* 245:175–195.

Fels, I. G., and Cheldelin, V. H., 1949a. Selenate inhibition studies. III. The role of sulfate in selenate toxicity in yeast, *Arch. Biochem. Biophys.* 22:402–405.

Finazzi Agro, A., Federici, G., Giovagnoli, C., Cannella, C., and Cavallini, D., 1972. Effect of sulfur binding on rhodanese fluorescence, *Eur. J. Biochem.* 28:89–93.

Flemming, R. W., and Alexander, M., 1972. Dimethylselenide and dimethyltelluride formation by a strain of *Penicillium*, *Appl. Microbiol.* 24:424–429.

Foyer, C. H. and Halliwell, B., 1976. The presence of glutathione and glutathione reductase in chloroplasts: A proposed role in ascorbic acid metabolism, *Planta*, 133:21–25.

Francis, A. J., Duxbury, J. M., and Alexander, M., 1974. Evolution of dimethylselenide from soils. *Appl. Microbiol.* 28:248–250.

Franke, K. W. and Moxon, A. L., 1936. A comparison of the minimum fatal doses of selenium, tellurium, arsenic and vanadium, *J. Pharmacol. Exptl. Therap.* 58:454–459.

Fuss, C. N., and Godwin, K. O., 1975. A comparison of the uptake of {^{75}Se} selenite, {^{75}Se} selenomethionine and {^{35}S} methionine in tissues of ewes and lambs, *Aust. J. Biol. Sci.* 28:239–249.

Galambos, J. T., and Green, I., 1964. Parallel labelling on nondialyzable components of rabbit urine following ^{75}SeO$_4$ and ^{35}SO$_3$ injections, *Biochim. Biophys. Acta* 83:204–208.

Ganther, H. E., 1966. Enzymic synthesis of dimethyl selenide from sodium selenite in mouse liver extracts, *Biochemistry* 5:1089–1098.

Ganther, H. E., 1968. Selenotrisulfides. Formation by the reaction of thiols with selenious acid, *Biochemistry* 7:2898–2905.

Ganther, H. E., 1971. Reduction of the selenotrisulfide derivative of glutathione to a persulfide analog by glutathione reductase, *Biochemistry* 10:4089–4098.

Ganther, H. E., and Baumann, C. A., 1962. Selenium metabolism. II. Modifying effects of sulfate, *J. Nutr.* 77:408–414.

Ganther, H. E., and Concoran, C., 1969. Selenotrisulfides. II. Cross-linking of reduced pancreatic ribonuclease with selenium, *Biochemistry* 8:2557–2563.

Ganther, H. E., Levander, O. A., and Baumann, C. A., 1966. Dietary control of selenium volatilization in the rat, *J. Nutr.*, 88:55–60.

Ganther, H. E., Oh, S. H., Chitharanjan, D., and Hoekstra, W. G., 1974. Studies on selenium in glutathione peroxidase, *Fed. Proc. Fed. Soc. Am. Exp. Biol.* 33:694.

Giasuddin, A. S. M., Caygill, C. P. J., Diplock, A. T., and Jeffrey, E. H., 1975. The dependence on vitamin E and selenium of drug dimethylation in rat liver microsomal fractions, *Biochem. J.* 146:339–350.

Gissel-Nielsen, G., 1970. Loss of selenium in drying and storage of agronomic plant species, *Plant Soil* 32:242–245.

Godwin, K. O., and Fuss, C. N., 1972. The entry of selenium into rabbit protein following the administration of $Na_2{}^{75}SeO_3$, *Aust. J. Biol. Sci.* 25:865–871.

Green, J., 1972. Vitamin E and the biological antioxidant theory, *Ann. N.Y. Acad. Sci.* 203:29–44.

Hahn, G. A., and Brown, J. W., 1967. Properties of a methionyl-t RNS synthetase from *Sarcina lutea, Biochim. Biophys. Acta* 146:264–271.

Halverson, A. W., and Monty, K. J., 1960. An effect of dietary sulfate on selenium poisoning in the rat, *J. Nutr.* 70:100–102.

Halverson, A. W., Guss, P. L., and Olson, O. E., 1962. Effect of sulfur salts on selenium poisoning in the rat, *J. Nutr.* 77:459–464.

Halverson, A. W., Hills, C. L., and Whitehead, E. I., 1964. Studies on selenium toxicity and chondroitin sulfate and taurine biosynthesis in the chick embryo, *Arch. Biochem. Biophys.* 107:88–91.

Hilz, H., and Lipmann, F., 1955. the enzymatic activation of sulfate, *Proc. Natl. Acad. Sci.* 41:880–890.

Hirooka, T., and Galambos, J. T., 1966. Selenium metabolism. I. Respiratory excretion, *Biochim. Biophys. Acta* 130:313–320.

Hoffman, J. L., McConnell, K. P., and Carpenter, D. R., 1969. Aminoacylation of *E. coli* methionine tRNA by selenomethionine, *Fed. Proc. Fed. Soc. Am. Exp. Biol.* 28:860.

Hofmeister, F., 1894. Ueber methylirung im theirkoerper, *Arch. Exp. Pathol. Pharmakol.* 33:198–215.

Holker, J. R., and Speakman, J. B., 1958. Action of selenium dioxide on wool, *J. Appl. Chem.* 8:1–3.

Hsieh, H. S., and Ganther, H. E., 1975. Acid-volatile selenium formation catalyzed by glutathione reductase, *Biochemistry* 14:1632–1636.

Huber, R. E., and Criddle, R. S., 1967. The isolation and properties of beta-galactosidase from Escherichia coli grown in sodium selenate, *Biochim. Biophys. Acta* 141:587–599.

Huber, R. E., Segel, I. H., and Criddle, R. S., 1967. Growth of *Escherichia coli* on selenate, *Biochim. Biophys. Acta* 141:573–586.

Hunter, F. E., Jr., Gebicki, J. M., Hoffsten, P. E., Weinstein, J., and Scott, A., 1963. Swelling and lysis of rat liver mitochondria induced by ferrous ions, *J. Biol. Chem.* 238:828–835.

Hurd-Karrer, A. M., 1937. Comparative toxicity of selenates and selenites to wheat, *Am. J. Bot.* 24:720–728.

Hurd-Karrer, A. M., 1938. Relation of sulphate to selenium absorption by plants, *Amer. J. Bot.* 25:666–675.

Jablonski, P. P., and Anderson, J. W., 1978. Light-dependent reduction of oxidized glutathione by ruptured chloroplasts, *Plant Physiol.* 61:221–225.

Jenkins, K. J., and Hidiroglow, M., 1971. Comparative uptake of selenium by low cystine and high cystine proteins, *Can J. Biochem.* 49:468–472.

Kemp, J. D., Atkinson, D. E., Ehrer, A., and Lazzarini, R. A., 1963. Evidence for the identity of the nicotinamide adenine dinucleotide phosphate-specific sulfite and nitrite reductases of *Escherichia coli, J. Biol. Chem.* 238:3466–3471.

Kumar, H. D., 1964. Adaptation of a blue-green alga to sodium selenate and chloramphenicol, *Plant Cell Physiol.* 5:465–472.

Kylin, A., 1967. The uptake and metabolism of sulfate in *Scenedesmus* as influenced by citrate, carbon dioxide, and metabolic inhibitors, *Physiol. Plant.* 20:139–148.

Lapage, S. P., and Bascomb, S., 1968. Use of Selenite reduction in bacterial classification, *J. Appl. Bact.* 31:568–580.

Leggett, J. E., and Epstein, E., 1956. Kinetics of sulfate absorption by barley roots, *Plant Physiol.* 31:222–226.

Leifson, E., 1936. New selenite enrichment media for the isolation of typhoid and parathyroid (salmonella) bacilli, *Amer. J. Hyg.* 24:423–432.

Letunova, S. V., 1970. Geochemical ecology of soil microorganisms, in C. F. Mills (ed.), *Trace Element Metabolism in Animals,* Livingstone, Edinburgh, pp. 432–437.

Levander, O. A., and Argrett, L. C., 1969. Effects of arsenic, mercury, thallium, and lead on selenium metabolism in rats, *Toxicol. Appl. Pharmacol.* 14:308–314.

Levander, O. A., Morris, V. C., and Higgs, D. J., 1973a. Acceleration of thiol-induced swelling of rat liver mitochondria by selenium, *Biochemistry* 12:4586–4590.

Levander, O. A., Morris, V. C., and Higgs, D. J., 1973b. Selenium as a catalyst for the reduction of cytochrome C by glutathione, *Biochemistry* 12:4591–4595.

Levander, O. A., Morris, V. C., and Higgs, D. J., 1974. Characterization of the selenium in rat liver mitochondria as glutathione peroxidase, *Biochem. Biophys. Res. Commun.* 58:1047–1052.

Levine, V. E., 1925. The reducing properties of microorganisms with special reference to selenium compounds, *J. Bacteriol.* 10:217–263.

Lewis, B., Johnson, C. M., and Delwiche, C. C., 1966. Release of volatile selenium compounds by plants: Collection procedures and preliminary observations, *J. Agr. Food Chem.* 14:638–640.

Lewis, B., Johnson, C. M., and Broyer, T. C., 1974. Volatile selenium in higher plants, *Plant Soil* 40:107–118.

Mahl, M. C., and Whitehead, E. I., 1961. Relationship between selenite and phosphate uptakes in respiring yeast cells, *Proc. S. Dakota Acad. Sci.* 40:93–97.

Massey, V., and Williams, Jr., C. H., 1965. On the mechanism of yeast glutathione reductase, *J. Biol. Chem.* 240:4470–4480.

Masukawa, T., and Iwata, H., 1977. Catalytic action of selenium in the reduction of methemoglobin by glutathione, *Life Sci.* 21:695–700.

Mautner, H. G., and Gunther, W. H. H., 1959. Selenopantethine, a functional analog of pantethine in the Lactobacillus helveticus system, *Biochim. Biophys. Acta* 36:561–562.

Mazelis, M., Levin, B., and Mallinson, N., 1965. Decomposition of methyl-methionine sulfonium salts by a bacterial enzyme, *Biochim. Biophys. Acta* 105:106–114.

McConnell, K. P., and Cho, G. J., 1965. Transmucosal movement of selenium, *Am. J. Physiol.* 208:1191–1195.

McConnell, K. P., and Cho, G. J., 1967. Active transport of L-seleno-methionine in the intestine, *Am. J. Physiol.* 213:150–156.

McConnell, K. P., and Hoffman, J. L., 1972. Methionine-selenomethionine parallels in rat liver polypeptide chain synthesis, *FEBS Lett.* 24:60–62.

McConnell, K. P., and Portman, O. W., 1952. Toxicity of dimethyl selenide in the rat and mouse, *Proc. Soc. Exp. Biol. Med.* 79:230–231.

McConnell, K. P., and Roth, D. M., 1966. Respiratory excretion of selenium, *Proc. Soc. Exp. Biol. Med.* 123:919–921.

McConnell, K. P., and Wabnitz, C. H., 1957. Studies on the fixation of radioselenium in proteins, *J. Biol. Chem.* 226:765–776.

McConnell, K. P., Hsu, J. M., Herrman, J. L., and Anthony, W. L., 1974. Parallelism between sulfur and selenium amino acids in protein synthesis in the skin of zinc deficient rats, *Proc. Soc. Exp. Biol. Med.* 145:970–974.

McCready, R. G. L., Campbell, J. N., and Payne, J. I., 1966. Selenite reduction by *Salmonella heidelberg, Can. J. Microbiol.* 1:703–714.

Millar, K. R., 1972. Distribution of Se^{75} in liver, kidney, and blood proteins of rats after intravenous injection of sodium selenite, *N.Z.J. Agr. Res.* 15:547–564.

Millar, K. R., and Shepphard, A. D., 1973. A comparison of the metabolism of methionine and selenomethionine in rats, *N.Z.J. Agr. Res.* 16:293–300.

Millar, K. R., Gardiner, M. A., and Shepphard, A. D., 1973. A comparison of the metabolism of intravenously injected sodium selenite, sodium selenate and selenomethionine in rats, *N. Z. J. Agr. Res.* 16:115–127.

Mudd, S. H., and Cantoni, G. L., 1957. Selenomethionine in enzymatic transmethylations, *Nature* 180:1052.

Mukai, K., Huang, J. J., and Kimura, T., 1974. Studies on adrenal steroid hydroxylases. Chemical and enzymatic properties of selenium derivatives of adrenal iron–sulfur protein, *Biochim. Biophys. Acta* 336:427–436.

Muth, O. H., Schubert, J. R., and Oldfield, J. E., 1961. White muscle disease (myopathy) in lambs and calves. VII. Etiology and prophylaxis, *Am. J. Vet. Res.* 22:466–469.

Ng, B. H., and Anderson, J. W., 1978a. Chloroplast cysteine synthases of *Trifolium repens* and *Pisum sativum, Phytochemistry* 17:879–885.

Ng, B. H., and Anderson, J. W., 1978b. Synthesis of selenocysteine syntheses from selenium accumulator and nonaccumulator plants, *Phytochemistry* 17:2069–2074.

Ng, B. H., and Anderson, J. W., 1979. Light-dependent incorporation of selenite and sulphite into selenocysteine and cysteine by isolated pea chloroplasts, *Phytochemistry* 18:573–580.

Nickerson, W. J., and Falcone, G. 1963. Enzymatic reduction of selenite, *J. Bacteriol.* 85:763–771.

Nigam, S. N., and McConnell, W. B., 1969. Seleno amino compounds from *Astragalus bisculcatus.* Isolation and identification of gamma-L-glutamyl-Se-methylseleno-L-cysteine and Se-methylseleno-L-cysteine, *Biochim. Biophys. Acta* 192:185–190.

Nigam, S. N., and McConnell, W. B., 1973. Biosynthesis of Se-methylselenocysteine, *Phytochemistry* 12:359–362.

Nisman, B., and Hirsch, M. L., 1958. Study of activation and incorporation of amino acids by enzymatic fractions of *Escherichia coli, Ann. Inst. Pasteur Paris* 95:615–636.

Nissen, P., and Benson, A., 1964. Absence of selenate esters and "selenolipid" in plants, *Biochem. Biophys. Acta* 83:400–402.

Obermeyer, B. D., Palmer, I. S., Olson, O. E., and Halverson, A. W., 1971. Toxicity of trimethylselenonium chloride in the rat with and without arsenite, *Toxicol. Appl. Pharmacol.* 20:135–145.

Olson, O. E., and Palmer, I. S., 1976. Selenoamino acids in tissues of rats administered inorganic selenium, *Metabolism* 25:299–306.

Olson, O. E., Schulte, B. M., Whitehead, E. I., and Halverson, A. W., 1963. Effect of arsenic on selenium metabolism in rats, *J. Agr. Food Chem.* 11:531–534.

Olson, O. E., Novacek, E. J., Whitehead, E. I., and Palmer, I. S., 1970. Investigations on selenium in wheat, *Phytochemistry* 9:1181–1188.

Opienska-Blauth, J., and Iwanowski, 1952. The effect of selenite on growth and conversion of carbohydrates in liquid cultures of *Escherichia coli, Acta Microbiol. Polon.* 1:273–359.

Orme-Johnson, W. H., Hansen, R. E., Beinert, H., Tsibris, J. C. M., Bartholomaus, R. C., and Gunsalus, I. C., 1968. On the sulfur components of iron-sulfur proteins. I. The number of acid-labile sulfur groups sharing an unpaired electron with iron, *Proc. Natl. Acad. Sci.* 60:368–372.

Painter, E. P., 1941. The chemistry and toxicity of selenium compounds with special reference to the selenium problem, *Chem. Rev.* 28:179–213.

Palmer, I. S., Fischer, D. D., Halverson, A. W., and Olson, O. E., 1969. Identification of a major selenium excretory product in rat urine, *Biochem. Biophys. Acta* 177:336–342.

Palmer, I. S., Gunsalus, R. P., Halverson, A. W., and Olson, O. E., 1970. Trimethylselenonium ion as a general excretory product from selenium metabolism in the rat, *Biochim. Biophys. Acta* 208:260–266.

Palmer, I. S., 1973. An example of the lack of parallelism in the metabolism of sulfur and selenium, *Proc. S. Dak. Acad. Sci.* 52:108–111.

Pan, F., Natori, Y., and Tarver, H., 1964. Studies on selenium compounds. II. Metabolism of selenomethionine and selenoethionine in rats, *Biochim. Biophys. Acta* 93:521–525.

Pan, F., and Tarrer, H., 1967. Comparative studies on methionine, selenomethionine, and their ethyl analogues as substrates for methionine adenosynetransferase from rat liver, *Arch. Biochem. Biophys.* 119:429–434.

Parizek, J., Ostadalova, I., Kalouskova, J., Babicky, A., and Benes, J., 1971. The detoxifying effects of selenium: Interrelations between compounds of selenium in certain metals, in W. Mertz and W. E. Cornatzer (eds.), *Newer Trace Elements in Nutrition*, Marcel Dekker, New York, pp. 85–122.

Patrick, H., Voitle, R. A., Hyre, H. M., and Martin, W. G., 1965. Incorporation of [32]phosphorous and [75]selenium in cock sperm, *Poult. Sci.* 19:587–591.

Paulson, G. D., Baumann, C. A., and Pope, A. L., 1966. Fate of a physiological dose of selenate in the lactating ewe: Effect of sulfate, *J. Anim. Sci.* 25:1054–1058.

Paulson, G. D., Baumann, C. A., and Pope, A. L., 1968. Metabolism of ^{75}Se-selenite, ^{75}Se-selenate, ^{75}Se-selenomethionine, and ^{35}S-sulfate by rumin microorganisms *in vitro, J. Anim. Sci.* 27:497–503.

Peterson, G., and Butler, G., 1962. The uptake and assimilation of selenite by higher plants, *Aust. J. Biol. Sci* 15:126–146.

Pope, A. L., Moir, R. J., Somers, M., Underwood, E. J., and White, C. L., 1979. The effect of sulphur on ^{75}Se absorption and retention in sheep, *J. Nutr.* 109:1448–1454.

Reamer, D. C., and Zoller, W. H., 1980. Selenium biomethylation products from soil and sewage sludge, *Science* 208:500–502.

Rhead, W. J., and Schrauzer, G. N., 1974. The selenium catalyzed reduction of methylene blue by thiols, *Bioinorg. Chem.* 3:225–242.

Rhead, W. J., Evans, G. A., and Schrauzer, G. N., 1974. Selenium in human plasma: Levels in blood proteins and behavior upon dialysis, acidificatin and reduction, *Bioinorg. Chem.* 3:217–223.

Robbins, P. W., and Lipmann, F., 1956. The enzymatic sequence in the biosynthesis of active sulfate, *J. Am. Chem. Soc.* 78:6409–6410.

Rosenfeld, I., 1962. Biochemical and chemical studies on *Astragalus* leaves and roots, *Univ. Wyoming Agr. Exp. Sta. Bull.* 385:1–43.

Rosenfeld, I., and Beath, O. A., 1948. Metabolism of sodium selenite by the tissues, *J. Biol. Chem.,* 172:333–341.

Rotruck, J. T., Pope, J. T., Ganther, H. E., Swanson, A. B., Hafeman, D. G., and Hoekstra, G. W., 1973. Selenium: Biochemical role as a component of glutathione peroxidase, *Science,* 179:588–590.

Said, A. K., and Hegsted, D. M., 1970. ^{75}Se-selenomethionine in the study of protein and amino acid metabolism of adult rats, *Proc. Soc. Exp. Biol. Med.* 133:1388–1391.

Sapozhnikov, D. I., 1937. The exchange of sulfur by selenium during the photoreduction of H_2CO_3 by purple sulfur bacteria, *Mikrobiologia (USSR)* 6:643–644.

Scala, J., and Williams, H. H., 1963. A comparison of selenite and tellurite toxicity in *Escherichia coli, Arch. Biochem. Biophys.* 101:319–324.

Schiff, J. A., and Hodson, R. C. 1973. Metabolism of sulfate, *Ann. Rev. Plant Physiol.* 24:381–414.

Schubert, J. R., Muth, O. H., Oldfield, J. E., and Remmert, L. F., 1961. Experimental results with selenium in white muscle disease of lambs and calves, *Fed. Proc. Fed. Am. Soc. Exp. Biol.* 20:689–694.

Schwarz, K., 1961. Development and status of experimental work on Factor 3-selenium, *Fed. Proc. Fed. Am. Soc. Exp. Biol.* 20:666–673.

Schwarz, K., and Sweeny, E., 1964. Selenite binding to sulfur amino acids, *Fed. Proc. Fed. Am. Soc. Exp. Biol.* 23:421.

Shepherd, L., and Huber, R. E., 1969. Some chemical and biochemical properties of selenomethionine, *Can J. Biochem.* 47:877–881.

Shrift, A., 1954a. Sulfur-selenium antagonism. I. Antimetabolite action of selenate on the growth of *chlorella vulgaris, Am. J. Bot.* 41:223–230.

Shrift, A., 1960. A role for methionine in division of *chlorella vulgaris, Plant Physiol.* 35:510–515.

Shrift, A., 1961. Biochemical interrelations between selenium and sulfur in plants and microorganisms, *Fed. Proc. Fed. Am. Soc. Exp. Biol.* 20:695–702.

Shrift, A., 1967. Microbial research with selenium, in *Selenium in biomedicine,* O. H. Muth (ed.), AVI Publishing Co., Westport, Connecticut, pp. 241–271.

Shrift, A., 1969. Aspects of selenium metabolism in higher plants, *Ann. Rev. Plant Physiol* 20:475–494.

Shrift, A., 1973. Selenium compounds in Nature and Medicine, in *Organic Selenium Compounds: Their Chemistry and Biology,* D. L. Klayman and W. H. H. Gunther (eds.), John Wiley, New York, pp. 763–814.

Shrift, A., and Kelley, E., 1962. Adaptation of *Escherichia coli* to selenate, *Nature (London)* 195:732–733.

Shrift, A., and Sproul, M., 1963. Nature of the stable adaptation induced by selenomethionine in *Chlorella vulgaris, Biochim. Biophys. Acta* 71:332–344.

Shrift, A., and Ulrich, J. M., 1969. Transport of selenate and selenite into *Astragalus roots, Plant Physiol.* 44:893–896.

Smith, M. I., Westfall, B. B., and Stohlman, E. F., 1938. Studies on the fate of selenium in the organism, *Pub. Health Rep.* 53:1199–1216.

Tappel, A. L., and Caldwell, K. A., 1967. Redox properties of selenium compounds related to biochemical function, in *Selenium in Biomedicine,* O. H. Muth, J. E. Oldfield, and P. H. Weswig (eds.), Avi Publishing Co., Westport, Connecticut, pp. 345–361.

Tappel, A. L., 1972. Vitamin E and free radical peroxidation of lipids, *Ann. N.Y. Acad. Sci.* 203:12–28.

Tilton, R. C., Gunner, H. B., and Litsky, W., 1967. Physiology of selenite reduction by enterococci. I. Influence of environmental variables, *Can. J. Microbiol.* 13:1175–1185.

Trelease, S. F., and Trelease, H. M., 1938. Selenium as a stimulating and possibly essential element for indicator plants, *Am. J. Bot.,* 25:372–380.

Tsay, D. T., Halverson, A. W., and Palmer, I. S., 1970. Inactivity of dietary trimethylselenonium chloride against the necrogenic syndrome of the rat, *Nutr. Rep. Int.* 2:203–207.

Tsibris, J. C. M., Namtvedt, M. J., and Gunsalus, I. C., 1968. Selenium as an acid labile sulfur replacement in putidaredoxin, *Biochem. Biophys. Res. Commun.* 30:323–327.

Tweedie, J. W., and Segal, I. H., 1970. Specificity of transport processes for sulfur, selenium and molybdenum anions by filamentous fungi, *Biochim. Biophys. Acta* 196:95–106.

Ulrich, J. M., and Shrift, A., 1968. Selenium absorption by excised *Astragalus* roots, *Plant Physiol.* 43:14–20.

Virupaksha, T., and Shrift, A., 1963. Biosynthesis of selenocystathionine from selenate in *Stanleya pinnata, Biochim. Biophys. Acta* 74:791–793.

Vlasakova, V., Benes, J., and Parizek, J., 1972. Application of gas chromatography for the analysis of trace amounts of volatile ^{75}Se metabolites in expired air, *Radiochem. Radioanal. Lett.* 10:251–258.

Weiss, K. F., Ayres, J. C., and Kraft, A. A., 1965. Inhibitory action of selenite on *Escherichia coli, Proteus vulgaris* and *Salmonella thompson, J. Bacteriol.* 90:857–862.

Whanger, P. D., Muth, O. H., Oldfield, J. E., and Weswig, P. H., 1969. Influence of sulfur on incidence of white muscle disease in lambs, *J. Nutr.* 97:553–561.

Whanger, P. D., Pedersen, N. D. and Weswig, P. H., 1973. Selenium proteins in ovine tissues. II. Spectral properties of a 10,000 molecular weight selenium protein, *Biochem. Biophys. Res. Commun.* 53:1031–1036.

White, C. L., 1980. Sulfur–selenium studies in sheep. II. Effect of a dietary sulfur deficiency on selenium and sulfur metabolism in sheep fed varying levels of selenomethionine, *Aust. J. Biol. Sci.* 33:699–707.

White, C. L., and Somers, M., 1977. Sulphur-selenium studies in sheep. I. The effects of varying dietary sulphate and selenomethionine on sulphur, Nitrogen and selenium metabolism in sheep, *Aust. J. Biol. Sci.* 30:47–56.

Widstrom, V. R., 1961. Effect of selenate ions on the growth of *Neurospora crassa* in the presence of various sulfur sources, *Proc. S. Dak. Acad. Sci.* 40:208–212.

Wilson, L. G., and Bandurski, R. S., 1958. Enzymatic reactions involving sulfate, sulfite, selenate and molybdate, *J. Biol. Chem.* 233:975–981.

Woolfolk, C. A., and Whiteley, H. R., 1962. Reduction of inorganic compounds with molecular hydrogen by *Micrococcus lactilyticus, J. Bacteriol.* 84:647–658.

Wright, P. L., and Mraz, F. R., 1965. Toxicity of sulfur-35, *Proc. Soc. Exp. Biol. Med.* 118:534–539.

Wu, M., and Wachsman, J. T., 1970. Effect of selenomethionine on growth of *Escherichia coli* and *Bacillus megaterium, J. Bacteriol.* 104:1393–1396.

Wu, M., and Wachsman, J. T., 1971. Selenomethionine, a methyl donor for bacterial nucleic acids, *J. Bacteriol.* 105:1222–1223.

Yamamoto, L. A., and Segal, I. H., 1966. The inorganic sulfate transport system of *Penicillium chrysogenum, Arch. Biochem. Biophys.* 114:523–538.

Yousef, M. K., and Johnson, H. D., 1970. ^{75}Se-selenomethionine turnover rate during growth and aging in rats, *Proc. Soc. Exp. Biol. Med.* 133:1351–1353.

Biological Interactions of Selenium with Other Substances

5

5.1 Cadmium

5.1.1 Pathological Effects

Single subcutaneous injections of cadmium chloride in amounts much below toxic levels selectively damage the testis of rats and other laboratory animals (Parizek, 1957). The site and mode of cadmium action have not been satisfactorily established even though there is general agreement that testicular injury is secondary to vascular changes. It is believed that cadmium specifically damages the testicular artery–pampiniform plexus complex and its countercurrent exchange mechanism (Gunn et al., 1963). Mason et al., (1964) believe that because of the sluggish blood flow through the intratesticular course of the testicular artery and its end-arterial type of capillary bed, cadmium might produce increased permeability of the capillary wall and perhaps may act directly on the parenchyma of the testis. This next produces an intertubular edema, increased intratesticular pressure, and interference with the vascular supply to the testis, which causes ischemia and necrosis of the tissue.

Hemorrhage, edema, and epithelial desquamation in proximal segments of the caput epididymis are also observed in cadmium injury (Gunn et al., 1963; Mason et al., 1964), but these manifestations are observed only when cadmium dosage has been three to six times the minimal for producing testicular damage. Since these types of damage occurred in the absence of the associated testis (Mason et al., 1964), they appear to result from cadmium acting upon a similar end-arterial termination of the superior epididymal branch of the testicular artery, in which blood flow is thought to be sluggish. The dense capsular portion of the first segments of the ductus epididymis, like the tunica of the testis, may also play a role. Other portions of the epididymis, supplied by the inferior epididymal artery which anstomoses with the artery of the vas deferens, are rarely involved. Both of the above interpretations were supported by the evidence of Waites and Setchell (1966). They found that within 3–12 hr after cadmium injection, there occurred a marked decrease in testicular blood flow as well as increased vascular permeability in the testis and caput epididymis but not in other portions of the epididymis.

Mason *et al.*, (1964) have designed experiments to determine the minimal amount of cadmium required to produce testicular damage in the rat. Mason and Young (1967) have also determined the minimal amounts of selenium and zinc required for the prevention of testicular damage. In these two reports, two types of histopathological change unlike those previously described were observed frequently in the excurrent duct system. One type of damage was characterized by partial or complete blockage of the ductuli efferentes, which was usually associated with variable degrees of distension and pressure degeneration in the seminiferous tubules of the testis. The other type of damage was marked by focal epithelial reactions and spermatocele formation in the ducts of the caput epididymis.

In addition to the testicular damage cadmium has been shown to interfere with a variety of biological processes. Chronic oral exposure in the adult animal produces kidney damage (Axelsson and Piscator, 1966), hypertension (Schroeder, 1965), anemia (Wilson *et al.*, 1941), tumors (Schroeder *et al.*, 1964), and in conjunction with some dietary deficiencies, osteomalacia (Nogawawa *et al.*, 1975). Cadmium also affects growing animals as well as the fetuses of pregnant animals by producing anemia and growth retardation (Webster, 1978). In addition, chronic respiratory exposure may cause emphysema (Friberg, 1950).

Acute exposure in the adult animal usually by injection, produced hemorrhage in sensory ganglia (Gabbiani, 1966) as well as the testicular necrosis which has been already described. If the animal is pregnant, acute exposure produces congenital anomalies (Ferm and Carpenter, 1968) or placental destruction and fetal death (Parizek, 1964). In the neonate cadmium injection causes hemorrhage in the central nervous system (Gabbiani *et al.*, 1967). Oral as well as respiratory exposure produces acute, local effects (Friberg *et al.*, 1974). Many of these toxic effects of cadmium can be prevented by prior or simultaneous exposure to other chemicals, in particular metals such as zinc and selenium, which also are able to interact with one another. Apparently cadmium, zinc, and selenium are able to compete with one another for binding sites for essential enzymes.

5.1.2 Cadmium–Zinc Interactions

Considerable interest has been shown in the interaction between cadmium and zinc since it has been demonstrated that prior or simultaneous administration of zinc salts had a marked protective effect against many of the acute lesions of cadmium toxicity. These have included the development of testicular damage and necrosis (Parizek, 1957), the development of a toxemialike condition in pregnant rats (Parizek *et al.*, 1969a), alterations in pancreatic and hepatic function (Merali and Singhal, 1976), protection against teratogenic effects (Ferm and Carpenter, 1968), and protection against placental destruction (Chiquoine, 1965). In laboratory animals zinc has also been shown to protect against some of the chronic effects of cadmium toxicity: zinc protects against cadmium-caused kidney damage (Vigliani, 1969); in the growing animal zinc is able to partially protect against anemia and growth retardation (Bunn and Matrone, 1966). Furthermore, many of the symptoms of chronic cadmium toxicity are quite similar to those of zinc deficiency (Underwood, 1977), with growth failure (Bunn and Matrone, 1966; Supplee, 1963), parakerotic lesions (Petering *et al.*, 1971; Powell *et al.*, 1964), and impaired glucose tolerance (Petering, 1974) occurring in both conditions. Because these symptoms could be eliminated by zinc supplementation, the toxicity of cadmium may result, in part, from metabolic disturbances of zinc. This would be consistent with the description of cadmium as an "antimetabolite of zinc" (Cotzias and Papavasiliou, 1964).

Cadmium-induced testicular damage may arise from competition between cadmium and zinc for binding sites on essential enzymes required for gametogenesis (Parizek, 1957). Hill and Matrone (1970) have suggested that those metals whose physical and chemical properties were similar would be antagonistic to each other biologically. Both zinc and cadmium are members of group IIB of the periodic table and have a similar tendency to form tetrahedral complexes. Evidence for this theory is supported by the inhibitory effect of cadmium *in vitro* on the activity of zinc-containing enzymes such as carboxypeptidase (Coleman and Vallee, 1961) and α-mannosidase (Snaith and Levvy, 1969), as well as the isomorphous replacement of zinc in metallothionein by cadmium (Kagi and Vallee, 1961). However, direct evidence for zinc displacement from metalloenzymes *in vivo* has yet to be found. Reductions have been reported in the activity *in vivo* of a zinc-containing enzyme, such as renal leucine aminopeptidase (Washko and Cousins, 1977). Claims have been made (Washko and Cousins, 1977) that this may be responsible for the proteinuria which develops in chronic cadmium poisoning. Even so, it is difficult to distinguish between reductions in enzyme activity as a result of impaired zinc supply and those arising from zinc displacement from the enzyme.

Even though their is evidence of an interaction between cadmium and zinc, it is difficult to assess its biological significance at exposure levels encountered by humans or animals. For example, the dietary intake in calves was raised to 640 mg/kg before obvious signs of zinc deficiency appeared in the form of growth failure and parakeratosis (Powell *et al.*, 1964). At much lower cadmium intakes, cadmium toxicosis can be induced if the diet contains inadequate amounts of zinc. This has been observed in turkey poults (Supplee, 1963) and in rats (Petering, 1974; Petering *et al.*, 1971). In the rat experiments, inclusion of 3.4 mg cadmium per liter of drinking water was sufficient to induce keratinization of the cornea, impair growth, and affect the peripheral circulation when the zinc intake was below requirement level, whereas no symptoms of toxicosis appeared when the zinc intake in the poults and rats was increased to more normal levels.

The occurrence of clinical signs of cadmium-induced zinc deficiency may only be artifactual, as some disturbances in zinc metabolism do occur at relatively low cadmium intakes. Concentrations of zinc in liver (Stonard and Webb, 1976), and to a lesser extent in kidney (Cousins *et al.*, 1973; Doyle and Pfander, 1975), are often increased in cadmium-treated animals. This is opposite of what would be expected if cadmium were displacing zinc from binding sites on proteins. Most of the additional zinc is probably incorporated into metallothionein (Stonard and Webb, 1976), which is the principal cadmium-binding protein in these tissues (Webb, 1975). There is significant variability between and within species, in the ratio of renal cadmium and zinc concentration, after long-term exposure to cadmium. These variations may reflect differences in the availability of zinc for enzymatic functions within the kidney. Attempts have been made to relate the increases in the cadmium : zinc ratio in kidneys to the incidence of hypertension in humans (Schroeder, 1965). However, at present this work is controversial and more investigation needs to be done.

Zinc concentrations in tissue other than liver and kidney from cadmium-treated animals have not been studied as extensively, but decreases in serum (Rice *et al.*, 1973), testes (Petering *et al.*, 1971), and in fetal tissue (Choudhury *et al.*, 1978) have been reported in rats and in other species.

The effects of dietary cadmium on zinc metabolism may originate from similar competitive processes which affect zinc absorption. Increased tissue retention and decreased fecal excretion of ^{65}Zn occurred in rats after a single injection of cadmium (Gunn *et al.*, 1962), but the opposite effect was observed in calves receiving high levels of cadmium in the diet (Roberts *et al.*, 1973). After the experiments were concluded, it was not possible to decide whether zinc metabolism was affected by cadmium at the intestinal or tissue level (Roberts *et al.*, 1973). The direct effect of a single large intragastric dose of cadmium on the absorption of ^{65}Zn has been demonstrated in rats (Evans *et al.*, 1974). Both transfer and mucosal uptake of ^{65}Zn were reduced when animals were supplemented with zinc. In contrast, only the mucosal uptake of ^{65}Zn was reduced in zinc-deficient animals. The experiment was complicated by giving the zinc-supplemented rats intraperitoneal injections of 200 μg zinc for two days and by fasting the rats for 18 hr prior to determination of ^{65}Zn absorption, as both of these procedures induce synthesis of zinc–thionein in liver and intestinal mucosa (Richards and Cousins, 1977). Owing to these additional experimental conditions, it is likely that abnormally high concentrations of this protein could have been present in the mucosa of the zinc-supplemented rats. These high-protein concentrations could have been significant in view of the involvement of metallothionein in the control of zinc absorption (Richards and Cousins, 1976) and in view of the greater affinity of cadmium for binding sites on the protein (Kagi and Vallee, 1961).

5.1.3 Cadmium–Selenium Interactions

Administration of selenium compounds has been shown to protect effectively against the toxicities induced by cadmium (Parizek *et al.*, 1968b). Selenium is also effective in protecting against the lethal doses of cadmium (Gunn *et al.*, 1968a). Therefore, selenium appears to have a general detoxifying effect on this element. The interactions of selenium and vitamin E with heavy metals are outlined in Table 5-1. Although selenium protects effectively against cadmium-induced testicular injury, it actually increases the cadmium content in the testis, indicating that selenium causes a redistribution of cadmium in this organ. Marked diversions of cadmium from 10,000- and 30,000-molecular-weight proteins to large-molecular-weight proteins were also observed (Chen *et al.*, 1975). Since the target of the cadmium-induced testicular injury appears to be the 30,000-molecular-weight protein, the diversion of cadmium by selenium appears to be one of the mechanisms involved in the protection against the cadmium-induced testicular injury.

Selenium was found to give a 22-fold increase in the cadmium content in the blood and doubled the uptake by the testis, while decreasing that in the liver by 48% and by 12% in the kidney (Chen *et al.*, 1975). The

Table 5-1. Relative Effectiveness of Selenium and Vitamin E against Heavy
Metal Effects

Metal	Influence of selenium and Vitamin E
Arsenic	Arsenic reduces selenium toxicity. Vitamin E has not been tested.
Bismuth	Selenium affects tissue distribution of bismuth. Vitamin E has not been tested.
Cadmium	Selenium is highly effective against cadmium damage. Vitamin E has not been tested.
Cobalt	Affects selenium absorption. Vitamin E has not been tested.
Copper	Affects selenium release indirectly. Vitamin E has not been tested.
Lead	Selenium has little effect on its toxicity. Vitamin E is highly protective.
Manganese	Deficiency parallels selenium deficiency. Vitamin E has not been tested.
Inorganic mercury	Selenium is effective against its toxicity. Vitamin E has very little influence on its toxicity.
Methylmercury	Selenium is highly effective against its toxicity. Vitamin E is effective but not as good as selenium.
Silver	Selenium is effective at excessively high levels. Vitamin E is highly effective against its toxicity.
Tellurium	Enhances selenium deficiency. Vitamin E has not been tested.
Thallium	Selenium protects against toxicity. Vitamin E has not been tested.
Vanadium	Enhances selenium deficiency. Vitamin E has not been tested.

diversion in the binding of cadmium in the soluble fraction to higher-molecular-weight proteins was also observed in the kidney and liver. This diversion may be a second mechanism involved in the protection of these organs against cadmium by selenium, in addition to a reductive effect of selenium on the tissue cadmium concentration. Similar to the testes, selenium was also found in these higher-molecular-weight cadmium-binding proteins. Based on a similar molecular weight of about 115,000, the cadmium-binding, selenium-containing proteins found in the kidney and liver appear to be similar. The diversion of cadmium from lower-molecular-weight proteins to larger ones was also found in the plasma. These cadmium-binding, selenium-containing proteins in plasma appear to be different from those of other organs since they have a larger molecular weight.

Since pretreatment by selenium increased cadmium content in the testis and all its subcellular fractions, the selenium protection against the cadmium-induced testicular injury could not be explained unless redistribution of cadmium occurred in at least one of the subcellular fractions. Even though a redistribution of cadmium in the subcellular organelles has not been demonstrated, the redistribution of cadmium by selenium in the soluble fraction was quite marked. Therefore, the diversion of cadmium from low-molecular-weight proteins, especially the unique 30,000-molecular-weight cadmium-binding protein (Chen *et al.*, 1975), to the larger-molecular-weight cadmium-binding protein appears to be at least one of the mechanism(s) responsible for the selenium protection against the cadmium-induced testicular damage. Support is given to this hypothesis by the fact that the soluble fraction was a major cadmium-binding component in the testis.

There are several possible mechanisms of the selenium diversion of cadmium binding which appear plausible: first, the selenium may be incorporated into metallothionein, the low-molecular-weight cadmium-binding protein, which may promote polymerization of this metalloprotein; secondly, a large-molecular-weight, preexisting protein could incorporate selenium into a sulfur–selenium bond which may have a higher affinity for cadmium; thirdly, selenium could be inducing a conformation change in this large-molecular-weight protein which might cause its affinity for cadmium to be greater. This latter process does not seem to be a major factor. Because of the time limitation, selenium does not appear to be inducing the synthesis of this large-molecular-weight protein. Thus one of the other two possibilities appear more plausible.

Gasiewicz and Smith (1978a) have administered simultaneous subcutaneous $Cd(Cl)_2$ and Na_2SeO_3 to rats. Evidence of a Cd–Se complex was detected in plasma by gel filtration chromatography. A similar complex was found in plasma after incubation of selenite, $Cd(Cl)_2$, rat erythrocytes, and plasma *in vitro,* and also after incubation of H_2Se, Cd, and plasma *in vitro*. When erythrocytes were not included in the *in vitro* incubation, no interaction of selenite, selenate, or selenodiglutathione with Cd was observed. These Cd–Se complexes were similar after they were characterized by gel filtration, ion exchange chromatography, affinity chromatography, and ammonium sulfate fractionation. These results support the hypothesis that H_2Se or a similarly reduced selenide is the product of selenite metabolism by rat erythrocytes. Hydrogen selenide also seemed to alter the distribution of inorganic mercury in rat plasma *in vitro* in such a way that the apparent molecular weights of the Cd–Se and the Se–Hg complexes associated with protein were similar. Incubation with Proteinase K showed that the stability of the Cd–Se complex in plasma depended upon the integrity of the native protein components. The properties of this complex suggested that it existed in a single form, but was reversibly associated with different plasma proteins under various conditions.

Using a pH 8.0 buffer for chromatography, Cd–Se peaks of 130,000 and 330,000 molecular weight appeared in plasma following the administration of Cd and selenite to rats. The Cd–Se peak at 330,000 appeared to be saturated at a concentration in plasma of approximately 30.0 nmol of Cd and Se per milliliter of plasma (Gasiewicz and Smith, 1976). When a pH 6.5 buffer was used, a higher concentration of Cd and Se (up to 0.13 μmol/ml of plasma) was associated with this peak. It seems unlikely that binding sites on the 330,000 species were able to accommodate 100 μmol of the complex through a pH change of less than 1.5 units. Although polymerization is another possible alternative, the complete polymerization of a molecule by a similar pH change also appears unlikely.

Gasiewicz and Smith (1976) also used blue sepharose and showed that the distribution of the complex could be altered by changing conditions other than pH. In this experiment, the formation of a stationary albumin phase resulted in an altered distribution and the subsequent association with albumin. Coupled with the ammonium sulfate precipitation data, these results imply that this complex exists in a single form. The nature and character of the complex, rather than the protein(s) it may be associated with, may play an important role not only in its *in vitro* properties, but its ultimate fate in intact biological systems.

Cadmium selenide (CdSe), like cadmium sulfide, is virtually insoluble in water (Seidell, 1952). When ^{75}Se-labeled H_2Se was bubbled into a buffer solution containing ^{109}Cd-labeled $CdCl_2$, over 90% of the radioactivity was associated with a precipitate following centrifugation at 2000 g for 10 min. The high affinity of Cd^{2+} for Se^{2-} and the stability of CdSe suggest that the complex forms even in the presence of plasma proteins. Several substances, including proteins, carbohydrates, and lipids, have been used to stabilize colloidal metal preparations. The formation of a colloidal Cd–Se complex whose stability is dependent upon proteins, seems to be a reasonable possibility.

The metabolism of ^{75}Se-labeled SeO_3^- and its conversion by intact rat erythrocytes *in vitro* to a form which complexes with Cd and plasma proteins were studied. When both excess SeO_3^- and N-ethylmaleimide were utilized, to lower erythrocyte-reduced glutathione (GSH) concentrations, it was shown that the uptake and metabolism of SeO_3^{2-} were GSH dependent. The probable intermediate was glutathione selenotrisulfide (GSSeSg) (Gasiewicz and Smith, 1978b). Secondary release of selenium by rat erythrocytes had no relation to the erythrocyte transport of oxidized glutathione (GSSG). It seems likely that SeO_3^{2-} follows the reductive metabolism as proposed by Ganther (1968):

$$H_2SeO_3 + 4GSH \rightarrow GSSeSG + GSSG + 3H_2O \qquad (5\text{-}1)$$

Sandholm (1973a) has postulated that selenium was released from erythrocytes as a glutathione complex. Jenkins and Hidiroglou (1972) suggested that most of the selenium released from bovine erythrocytes *in vitro* was in the form of GSSeSG. They proposed that the release mechanism was similar to the transport of GSSG. However, Gasiewicz and Smith (1978b) concluded that the release of selenium is not related to the release of GSSG and that the selenium form released was not GSSeSG. In their experiment, fluoride, an inhibitor of GSSG release, had no effect on selenium release from erythrocytes. On the other hand, chromate, an inhibitor of glutathione reductase (Srivastava, 1969), in spite of increasing the GSSG levels of both plasma and red cells, can bring about complete inhibition of selenium release from red cells. The release of selenium appeared to be secondary to a reaction catalyzed by glutathione reductase. The similarity of I_{50} values for chromates inhibition of glutathione reductase and for the inhibition of selenium release further suggested that these two events are interrelated. Hydrogen selenide or a similar product of GSSeSG reduction was thought to be the active product of SeO_3^{2-} metabolism by rat erythrocytes. The metabolic path from SeO_3^{2-} to H_2Se has been documented by Hsieh and Ganther (1975). The mechanism seems to exist both *in vitro* and *in vivo*, and may be the one whereby the tissue distribution of cadmium is altered and its toxicity reduced. The efficacy of SeO_3^{2-} in protecting against cadmium toxicity is much greater than that of SeO_4^{2-} and the seleno-amino acids. Presumably these compounds should first be converted to SeO_3^{2-} or go through an alternate pathway to H_2Se.

The fate of the Cd–Se complex has not been studied in detail. The amount of the complex is about half its peak value after 24 hr. The excretion of cadmium is not significantly increased and cadmium inhibits the urinary excretion of selenium, even though six times as much selenium as cadmium appears in the excreta by 48 hr. It is possible that the cadmium may deposit slowly in other tissues as the complex breaks down. The liver burden of cadmium is less when selenium is given simultaneously, therefore other tissues must have higher levels. The slow release rate of cadmium may allow secondary protection such as is afforded by the synthesis of metallothionein. Selenium itself had no effect on metallothionein levels induced in rat liver and kidney by cadmium (Piotrowski *et al.*, 1977).

Simultaneous or prior injection of selenium compounds has prevented the development of necrosis of the placentae (Parizek *et al.*, 1968a) in rats receiving a single injection of cadmium. Selenium has also given protection against injury to pancreatic β cells in subacute cadmium toxicity (Merali and Singhal, 1975a). Reddy *et al.* (1978) have observed enhanced lung toxicity of intratracheally instilled cadmium chloride in selenium-deficient rats. Selenium also protects against cadmium-induced hemorrhagic necrosis of nonovulating ovaries (Parizek *et al.*, 1968b), damage to the lactating mammary gland (Parizek *et al.*, 1968), teratogenicity (Holmburg and Ferm, 1969), and lethality (Parizek *et al.*, 1968b). It also has been shown (Perry *et al.*, 1974) that inclusion of relatively high Se concentrations (0.9 and 3.5 mg/liter) in the drinking water could prevent the increase in systolic blood pressure which occurred in rats receiving cadmium for up to one year.

In rats, selenite, when given as daily im injections of SeO_2 (2.0 mg/kg) for seven days, also prevents the depression of pancreatic function and the elevation of the activities of the hepatic gluconeogenic enzymes (phosphoenolpyruvate carboxykinase, pyruvate decarboxylase, fructose-1:6-diphosphatase and glucose-6-phosphatase) that result from the repeated administration of subacute doses of Cd^{2+} (1 mg $CdCl_2$/kg, sc, twice daily for seven days). On the other hand there was no significant effect on the Cd^{2+}-induced increase in the hepatic concentration of cyclic AMP (Merali and Singhal, 1975).

5.1.4 Effect on Drug Response

Cadmium treatment can potentiate drug response and inhibit hepatic microsomal drug metabolism in male rats (Schnell, 1978). The effect of cadmium on the inhibition of the monooxygenase system is primarily mediated by a decrease in the concentration of hepatic microsomal cytochrome P-450 (Means *et al.*, 1979). Studies by Burk *et al.* (1974a) have attempted to clarily the role of selenium on this hepatic mono-oxygenase system. Selenium deficiencies did not alter the activity of the mono-oxygenase enzymes, the levels of the cytochromes P-450, b_5 or NADPH–cytochrome reductase activity (Burk and Masters, 1975).

Early and Schnell (1981) have examined the effect of selenium on cadmium-induced inhibition of drug metabolism in male, Sprague–Dawley rats. Prior administration of sodium selenite (1.6 mg Se/kg, ip) blocked the cadmium-induced (0.84 mg Cd/kg, ip) increase of hexobarbital-induced hypnosis and inhibition of hepatic microsomal biotransformation of ethylmorphine or aniline. Selenium also blocked cadmium-induced reduction in microsomal cytochrome P-450 content and the microsomal binding of both ethylmorphine and aniline. However, pretreatment of rats with selenium did not prevent the inhibitory effect of cadmium $(10^{-6}–10^{-3}$ $M)$ when added *in vitro* on either ethylmorphine or aniline biotransformation. Moreover, the reduction in biotransformation of both substrates following *in vivo* cadmium administration was not reversed after the *in vitro* administration of selenium. Instead selenium produced further concentration decreases in drug metabolism. In another *in vitro* experiment, it was found that the inhibition in drug metabolism induced by *in vitro* additions of cadmium was not affected by additions of selenium to the incubation either before or after the cadmium. Therefore, for selenium to prevent the cadmium-induced inhibition of hepatic drug metabolism, the *in vivo* administration of selenium is required. The results of this investigation certainly argue against the formation of a nontoxic Se : Cd complex. The *in vitro* studies showed that both cadmium and selenium in sufficient concentrations could inhibit the mono-oxygenase enzyme system. Thus, if a chemical complexation of the two elements were the mechanism of detoxification, the inhibitory effect of one element should be decreased in the presence of the other. Early and Schnell (1981) found in fact, that the inhibitory effect of each element was enhanced in the presence of the other.

5.2 Arsenic

Arsenic was first found to protect against the chronic toxicity of selniferous grains or of selenite in rats (Dubois *et al.*, 1940). Arsenic also diminished the chronic toxicity of seleniferous grains or of selenite in dogs (Rhian and Moxon, 1943), cattle (Moxon *et al.*, 1944), and swine (Wahlstrom *et al.*, 1955). In poultry, arsenic not only decreased the growth inhibition caused by excess selenium but also improved the poor hatchability of eggs from selenized birds (Moxon and Wilson, 1944), apparently by reducing the amount of the element that is incorporated into the egg (Krista *et al.*, 1961). The simultaneous injection of arsenite also protects against single toxic doses of selenite in rats (Palmer and Bonhorst, 1957). At levels of chronic toxicity the absorption and retention of selenium did not seem to be influenced by protective amounts of arsenite (Klug *et al.*, 1950a. After injection in rats with single subtoxic doses of selenite, arsenic has been reported to increase the concentration of selenium in blood and to decrease it in the liver (Palmer and Bonhorst, 1957). Moreover, Kamstra and Bonhorst (1953) have observed that the amount of selenium exhaled as volatile compounds following the injection of selenite was greatly decreased when selenite was present. Since the formation of volatile selenium compounds has long been regarded as a detoxification process, this latter effect of arsenite would be expected to increase rather than to decrease the symptoms of selenium toxicity unless compensatory mechanisms were operative.

Nonetheless, the major pathway of selenium elimination could be the exhalation of volatile selenium especially when large amounts of selenium are administered. Schultz and Lewis (1940) have reported that when adult rats were injected with 2.5–3.5 mg of selenium/kg as selenite, the rats exhaled 17%–52% of the dose within 8 hr as volatile selenium (McConnell, 1942). The compound exhaled is thought to be mainly dimethyl selenide (McConnell and Portman, 1952a), which implies that this mode of selenium elimination involves a very active reduction of selenium. The biosynthesis of the volatile product, dimethyl selenide, from selenite has been studied extensively at the subcellular level by Ganther and Hsieh (1974). The reaction pathway seems to consist of reduction of the selenite to the selenide oxidation state followed by methylation by methyl transferase enzymes. The microsomal fraction of the liver contains the methyl transferase which is very sensitive to arsenite. This sensitivity may account for the ability of arsenic to inhibit the production of volatile selenium compounds. Oral administration of arsenic has been shown to detoxify selenium regardless of whether the arsenic was given in the diet or in the drinking water (DuBois et al., 1940). This report suggested that arsenic might decrease the toxicity of selenium by combining with it in the gastrointestinal tract, thereby decreasing the absorption of the element. But it was later shown that arsenic could prevent selenium poisoning, even when compounds of both elements were simulatenously injected. This finding strongly suggested that arsenic did not interfere with the gastrointestinal absorption of selenium. Ganther and Baumann (1962) have carried out total metabolic studies on rats injected with subacute doses of arsenic and selenium and found that, in addition to blocking the production of volatile selenium compounds, arsenic markedly decreased the retention of selenium in the liver and also increased the levels of selenium appearing in the gastrointestinal tract. Levander and Baumann (1966a) showed that as the dose of arsenic given to the rat was varied, there was an inverse relationship found between the amount of selenium retained in the liver and the amount which appeared in the gastrointestinal tract. This inverse relationship suggested that arsenic might act by promoting the biliary excretion of selenium. Experiments in rats or guinea pigs with cannulated bile ducts demonstrated that this was the case (Levander and Baumann, 1966b). Animals which were injected with both arsenic and selenium excreted ten times as much selenium into the bile during a 3-hr collection period as animals injected with selenium alone. Much less selenium was retained in the livers of the arsenic-treated rats versus the controls. On the other hand, there was no difference in the amount of selenium appearing in the gastrointestinal contents between these two groups of animals whose bile ducts were cannulated. The increased volume of bile excreted by the arsenic-treated rats could not account for the increased level of selenium in the bile of these rats (Levander, 1977).

This arsenic effect in stimulating the biliary excretion of selenium was seen over a wide range of dosages and under different experimental conditions. The most active form of arsenic for enhancing the biliary excretion of selenium was sodium arsenite even though sodium arsenate was also reasonably effective. Various organic arsenicals, such as 3-nitro-4-hydroxyphenylarsonic acid or arsanilic acid were much less effective. Arsenite also increased the biliary excretion of selenium when selenate was administered, but had no effect on the biliary excretion of sulfur given as sulfate. Selenite stimulated the excretion of arsenic into the bile, just as arsenite stimulated the excretion of selenium into the bile. Even though both mercury and thallium blocked the formation of volatile selenium compounds, only arsenic was quite specific in increasing the biliary excretion of selenium. Other heavy metals such as mercury, thallium, and lead had no effect in this regard. Since only 37% of the biliary selenium was dialyzable against buffered saline, the majority of the selenium in the bile from arsenic treated rats was loosely bound to macromolecules. Addition of 10^{-3} M glutathione to the dialysis medium increased the dialyzable fraction to 73% (Levander and Baumann, 1966b).

It appears that the increased biliary excretion of selenium caused by arsenic might provide a reasonable explanation for the ability of arsenic to counteract the toxicity of selenium. Some investigators have been unable to find any effect of arsenic on the fecal or urinary excretion of selenium when both arsenic and selenium were given at low dosages (Olson et al., 1963). Other investigators have shown the selenium levels were decreased in the livers of animals chronically poisoned with selenium and treated with arsenic, as compared to control animals who were given selenium only (Levander and Argrett, 1969). These latter observations are consistent with the hypothesis that arsenic clears selenium from the liver, which in many species is the primary target organ of selenium poisoning.

Although there are several possibilities, the precise chemical mechanism by which arsenic detoxifies selenium is not known. One possible explanation is that selenium and arsenic react in the liver to form a detoxification conjugate which is then excreted into the bile. This explanation would seem to be consistent with the fact that arsenic and selenium both increase each other's biliary excretion. It is not known whether this mechanism is related to the inhibitory effect of arsenite on the methyl-transferase which forms dimethyl selenide. If the methyltransferase were blocked, large amounts of hydrogen selenite could be generated in the liver. The excess hydrogen selenide could react with any arsenite which might be present, in a similar way to the reaction between thiols and arsenite. The detoxification conjugate might be a selenoarsenite which is excreted into the bile. Other metabolic interactions between selenium and arsenic may also be possible. Klug *et al.*, (1950b) showed that hepatic succinic dehydrogenase levels were markedly depressed in rats poisoned with selenium, but this depression was markedly reduced in animals which were also treated with arsenic. In addition, Levander *et al.*, (1973a) demonstrated that the glutathione-induced swelling of rat liver mitochondria, stimulated by selenite, could be markedly inhibited by arsenic, cadmium, or mercury. Unlike cadmium or mercury, arsenic had little or no inhibitory effect on the selenium-catalyzed reduction of cytochrome c by glutathione in a chemically defined model system (Levander *et al.*, 1973b). These results suggest that the arsenite acted in the mitochondrial system, but not the chemical system, by reacting with a unique grouping of ligands that were not present in the chemical model system. Levander *et al.* (1973b) have suggested that the unique grouping might be a selenopersulfide in close proximity to a sulfhydryl ligand. It is possible that an inhibitory complex might be formed between the arsenite and the selenopersulfide.

Other possible arsenic–selenium interactions of metabolic significance are based on the chemical oxyanion concepts developed by Matrone (1974). The most important parameters to be considered are the number of π–d bonds and the anion orbital configuration. On this basis, it was predicted and verified experimentally that selenate could at least partially prevent the uncoupling of oxidative phosphorylation caused by arsenate (Hill, 1975). Perhaps this metabolic antagonism may be explained by the reciprocal inhibition of uptake of these two anions by mitochondria. In yeast Bonhorst (1955) has carried out similar experiments in an attempt to use anion antagonists as possible indicators of the mechanism of selenium toxicity. More work is needed to establish whether or not these *in vitro* phenomena contribute to the protective effect of arsenic in selenium poisoning.

Although most of the evidence clearly demonstrates that arsenic decreases selenium toxicity under most experimental conditions, there are certain specific situations in which arsenic increases the toxicity of selenium. Poisoning by trimethylselenonium chloride has been shown by Obermeyer *et al.* (1971) to be markedly increased by simultaneous injection with arsenite. Ordinarily trimethylselenonium ion is a compound of relatively low toxicity. Dimethylselenide, which is generally considered a detoxification product of selenium metabolism (McConnell and Portman, 1952b) and a rather innocuous selenium compound, also was more toxic when arsenite was injected along with the dimethylselenide. These two reports of the highly synergistic toxicity between arsenic and methylated selenium is similar to that reported by Parizek et al. (1971) between dimethylselenide and mercury. [The synergistic mechanism of these two cases is unknown, but because selenium is likely methylated in the environment (Francis et al., 1974), further studies might be important in elucidating these metabolic interrelationships.]

Another example of the arsenic–selenium antagonism has been observed by Holmberg and Fern (1969), who demonstrated that selenium decreased the teratogenic toxicity of arsenic in hamsters when salts of arsenic and selenium were injected at the same time. Palmer et al. (1973) have found that arsenite decreased the toxicity of several selenium compounds to chick embryos. The toxicity of trimethylselenonium ion to chick embryos was even decreased by arsenite, even though these two compounds have a marked synergistic toxicity in rats.

The mechanism of how selenium decreases the toxicity of arsenic in the embryo or how arsenic decreases the toxicity of selenium is not known but some of the metabolic relationships between arsenic and selenium which were previously described might be involved. In contrast, Walker and Bradley (1969) have found a direct synergistic effect of sodium arsenate and selenocystine on chromosomal crossing over in fruit flies. They related this effect to the possible incorporation of arsenate and selenocystine into DNA and chromosomal protein. Because this effect was a direct one, their finding does not explain the arsenic–selenium antagonism observed by Palmer et al. (1973) had Holmberg and Fern (1969).

Because arsenic has been found to decrease selenium accumulation in the tissues, one would expect that arsenic could increase the cases of nutritional deficiency of selenium in animals fed low-selenium diets. However, all experiments trying to show that feeding arsenic compounds would promote selenium deficiency have been unsuccessful. The incidence or severity of gizzard myopathy in turkey poults (Scott et al., 1967) fed low-selenium diets was not increased by the addition of arsanilic acid or p-ureidobenzenearsonic acid to the diet. Neither sodium arsenite (Whanger, 1976) nor arsanilic acid (Bunyan et al., 1968) increased the development of liver necrosis in rats fed diets deficient in selenium and vitamin E. Weanling rats fed low levels of selenium were also not affected by arsanilic acid (Halverson and Palmer, 1975). Finally, when arsenic trioxide or sodium arsenate were added to a selenium-deficient diet, there was no significant effect on the induction of white muscle disease in lambs (Whanger *et al.*, 1976). There was also no significant effect on the elevated activities of several plasma enzymes associated with this selenium deficiency.

The known metabolic relationships between arsenic and selenium have led some workers to test arsenic in order to see if arsenic would prevent selenium deficiency diseases in animals. Schwarz and Foltz (1957) have reported that liver necrosis in rats was not affected by sodium arsenate. In addition, exudative diathesis in chicks was not affected by combinations of arsenate arsenite. One report, which claimed that the incidence of selenium deficiency induced myopathy in lambs was reduced by sodium arsenate (Muth *et al.*, 1971), has not been confirmed (Whanger *et al.*, 1976).

Olson (1960) has discussed the use of organic arsenicals as a practical means to cure or prevent selenium poisoning in farm animals. Both arsanilic acid and 3-nitro-4-hydroxyphenylarsonic acid are very attractive possibilities since they were already used as feed additives to stimulate the growth of swine and poultry. At this time, this does not seem to be a feasible way of controlling selenium poisoning in livestock.

When the use of arsenicals in humans as antisyphilitic drugs was still an acceptable practice, Amor and Pringle (1945) suggested that a tonic containing arsenic should be used as a prophylactic against selenium poisoning. It is possible from the industrial hygienist's point of view that arsenic could be used as an antidote against selenium poisoning or selenium could be used as an antidote against arsenic poisoning.

5.3 Copper

Adding high dietary levels of copper to a chick diet marginal in selenium has resulted in selenium deficiency signs (Jensen, 1974). In addition, both exudative diathesis and muscular dystrophy were observed in chicks fed a high copper level (Witting and Horwitt, 1964). Because selenium is an integral part of the enzyme glutathione peroxidase, high dietary levels of copper probably reduce the availability of selenium in tissue for synthesis of glutathione peroxidase by interfering with absorption and/or formation of insoluble intracelluar selenium compounds. In a similar way, when chicks were fed selenium dioxide, the toxic effects of selenium, as measured by growth retardation and mortality could be partially alleviated by the inclusion of cupric sulfate in the diet. Jensen (1975) has also reported that dietary copper has counteracted the growth depression and prevented mortality when toxic amounts of selenium were included in the diet.

In another experiment, 34 adult ponies were used to determine the effects of single oral doses of copper supplements (0, 20, and 40 mg of Cu/kg of body weight) on the toxicity of oral doses of selenium supplements (0, 2, 4, 6, and 8 mg of Se/kg of body weight), which were administered 24 hr after the copper was given. Selenium toxicosis signs, i.e., sweating, diarrhea, tachycardia, tachypnea, mild pyrexia, lethargy, and colic, developed in ponies given 6 and 8 mg of Se/kg of body weight with no copper pretreatment (Stowe, 1980). Two of four ponies given 6 mg of selenium/kg and both ponies given 8 mg of Se/kg with no copper pretreatment died within 36 hr after being given the selenium. Ponies which had been given either 20 or 40 mg of Cu/kg were not affected by the subsequent selenium supplement, regardless of dosage. The pretreatment by copper did not seem to inhibit the absorption of selenium, as serum selenium concentrations were similar, but copper pretreatment hastened the disappearance of the selenium from the serum.

In another type of experiment, rats receiving a low intake of dietary selenium have a decreased threshold of copper toxicity, as compared with selenium-supplemented controls (Godwin et al., 1977). When a dosage between 0.16 and 0.24 mg Cu/100 g body weight was given intraperitoneally to selenium-deficient rats, deaths occurred, with an accompanying hemolytic crisis. The mode of administration apparently makes a difference because if the same amounts of copper were given by oral means, similar though less severe changes occurred. The toxic effect is likely due to increased fragility of the erythrocyte membranes following the low selenium intake. Copper can induce lipid peroxidation in hepatocytes (Lindquist, 1968) and oxidative damage to red cells (Metz, 1969), which are both among the kind of lesions selenium is supposed to counteract. Thus, it would be tempting to hypothesize that at least one role for selenium is the defense against manifestations of copper toxicity.

In lambs with suboptimal copper status, Thomson and Lawson (1970) have observed a significant increase of liver copper concentrations after selenium administration, whereas no effect was observed in copper-supplemented lambs. Fehrs et al. (1981) have observed that feeding 1 ppm of selenium as sodium selenite, along with either 0 or 100 ppm of supplemental copper to Holstein calves, increased tissue copper in the calves fed high-selenium–high-copper diets. Copper was also greater in the heart tissue of both groups fed high selenium than in the controls. In contrast, Lee and Jones (1976) found small or no effects of selenium on liver copper concentrations in sheep. In both of these studies, the levels were within 150–460 mg Cu/g dry matter or about 45–140 mg Cu/g wet weight. Silvertsen et al. (1978) have studied the liver concentrations of selenium in 88 normal sheep and in 45 cases of chronic copper poisoning, at different intervals of liver copper concentrations. Their study did not reveal any systemic relationships between copper and selenium concentration in sheep livers fed with normal to high copper levels. Similar results have been reported by McGuire et al. (1981). In calves 1 ppm of added dietary selenium increased copper retention, but high copper had little effect on tissue ^{75}Se retention. In another experiment, orally given selenite (5.7 or 57 μmol/100 g food) has increased the kidney levels of copper in rats (Alexander and Aaseth, 1980). In the same experiment, selenite (5 μmol/kg) was shown to decrease the copper excretion in bile, but glutathione levels were not significantly affected. In calves, accumulation of copper in liver prevented an elevation of copper in the plasma when 100 or 200 μg/g was fed (Amer et al., 1973). Neutrophils were isolated from Friesian steers fed diets deficient in either Se or Cu or both. The neutrophils from the steers fed the Se-, Cu-, and Se–Cu-deficient diets were less able to kill ingested C. albecans than were the neutrophils from the steers fed a diet with Se and Cu supplements. The effect of the combined Se–Cu deficiency did not appear to be additive (Boyne and Arthur, 1981).

No biochemical basis for a Cu–Se interaction has been reported Rafter (1980) had found that copper inhibited yeast glutathione reductase. Glutathione reductase is thought to be necessary for the release of reduced selenium from erythrocytes (Gasiewicz and Smith, 1978b), an important step in the metabolism of intravenously injected selenite (Sandholm, 1973).

5.4 Silver

Administration of silver nitrate or silver lactate in the drinking water of rats fed a vitamin E-low diet resulted in muscular dystrophy, necrotic degeneration of the liver, and increased mortality (Shaver and Mason, 1951). Similar effects on selenium toxicity by silver in rats have been observed by Cabe et al. (1979). A proexudative effect of silver acetate when fed at a level of 20 ppm silver to vitamin E-deficient chicks has been reported by Dam et al. (1958). Diplock et al. (1967) have found a high incidence of liver necrosis and death in weanling rats given a low-casein diet and silver acetate (130 ppm in the diet or 1500 ppm in the drinking water). Even though both vitamin E and selenium prevented the disease, it was found that the silver treatment did not increase the metabolism of small doses of radioactive α-tocopherol which had been previously administered. Diplock et al. (1967) also found that the silver-treated animals actually contained more α-tocopherol in the liver than in the controls. Van Vleet et al. (1981b) has observed that pigs fed Ag (3000 mg/kg of feed as acetate) induced lesions characteristic of selenium–vitamin

E deficiency with accumulations of serous transudates in body cavities and hepatic and cardiac necrosis. Blood glutathione peroxidase activities decreased to low levels several weeks before the pigs died with lesions of selenium–vitamin E deficiency. Even though silver toxicity is more complex in the chick than in the rat, Bunyan et al. (1968) also induced a silver deficiency.

Jensen et al. (1974) have added silver acetate or silver nitrate to the diets of large white turkey poults. The highest level of silver (900 ppm) depressed growth rate, reduced packed cell volume, slightly reduced hemoglobin level, and also caused cardiac enlargement. A varying incidence of degeneration of the gizzard musculature was induced by the high silver level. The degeneration of the gizzard musculature was completely prevented by adding 1 ppm selenium or partially prevented by adding 50 IU vitamin E/kg diet. Their results indicated that a high level of silver induced a selenium and also a copper deficiency when added to practical diets for turkey poults. Jensen (1975) has also studied the effect of dietary silver (1000 ppm) on the effect of a toxic 5 and 10 ppm of dietary selenium. The dietary silver counteracted the growth depressions and prevented mortality at the higher levels of selenium. Hepatic selenium of chicks fed silver was less than that of the control chicks when diets containing 5 or 10 ppm of selenium were fed. Results of an experiment to determine the effects of dietary silver on the distribution of ^{75}Se, administered either orally or intramuscularly, showed that silver interfered with absorption of selenium. The results of these experiments suggest that silver modifies the toxicity of selenium both by interfering with selenium absorption and by causing the accumulation of nontoxic selenium compounds in the tissues. Ducklings fed Ag (3000 mg/kg of feed as acetate) developed lesions characteristic of selenium–E deficiency, such as necrosis of skeletal and cardiac muscle and of smooth muscle of the gizzard and intestine (Van Vleet, 1981a).

Wagner et al. (1975) found that selenium (0.5 ppm, as sodium selenite) prevented the growth depression observed with 75 ppm silver, and markedly improved growth and survival of rats given 750 ppm of silver even though the silver concentration of liver and kidney were greater. Numerous other experiments with 750 ppm silver showed that either 0.5 ppm of selenium or 100 IU/kg of vitamin E largely prevented the growth depression. Liver glutathione peroxidase activity in rats given 0.5 ppm selenium plus 75 ppm silver for 52 days was only about 30% of that for rats given 0.5 ppm selenium without silver. When the rats were given 750 ppm silver and 0.5 ppm selenium for 52 days, glutathione peroxidase was practically undetectable. Dietary silver also decreased glutathione peroxidase in erythrocytes and kidney. Black et al. (1979) has reported that silver markedly decreased the activity of glutathione peroxidase in the liver and kidneys of rats. Tissue selenium levels were also decreased by silver. Silver had no effect upon the glutathione transferase in liver and testes.

A time study of the liver levels of glutathione peroxidase, silver, and selenium in rats given 750 ppm silver plus 0.5 ppm of selenium showed that there was an initial rise in liver selenium concentration during the first week, paralled by a rise in liver silver. After this time, liver selenium and silver concentrations fell to a minimum at 22 days but rose slowly thereafter. Throughout the 22-day period liver glutathione peroxidase activity fell progressively and remained at low levels. This study suggests that the liver accumulated a biologically inactive form of selenium which might be complexed to silver, during the first week. However, by increasing dietary selenium from 0.5 to 5 ppm, the effect on liver glutathione peroxidase was partially overcome. The effect of silver on liver glutathione peroxidase was reversible; in those rats which received 0.5 ppm selenium, enzyme activity returned to normal levels within five weeks when the administration of silver was stopped. These results suggest that silver induces a conditional deficiency of selenium and makes selenium unavailable for glutathione peroxidase activity. The effectiveness of low levels of selenium against silver toxicity might be due to its ability to overcome the conditioned deficiency of selenium. It is not necessary that the significant fraction of the total silver be complexed by selenium, nor is it necessary for selenium to decrease tissue levels of silver.

Vitamin E was as effective as selenium in overcoming the growth depression produced by silver, but did not prevent the depression of glutathione peroxidase. Therefore, it appears that the growth depression is not related to the silver-induced decrease in glutathione peroxidase.

5.5 Cobalt

Evidence for a possible interaction between cobalt and selenium in sheep nutrition has accumulated in Western Australia in high-rainfall districts where both trace elements may be present in soils and plants in marginal or deficient amounts. Cobalt deficiency causes sheep disease syndromes ranging from suboptimal growth and production to severe wasting and death. These symptoms were described by Bennetts (1959) in many parts of the western and southern littorals and in other high-rainfall areas of southwest Western Australia.

Gardiner (1961) first recognized selenium deficiency in Western Australia. Deficiency areas were rapidly outlined and within the decade were fairly well known with respect to rainfall patterns, soil types, and agricultural practices (Gardiner, 1969).

The only significant selenium-responsive condition so far found in Western Australia is white muscle disease in young lambs. The disease seemed to vary with the season. It appeared in young lambs during the spring flush of pasture growth and in weaners during the dry summer months. These conditions and those resulting from a deficiency of cobalt occur sporadically throughout the same large area west and south of the 20-in. isohyet and are both apparently influenced by the same climatic, seasonal, and geological factors. These sporadic deficiencies that appear may also be dependent on the supplementary use of the trace elements, of which has become widespread in this area.

An interrelationship between selenium and cobalt was suggested as a factor in the complex cause of Clover disease in sheep in Western Australia (Gardiner and Nairn, 1969) and in the susceptibility of sheep to chronic selenium toxicity (Gardiner, 1966).

Gardiner and Nicol (1971) have investigated the effect on rats of prolonged feeding of each trace element on their subsequent concentrations in heart, skeletal muscle, liver, and kidney, which are tissues that are presumed to be metabolically important in the pathogenesis of conditions associated either with deficiencies or excesses of cobalt and/or selenium. Gardiner and Nicol (1971) found that the levels of selenium in the heart and skeletal muscle were positively related to dietary selenium and negatively related to dietary cobalt. They also found that the levels of selenium in liver and kidneys were directly related to dietary selenium and negatively related to an interaction between dietary selenium and cobalt. Dietary selenium did not influence the heart muscle cobalt level. Gains in body weight and histological appearance of the major tissues of rats fed either or both of the trace elements in concentrations as high as 8.6 ppm in the ratio were not affected except for some minor liver changes in those fed the ration containing 8.6 ppm of cobalt and low levels of selenium.

Grice et al. (1969) have induced a cardiomyopathy in rats given oral doses of cobalt as low as 3 mg/kg daily for 180 days. Grice et al. (1969) have also compared the cardiomyopathy induced by excess cobalt to the cardiomyopathies observed in humans induced by adding cobalt to beer. The cobalt intake in the rat experiment of Grice et al. (1969) is about half that of the young rats of Gardiner and Nicol (1971), who fed a 8.6 ppm cobalt ration but could detect no cobalt cardiopathy. Because the heart is indirectly affected in the selenium-deficiency-caused white muscle disease, it is possible that an excess of cobalt might have a synergistic effect on the selenium-induced heart lesions. It is likely that the cobalt–selenium interactions reported by Gardiner and Nicol (1971) probably occurred in the gut, with effects on absorption rather than in the tissues studied. Macroscopic lesions were not observed in pigs fed cobalt

(500 mg/kg, as chloride). However, evidence of selenium–vitamin E deficiency was indicated by microscopically detected necrosis of cardiac and skeletal muscle in 50% of the pigs fed cobalt (Van Vleet et al., 1981b). Ducklings fed cobalt (200 or 500 mg/kg, as chloride) developed lesions characteristic of selenium–vitamin E deficiency, such as necrosis of skeletal and cardiac muscle, and of the smooth muscle of the gizzard and intestine (Van Vleet, 1981a). Complete protection was afforded by 2 mg/kg of selenite.

5.6 Manganese

The essentiality of manganese in animal nutrition was shown by Kemmerer et al. (1931), who showed that this element was necessary for growth in mice. In the same year Orent and McCollum (1931) demonstrated that manganese was necessary to prevent testicular degeneration in male rats as well as the maintenance of normal ovarian function in female rats. A dietary deficiency of manganese in poultry results in two disease entities: (1) perosis or slipped tendon (Wilgus et al., 1936) and (2) chondrodystrophy (Lyons and Insko, 1937). Leach et al. (1969) showed that two enzymes involved in chondroitin sulfate synthesis have decreased activity along with manganese deficiency. These enzymes, which are found in the microsomal fractions of epiphyseal cartilage homogenate, are (1) polysaccharide polymerase, which catalyzes polysaccharide synthesis from UDP–N-acetylgalactosamine and UDP–glucuronic acid, and (2) galactotransferase, which catalyzes the incorporation of galactose from UDP–galactose into the galactose–galactose–xylose trisaccharide. This latter reaction serves to link the mucopolysaccharide and protein. The increases in enzyme may explain the chondrodystrophy and perosis found in poultry. There has been one case of manganese deficiency reported in humans (Doisy, 1973). Manganese salts in this case were accidently omitted from a purified diet being used to study vitamin K deficiency in an adult volunteer.

Manganese has been suggested as a necessary cofactor for many enzymatic reactions of which the best known is hepatic arginase (Underwood, 1971). One known manganese metalloenzyme is pyruvate carboxylase from chicken liver (Scrutton et al., 1966). This metalloenzyme contains 4 g-atoms of firmly bound manganese in constant proportions to 4 mol of bound biotin. Pyruvate carboxylase catalyzes the fixation of CO_2 by pyruvate, to form oxaloacetate. In addition, manganese may be involved in the synthesis of protein (Weser and Koolman, 1970), DNA (Wiberg and Neuman, 1957), and RNA synthesis (Windall and Tata, 1966).

Only one experiment studying the interrelationship of manganese and selenium has been attempted (Burch et al., 1975). Pigs received a diet containing 0.59 ppm of manganese. Pigs which served as controls received the same quantity of the same diet which was supplemented with 22 ppm of manganese. After six weeks, pigs fed the low-manganese diet had a decrease in tissue manganese levels as well as a decrease in tissue selenium content. The dietary regimen did not affect the tissue levels of zinc, copper, calcium, and magnesium. The experimental animals were probably manganese deficient as evidenced by the fact that there was a statistically significant decrease in hepatic arginase activity that could be enhanced by the addition of manganese. There was no significant differences in other hepatic enzymes, glutamic dehydrogenase, ornithine transcarbamylase, leucine aminopeptide, and isocitric dehydrogenase. These studies indicate that there may be a interaction between tissue selenium and manganese. Tissue selenium levels were decreased to a statistically significant degree in all examined tissues, except the kidney, in the manganese-deficient animals. Even though the decrease in the kidney was statistically insignificant, there was a decrease in renal selenium which represents a 25% lower value than that found in the control kidneys. In all other tissues the decrease varied from a low of 52% found in spleen, to a high of 76% found in muscle.

It is not known what the relationship is between the low-manganese diet and the low-tissue-selenium levels. Furthermore, it is also unknown how the low-tissue-selenium levels occur. However, it would appear that either decreased absorption and/or increased excretion of selenium must be involved instead of a redistribution of total body selenium, since all tissues studied had low selenium levels.

5.7 Lead

Chronic lead poisoning can affect the gastrointestinal, hematopoetic, or central nervous system, producing clinical signs accordingly. Autopsies of animals subjected to either chronic or acute lead toxicities showed central nervous system (CNS) and/or gastrointestinal lesions. In addition, edema of the kidneys as well as abnormal erythrocyte metabolism and morphology has been demonstrated (Lynch et al., 1976).

There has been some success in regard to attempts to reduce absorption and/or increase the excretion of absorbed lead. Feeding excess calcium to swine has a protective effect against lead toxicity, but an excess of dietary zinc increases the problem (Hsu et al., 1975). Chelating agents have been widely utilized to remove ingested lead, but these agents sometimes have an adverse effect on the metabolism of essential minerals (Lilis and Fischbein, 1976). Both selenium and vitamin E have both been shown to be involved in decreasing the toxic effects of lead in rats.

However, vitamin E status appeared to be more important than selenium (Levander *et al.*, 1977c).

In the experiments of Levander *et al.* (1977c), weanling male rats were fed for three months a Torula yeast diet which was supplemented with selenium and vitamin E. Of the rats fed each diet, one group received 250 ppm lead in the drinking water whereas another group did not. In rats not poisoned with lead neither vitamin E or selenium deficiency affected spleen weight, hematocrit value, or erythrocyte mechanical fragility. Deficiency of vitamin E increased the splenomegaly, anemia, and the mechanical fragility of the red cells of lead-poisoned rats. But excess levels of selenium (2.5 and 5 ppm) in the vitamin E-deficient diet had little or no effect on spleen size or hematocrit of rats not receiving lead, but partially prevented the splenomegaly and anemia of lead-poisoned rats. An excess of selenium delayed the decrease in filterability of red cells from either nonpoisoned or lead-poisoned vitamin E-deficient rats, but was not as effective as vitamin E. These results show that vitamin E status of rats is more important than the selenium status in determining response to toxic levels of lead. Excess dietary selenium did partially protect against lead poisoning in vitamin E-deficient rats, but the levels of selenium used were also toxic.

The greater effect by vitamin E is consistant with previous studies showing a relationship between lead poisoning and vitamin E. Lead poisoning causes a marked anemia in vitamin E-deficient rats (Levander *et al.*, 1975). Erythocytes from these animals have an increased mechanical fragility (Levander *et al.*, 1975), decreased peroxidative fragility and osmotic fragility (Levander *et al.*, 1977a), and decreased filterability (Levander *et al.*, 1977b). Erythrocyte filterability from lead-poisoned, vitamin E-deficient rats, was found to be negatively correlated with the degree of lipid peroxidation in the red cells after incubation *in vitro* (Levander *et al.*, 1977b). Levander *et al.* (1977b) have demonstrated that a synthetic antioxidant, N,N'-diphenyl-p-phenylenediamine, is as active as vitamin E in protecting erythrocytes of vitamin E-deficient rats against the decreased filterability caused by lead. These experimental results suggest that lead may exert its harmful effect in the interior, hydrophobic region of the red cell membrane, where the lipid soluble antioxidants can penetrate. In contrast, selenium as part of glutathione perioxidase occurs mainly in the aqueous phase of the cell since the enzyme is found largely in the cytoplasm (Ganther *et al.*, 1976).

Selenium supplementation of 0.5 ppm to vitamin E-supplemented, lead-poisoned rats had a slightly harmful effect, since such treatment caused a slight increase in spleen size and decrease in hematocrit (Levander *et al.*, 1977b). Similar selenium supplementation also increased the urinary excretion of δ-aminolevulinic acid (ALA) by lead-poisoned rats also receiving vitamin E. Similarly, Cerklewski and Forbes (1976) observed a deleterious effect of 1 ppm of selenium on the urinary excretion of ALA by lead-poisoned, vitamin E-supplemented rats, even though 0.5 ppm was beneficial. Stone and Soares (1976) found no protective effect of 1 ppm of selenium against the inhibition of red cell δ-amino-levulinic acid dehydratase in Japanese quail which had been poisoned with lead for 12 weeks.

Rastogi et al. (1976) have observed that toxic levels of selenium counterbalanced toxic levels of lead in rats as evidenced by growth rate, food consumption, δ-amino-levulinic acid dehydratase, and P-450 enzymic activities. Levander and Argrett (1969) have demonstrated that injecting rats with lead acetate had no effect on the short-term metabolism of an injected dose of sodium selenite. Bell et al. (1978) have reported that feeding 250 ppm lead had only minor effects on chicks with some indication that lead aggravated signs of selenium deficiency.

Little is known about the interactions between organic lead and Se or vitamin E. Rastogi et al. (1976) have demonstrated some protective effects of Se against lead naphthenate toxicity. Skilleter (1975) has provided evidence for changes in membrane permeability following triethyllead poisoning. Ramstoeck et al. (1980) have studied the ability of trialkyllead compounds to induce lipid peroxidation (measured by ethane production in vivo) in rats fed a diet deficient in vitamin E and selenium or supplemented with 200 IU of D,L-d-tocopherol or 0.6 ppm selenium supplied as sodium selenite. In vitamin E- and selenium-deficient animals, trimethyllead induced lipid peroxidation in vivo to the greates extent. Vitamin E and, to a lesser extent selenium, offered protection against this effect of trimethyllead. In the case of triethyllead, metabolism of the chemical caused the release of large amounts of ethane and ethylene; therefore, the use of ethane production as an index of lipid peroxidation was prevented.

Even though current data do not allow conclusions with regard to the scope of the interaction of lead and selenium, present data suggest that the nature of the interaction is not nearly as complex and as extensive as those of mercury and selenium, cadmium and selenium, or arsenic and selenium.

5.8 Mercury

5.8.I. Inorganic and Organic Mercury

Parizek and Ostadalova (1967) first demonstrated that selenite and to a lesser extent selenomethionine, markedly decreased the acute nephrotoxicity of mercuric mercury in rats, as long as the selenium compound was given after the mercury compound. If, on the other hand, selenium was given before mercuric mercury, males showed an increased mortality. Similar results were observed if mercuric mercury and dimethylselenide or trimethylselenonium ion were administered (Parizek *et al.* 1971). In chronic exposure in rats, selenium reduced the toxicity of mercuric mercury (Groth, 1973). The protective effect of selenium against mercury poisoning in the rat has been confirmed by several investigators. Burk *et al.* 1974b) made similar observations of the antagonism of selenium on mercuric-chloride-poisoned animals. On the other hand, mercuric mercury did not affect the toxicity of selenite (Levander and Argrett, 1969). In a similar way, selenium has been observed to reduce the toxicity of methyl mercury when administered at the same time (E1-Begearmi *et al.*, 1977).

The mechanism by which selenium reduces the toxicity of the mercury compounds are not well understood and appear to be complex. Simultaneous administration of selenium increases whole-body retention of mercury (El-Begearmi *et al.*, 1977). In general, the increased retention of mercury occurs in the blood, liver, and the spleen, while there is a decrease in the kidneys (Parizek *et al.*, 1969b). On the other hand, selenite may not influence the accumulation of mercury in rat organs. The amounts of mercury found in the tissues depended on the chemical form and dose of mercury and the duration of the experiment. The methyl mercury content of the brains, in methyl mercury- and selenium-treated animals, increased but neurological disturbances did not appear. Selenium is first associated with protein and then next, retains mercury. In this way mercury is prevented from reaching the target structures (Burk *et al.*, 1974b). After this occurs selenium may change the distribution of mercury among the soluble proteins in liver or kidney (Mengel and Karlog, 1980). Selenium was found to be retained especially in the kidneys, when reducing the toxicity of mercury compounds (Burk *et al.*, 1974b). Mercury significantly depressed the activity of glutathione peroxidase (Black *et al.*, 1979). The most pronounced inhibition of glutathione peroxidase was found in the liver and kidneys. In the same experiment glutathione transferase was initially increased in the kidneys.

In red blood cells, selenite accelerated the uptake of inorganic mercury (Imura and Naganuma, 1978). The major fractions of mercury and selenium were eluted in a high-molecular-weight fraction on gel filtrations of rat plasma (Burk *et al.*, 1974b), as well as rabbit stroma-free hemolysate (Imura and Naganuma; 1978) following simultaneous administration of mercuric chloride and selenite.

Burk *et al.* (1974b) have suggested that one action of selenite is to stimulate the removal of inorganic mercury by metallothionein. However,

Chmielnicka and Brzeznicka (1978) have found that selenite interfered with the stimulation of metallothionein synthesis by a variety of inorganic and organic mercurials. Potter and Matrone (1977) have suggested that selenium protects against mercury toxicity by a direct interaction with methyl mercury, thus implying that the product is less toxic than methylmercury alone.

5.8.2 Tissue Distribution

Kari and Kauranen (1978) have reported that the mercury and selenium contents in livers of seals from fresh and brackish waters in Finland were strongly correlated, with the Hg/Se ratio being 1.16:1. There were no correlations found in either the kidneys or the muscles. Tamura et al. (1975) have observed in fish that the molar concentration levels of selenium exceeded those of mercury in all tissues, but no constant ratio could be found. Administration of selenite prevented mercury inhibition of kidney glutathione peroxidase in mice (Wada et al., 1976). In this experiment, they found a Hg/Se ratio which ranged from 1:1 in the liver to 2.2:1 in the kidneys. Burk et al. (1974a) have reported that there is a 1:1 molar ratio of Hg/Se in soluble plasma proteins. This ratio has also been looked at by several others. Koeman et al. (1975) have found that in the livers of marine mammals the molar Hg/Se ratio was also 1:1. However, they failed to find a constant ratio among marine birds. Chen et al. (1975) have reported that the mercury in the soluble fractions of the liver, kidney, spleen, and testicles of rats administered only mercuric chloride was found mainly in the low-molecular-weight fractions (about 10,000 MW) when eluted from the gel filtration column. If the rats were pretreated with selenite before mercuric chloride was administered, most of the mercury instead was bound to higher-molecular-weight fractions (65,000 to about 200,000 MW).

Cappon and Smith (1981a) have studied the content, chemical form, and distribution of mercury and selenium for selected human and animal tissue samples by gas chromatography. Methylmercury averaged for human brain, heart, spleen, liver, kidney, and placenta, 38.7%, 40.2%, 57.0%, 39.6%, 6.0% and 57.1%, respectively, of the total mercury content. Similar results were obtained for the heart and liver of a whitetail deer. The amounts of selenium paralleled the amounts of mercury. For all samples except kidney, liver, and deer meat, 55%–76% of the total selenium content was water extractable. Selenate [Se(VI)] was more extractable on a percentage basis than selenite [Se(IV)] and selenide [Se(II)]. For all samples, a significant portion of the total selenium content, averaging 27%, was present as selenate.

5.8.3 Properties of the Mercury–Selenium Complex

Naganuma and Imura (1980) have reported that bis(methylmercuric) selenite (BMS), $(CH_3Hg_2)Se$, was formed from methylmercury and selenite in rabbit blood. BMS was also formed in the reaction of methylmercury and selenite in the presence of reduced glutathione (GHS) (Naganuma and Imura, 1980). Since GSH exists extensively in animal tissues at high concentrations, there is also a possibility that BMS is formed in animal tissues from methylmercury and selenium. The formation of BMS could play an important role in the modifying effect of selenium on methylmercury toxicity. Naganuma et al. (1980) has studied the in vitro and in vivo formation and decomposition of BMS in mice. When methylmercury and selenite were added in vitro the homogenates of the soluble fraction or insoluble fraction of mouse liver, kidney, spleen, and brain, substantial amounts of BMS were formed. BMS was also formed by the addition of methylmercury into the soluble fraction of liver or kidney obtained from the mouse pretreated with selenite. However, in the tissues of mice injected intravenously with both methylmercury and selenite or even with BMS itself, BMS was hardly detected. These experiments, together with the rapid decomposition of BMS observed in vitro, suggest that the cycle of formation and decomposition of BMS may occur rapidly in vivo. Naganuma et al. (1980) also found that the concentrations of mercury and selenium in the brain of the mice receiving BMS were significantly higher than those of the mice administered methylmercury and/ or selenite. This observation suggests a possibility that the increase of brain-mercury concentration by the administration of selenite reported by several investigators is due to the formation of BMS from methylmercury and selenite.

5.8.4 Teratogenicity

Methylmercury has been shown to be a powerful environmental contaminent, which differs from many other mercurials in that it readily crosses the placental barrier and accumulates in the fetal tissues (Ukita et al, 1967). Several studies have established the teratogenicity of methylmercury in frogs (Dial, 1976), fish (Dial, 1978), mice (Su and Okita, 1976) rats (Snell et al., 1977), hamsters (Harris et al., 1972), cats (Khera, 1973), and humans (Harada et al., 1975). Su and Okita (1976) have observed that cleft palate is a prominant feature of methylmercury teratogenicity. Under appropriate conditions of administration, the incidence of cleft palate may be as high as 97% among the surviving fetuses of mice.

Lee *et al.* (1979) have examined the effects of simultaneous administration of sodium selenite on the teratogenic action of methylmercury. Simultaneous administration of selenium at concentrations of 0.0625–3.5 mg/kg/day did not reduce the incidence of cleft palate. However, concentrations of 0.5 and 1.0 mg/kg selenium appeared to increase the maternal toxicity and the teratogenicity of methylmercury. Perhaps bis(methylmercuric) selenite is formed and is responsible for the latter increase. Nobunaga *et al.* (1979) have performed a similar experiment using female IVCS mice. Embryolethality was decreased, but the incidence of cleft palate was increased in the group receiving the high dose of selenite combined with the high dose of methylmercury.

Interactions between dietary mercury and selenium have been observed in chickens whereby significant amounts of both elements are deposited in the eggs (Sell *et al.*, 1974). When mercury was included in the diets (20 ppm as methylmercury) for laying hens, dietary selenium (8 ppm) was shown to increase the mercury content in egg white (Magat and Sell, 1979). There was a simultaneous reduction of the mercury content in egg yolk. About 97% of the total mercury in egg white was associated with ovalbumin and selenium had very little influence on this. The largest proportion of selenium was found in the globulin of the egg white. It is possible that this preferential binding of mercury by ovalbumin and of selenium by globulin, may be the reason for essentially no interaction of these two elements in egg whites. Latshaw (1975) has suggested that the different affinity of selenium for egg white and yolk may be due to the origin of the corresponding proteins from the oviduct and liver, respectively. The liver is a major site of selenium metabolism, whereas the lower concentration in the oviduct tissue could explain the lower selenium concentration in the egg white.

Cappon and Smith (1981) have studied the distribution of mercury and selenium in the whites and yolks of eggs from chickens fed seed grain treated with a mercurial fungicide. The majority of the mercury was associated with the white portion of the egg, whereas selenium was mainly associated with the yolks. The majority of the egg white mercury and selenium from the eggs was water extractable. Both elements appear to be either unbound or bound to polar constituents such as simple amino acids, peptides, and low-molecular-weight proteins (MW < 100,000). The existence of unbound mercury as CH_3Hg^+ or Hg^{2+} and selenium as SeO_3^2 or SeO_4^2 is highly unlikely, since these species have a strong affinity for sulfhydryl (SH) binding sites in tissue. Therefore, the association of both elements with water-soluble (polar) proteins is most likely in egg white. Cappon and Smith (1981) also suggested that both mercury and selenium have a slightly greater affinity for higher-molecular-weight (nonpolar) constituents such as lipoproteins or phospholipids in the yolk. In addition to the whole animal studies, a tissue culture model for mercury–selenium interactions has been described (Potter and Matrone, 1977). Chang's liver cells and 3T3 cells were protected against mercury toxicity by selenite in these cultures.

5.9 Thallium

Hollo and Zlatarov (1960) have reported that death in rats due to thallium poisoning could be prevented by the parental administration of selenate. This observation was later confirmed by Rusiecki and Brzezinski (1966), who showed that oral administration of selenate prevented the toxicity of thallium. The content of thallium in liver, kidneys, and bones was markedly increased by the selenate treatment. Conversely, the subcutaneous injection of thallium acetate increased the retention of selenium in the liver and kidney and also diminished pulmonary and urinary excretion of selenium (Levander and Argrett, 1969). In vitamin E- and selenium-deficient rats, 10 ppm of dietary thallium did not promote liver necrosis.

5.10 Tellurium

In a study in ducklings, tellurium toxicosis resulted in the development of cardiac lesions that resembled those of selenium–vitamin E deficiency (Carlton and Kelly, 1967). Ducklings fed 500 ppm tellurium tetrachloride developed characteristic clinical signs and pathologic alterations of selenium–vitamin E deficiency, in spite of being fed diets adequate in selenium and vitamin E. Affected birds had anorexia, slowed growth, a reluctance to stand, and eventually many died. Lesions included myopathy of gizzard, intestine, skeletal muscles, heart, and the hydropericardium (Van Vleet, 1977; Van Vleet et al., 1981a). The birds fed tellurium had marked vascular injury with prominent epicardial and myocardial congestion, edema, and hemorrhage. Hydropericardium was frequently severe. Myocardial necrosis involved all chambers and tended to involve the full wall. The damaged hearts in these ducks fed tellurium appeared similar to the red "mulberry hearts" described in pigs with selenium–vitamin E deficiency (Grant, 1961). Most other animal species with selenium–vitamin E deficiency have pale hearts. Focal cerebral malacia was also observed in the ducks fed tellurium. Clinical signs and lesions were prevented completely in ducklings fed tellurium with supplements of selenium at 5.0

ppm as sodium selenite or vitamin E as α-tocopherol at 200 IU/kg of ration. Clinical signs and lesions were partially protected by addition of selenium at 1.0 or 0.1 ppm as sodium selenite, vitamin E as α-tocopherol at 30 IU/kg of feed, or 0.3% ethoxyquin. No protection was afforded by the addition of 0.4% methionine.

Six weanling pigs were fed for ten weeks a commercial ration that was adequate in selenium and vitamin E content. These six were compared to another six pigs, who in addition to the adequate selenium and vitamin E diet were also fed tellurium tetrachloride (500 mg/kg of diet) (Van Vleet et al., 1981b). Evidence of selenium–vitamin E deficiency, as indicated by microscopically detected necrosis of cardiac and skeletal muscle, was present in 65% of the tellurium fed pigs. The pigs fed tellurium had a marked decrease of blood glutathione peroxidase activity over the last six weeks of the feeding period.

5.11 Vanadium

Vanadium added to a low-selenium–vitamin E basal diet enhanced the development of selenium–vitamin E deficiency in rats (Whanger and Weswig, 1978), but white Pekin ducklings fed vanadium (100 mg/kg, as vanandate) for 15–28 days did not develop lesions characteristic of selenium–vitamin E deficiency (Van Vleet et al., 1981a). Macroscopic lesions of selenium–vitamin E deficiency were not found in pigs fed vanadium (200 mg/kg, as vanadate) (Van Vleet, 1981b). However, evidence of selenium–vitamin E deficiency, as indicated by microscopically detected necrosis of cardiac and skeletal muscle, was present occasionally in pigs fed vanadium supplements.

5.12 Bismuth

Subcutaneous administration of bismuth in single or multiple administrations resulted in the deposition of this metal mainly in the kidneys, which contained over 50% of the available pool of bismuth. The kidneys of the rats who were given bismuth contained bismuth in the soluble fraction in which it was complexed with a protein of molecular weight of about 7000 (Szymanska et al., 1978). Multiple administration of bismuth increased the level of this protein. When selenite was administered there was an increase in the accessible pool of bismuth, possibly due to a drop in excretion, and also changes in the organ distribution of this metal. Selenite administration decreased the retention of bismuth in the kidney while the bismuth content of the liver and in the other tissues were increased. These changes were accompanied by a change in the chemical form of bismuth present in the kidneys as evidenced by the total disappearance of the protein complex of molecular weight of 7000. The bismuth binding by the soluble fraction of the kidney is conditioned by formation of a bismuth complex with a low-molecular-weight protein which seemed to belong to metallothioneinlike proteins (Szymanska *et al.*, 1977). The increased synthesis of this protein due to bismuth administration was not abolished completely. Strong analogies in the behavior of bismuth and mercury become evident on the basis of the presented studies. In the organism both of these metals are bound mainly in the kidneys, chiefly as a complex with the low-molecular-weight inducible protein. In both the case of bismuth and mercury, selenium abolished or diminished affinity of the metals for kidneys, thereby eliminating the role of the low-molecular-weight renal protein in the binding of these metals. In both cases the effect of selenium is expressed in a distribution change of the metal. The role of the liver is increased and there is also a shift of the metal to heavy subcellular fractions, along with a tendency of the metal binding on proteins of highest molecular weights.

5.13 Other Substances

The organophosphate tri-*o*-cresyl phosphate (TOCP) has been reported to induce creatinurea and muscular dystrophy in rabbits (Bloch and Hottinger, 1943), leg weakness in lambs (Draper *et al.*, 1952), liver necrosis and lung hemorrhage in rats (Hove, 1955), and testicular atrophy, bronchopneumonia, and muscular disorders in dogs (Carpentar *et al.*, 1959). The administration of vitamin E effectively prevented these symptoms. However, vitamin E does not completely correct the severely depressed growth rate resulting from the administration of TOCP (Hove, 1955).

Shull and Cheeke (1973) have administered TOCP to low-selenium-low-vitamin-E, Torula-yeast-based diets to both rats and Japanese quail. The diets resulted in growth cessation and mortality. Supplementation of the rat diets with various levels of either selenium or vitamin E prevented both growth cessation and mortality. Rats which had stopped growing following TOCP administration responded with renewed growth when the diet was supplemented with either vitamin E or selenium. The response was greater with selenium than vitamin E, suggesting that TOCP may be a selenium antagonist. With TOCP-fed Janpanese quail, the gains were significantly greater and the mortality lower when selenium was supplemented than when vitamin E was added.

A metabolic interaction between selenium and vitamin E has been well established (Hoekstra, 1974). In many cases selenium had a greater effect on heavy metal toxicity than vitamin E. However, vitamin E had a greater effect against lead toxicity than selenium. Other dietary factors include cystine levels, level of protein, and the kind of protein. Stillings *et al.* (1974) found that the combination of selenium and cystine produced a greater additive effect against methylmercury toxicity in rats than either one of these alone (Stillings *et al.*, 1974). When fish protein replaced casein in the basal diet, toxicity signs were reduced. In addition, a 20% protein level from either source reduced toxicity symptoms as compared to a 10% protein level.

References

Alexander, J., and Aaseth, J., 1980. Biliary excretion of copper and zinc in the rats as influenced by diethylmaleate, selenite, and diethyldithiocarbamate, *Biochem. Pharmacol.* 29:2129–2133.

Amer, M. A., St. Laurent, G. J. and Brisson, G. J., 1973. Supplemental copper and selenium for calves: Effects upon ceruloplasmin activity and liver copper concentration, *Can. J. Physiol. Pharmacol.* 51:649–653.

Amor, A. J., and Pringle, P., 1945. A review of selenium as an industrial hazard, *Bull. Hyg.* 20:239–241.

Axelsson, B., and Piscator, M., 1966. Renal damage after prolonged exposure to cadmium. An experimental study, *Arch. Environ. Health* 12:360–373.

Bell, M. C., Bacon, J. A., Bratton, G. R., and Wilkinson, J. E., 1978. Effects of dietary selenium and lead on selected tissues of chicks, in *Trace Element Metabolism in Animals*, Vol. 3, M. Kirchgessner (ed.), Freising-Weihenstephan, West Germany, pp. 604–607.

Bennets, H. W., 1959. Copper and cobalt deficiency of livestock in Western Australia, *J. Agric. West. Aust.* 8(3rd series):631–636.

Bloch, H., and Hottinger, A., 1943. Creatinuria in poisoning by tri-*o*-cresylphosphate and the influence of vitamin E upon it, *Z. Vitaminforsch* 13:9–18.

Black, R. S., Whanger, P. D., and Tripp, M. J., 1979. Influence of silver, mercury, lead, cadmium, and selenium on glutathione peroxidase and transferase activity in rats, *Biol. Trace Element Res.* 1:313–324.

Bonhorst, C. W., 1955. Selenium poisoning. Anion antagonists in yeast as indicators of the mechanism of selenium toxicity, *J. Agr. Food Chem.* 3:700–703.

Boyne, R., and Arthur, J. R., 1981. Effects of selenium and copper deficiency on neutrophil function in cattle, *J. Comp. Path.* 91:271–276.

Bunn, C. R., and Matrone, G., 1966. *In vivo* interactions of cadmium, copper, zinc, and iron in the mouse and rat, *J. Nutr.* 90:395–399.

Bunyan, J., Diplock, A. T., Cawthrone, M. A., and Green, J., 1968. Vitamin E and stress 8. Nutritional effects of dietary stress with silver in vitamin E-deficient chicks and rats, *Brit. J. Nutr.* 22:165–182.

Burch, R. E., Williams, R. V., Hahn, H. K. J., Jetton, M. M., and Sullivan, J. F., 1975. Tissue trace element and enzyme content in pigs fed a low manganese diet. I. A relationship between manganese and selenium, *J. Lab. Clin. Med.* 86:132–139.

Burk, R. F., and Master, B. S. S., 1975. Some effects of selenium deficiency on the hepatic microsomal cytochrone P-450 system in the rat, *Arch. Biochem. Biophys.* 170:124–131.

Burk, R. F., Foster, K. A., Greenfield, P. M., and Kiker, K. W., 1974. Binding of simultaneously administered inorganic selenium and mercury to a rat plasma protein, *Proc. Soc. Exp. Biol. Med.* 145:782–785.

Cabe, P. A., Carmichael, N. G., and Tilson, H. A., 1979. Effects of selenium, alone and in combination with silver or arsenic in rats, *Neurobehav. Toxicol.* 1:275–278.

Cappon, C. J., and Smith, J. C., 1981. Chemical form and distribution of mercury and selenium in eggs from chickens fed mercury-contaminated grain, *Bull. Environ. Contam. Toxicol.* 26:472–478.

Carlton, W. W., and Kelly, W. A., 1967. Tellurium toxicosis in Pekin ducks, *Toxicol. Appl. Pharmacol.* 11:203–214.

Carpentar, H. M., Jenden, D. J., Shulman, N. R., and Tureman, I. R., 1959. Toxicology of a triarylphosphate oil. I. Experimental toxicology, *A.M.A. Arch. Ind. Health* 20:234–252.

Cerklewski, F. L., and Forbes, R. M., 1976. Influence of dietary selenium on lead toxicity in the rat, *J. Nutr.* 106:778–783.

Chen, R. W., Whanger, P. D., and Weswig, P. H., 1975. Selenium-induced redistribution of cadmium binding to tissue proteins: A possible mechanism of protection against cadmium toxicity, *Bioinorg. Chem.* 4:125–133.

Chiquoine, A. D., 1965. Effect of cadmium chloride on the pregnant albino mouse, *J. Reprod. Fertil.* 10:263–265.

Chmielnicka, J., and Brezeznicka, E. A., 1978. The influence of selenium on the level of mercury and metallothionein in rat kidneys in prolonged exposure to different mercury compounds, *Bull. Environ. Contamin. Toxicol.* 19:183–190.

Choudhury, H., Hasting, L., Menden, E., Brockman, D., Cooper, G. P., and Petering, H. G., 1978. Effects of low level prenatal cadmium exposure on trace metal body burden and behavior in Sprague-Dawley rats, in *Proceedings of the 3rd International Symposium on Trace Element Metabolism in Man and Animals*, M. Kirchgessner (ed.) Freising-Weihenstephan, West Germany, pp. 549–552.

Coleman, J. E., and Vallee, B. L., 1961. Metallocarboxypeptidases: Stability constants and enzymatic characteristics, *J. Biol. Chem.* 236:2244–2249.

Cotzias, G. C., and Papavasiliou, P. S., 1964. Specificity of zinc pathway through the body: homeostatic considerations, *Am. J. Physiol.* 206:787–792.

Dam, H., Nielsen, G. K., Prange, I., and Sondergaard, E., 1958. Exudative diathesis produced by vitamin E-deficient diets without polyenoic fatty acids, *Experientia* 14:291–294.

Dial, N.A., 1976. Methylmercury: Teratogenic and lethal effects in frog embryos, *Tetratology* 13:327–333.

Dial, N. A., 1978. Methylmercury: Some effects on embryogenesis in the Japanese medaka, *Oryzias latipes*, *Teratology* 17:83–92.

Diplock, A. T., Green, J., Bunyan, J., McHale, D., and Muthy, I. R., 1967. Vitamin E and stress 3. The metabolism of D-α-tocopherol in the rat under dietary stress with silver, *Brit. J. Nutr.* 21:115–125.

Doisy, E. A., 1973. Micronutrient control of biosynthesis of clotting proteins and cholesterol, in *Trace Substances in Environmental Health*, Vol. 6, D. D. Hemphill (ed.), University of Missouri Press, Columbia, Missouri, pp. 193–199.

Draper, H. H., James, M. F., and Johnson, B. C., 1952. Trio-*o*-cresylphosphate as a vitamin E antagonist for the rat and lamb, *J. Nutr.* 47:583–599.

Dubois, K. P., Moxon, A. L., and Olson, O. E., 1940. Further studies on the effectiveness of arsenic in preventing selenium poisoning, *J. Nutr.* 19:477–482.

Early, J. L., and Schnell, R. C., 1981. Selenium antagonism of cadmium-induced inhibition of hepatic drug metabolism in the male rat, *Toxicol. Appl. Pharmacol.* 58:57–66.

El-Begearmi, M. M., Sunde, M. L., and Ganther, H. E., 1977. A mutual protective effect of mercury and selenium in Japanese quail, *Poult. Sci.* 56:313–322.

Evans, G. W., Grace, C. I., and Hahn, C., 1974. The effect of copper and cadmium on ^{65}Zn absorption in zinc-deficient and zinc-supplemented, *Bioinorganic Chem.* 3:115–120.

Fehrs, M. S., Miller, W. J., Gentry, R. P., Neathery, M. W., Blackmon, D. M., and Heinmiller, S. R., 1981. Effect of high but nontoxic dietary intake of copper and selenium on metabolism in calves, *J. Dairy Sci.* 64:1700–1706.

Ferm, V. H., and Carpenter, S. J., 1968. The relationship of cadmium and zinc in experimental mammalian teratogenesis, *Lab. Invest.* 18:429–432.

Francis, A. J., Duxbury, J. M., and Alexander, M., 1974. Evolution of dimethylselenide from soils, *Appl. Microbiol.* 28:248–250.

Friberg, L., 1950. Health hazards in the manufacture of alkaline accumulators with special reference to chronic cadmium poisoning, *Acta Med. Scand.* 138, Suppl. 240.

Friberg, L. Piscator, M., Nordberg, G. F., and Kjellstrom, 1974. *Cadmium in the Environment*, 2nd ed., CRC Press Inc., Cleveland, Ohio, pp. 93–202.

Gabbiani, G., 1966. Action of cadmium chloride on sensory ganglia, *Experientia* 22:261–264.

Gabbiani, G. Baic, D., and Deziel, C., 1967 Toxicity of cadmium for the central nervous system, *Exp. Neurol.* 18:154–160.

Ganther, H. E., 1968. Selenotrisulfides: Formation by the reaction of thiols with selenious acid, *Biochemistry* 7:2898–2905.

Ganther, H. E., and Baumann, C. A., 1962. Selenium metabolism. I. Effects of diet, arsenic and cadmium, *J. Nutr.* 77:210–216.

Ganther, H. E., and Hsieh, H. S., 1974. Mechanisms for the conversion of selenite to selenides in mammalian tissues, in *Trace Element Metabolism in Animals*, Vol. 2., W. G. Hoekstra, J. W. Suttie, H. E. Ganther, and W. Mertz (eds.), University Park Press, Baltimore, pp. 339–353.

Ganther, H. E., Hafeman, D. G., Lawrence, R. A., Serfass, R. E., and Hoekstra, W. G., 1976. Selenium and glutathione peroxidase in health and disease: A review. In *Trace Elements in Human Health and Disease*, Vol. II, A. S. Prasad (ed.), Academic Press, New York, pp. 165–234.

Gardiner, M. R., 1961. White muscle disease of sheep, *J. Agric. West Aust.* 2(4th series):497–501.

Gardiner, M. R., 1966. Chronic selenium toxicity studies in sheep, *Aust. Vet. J.* 42:442–448.

Gardiner, M. R., 1969. Selenium in animal nutrition, *Outlook in Agriculture* 6:19–28.

Gardiner, M. R., and Nairn, M. E., 1969. Studies on the effect of cobalt and selenium in clover disease of ewes, *Aust. Vet. J.* 45:215–222.

Gardiner, M. R., and Nicol, H., 1971. Cobalt–selenium interactions in the nutrition of the rat, *Austral. J. Exp. Biol. Med. Sci.* 49:291–296.

Gasiewicz, T. A., and Smith, J. C., 1976. Interactions of cadmium and selenium in rat plasma *in vivo* and *in vitro*, *Biochim. Biophys. Acta* 428:113–122.

Gasiewicz, T. A., and Smith, J. C., 1978a. Properties of the cadmium and selenium complex formed in rat plasma *in vivo* and in vitro, *Chem. Biol. Interact* 23:171–183.

Gasiewicz, T. A., and Smith, J. C., 1978b. The metabolism of selenite by intact rat erythrocytes in vitro, *Chem. Biol. Interact.* 21:299–313.

Godwin, K. O., Partick, E. J., and Fuss, C. N., 1977. Adverse effects of copper, and to a lesser extent iron, when administered to selenium-deficient rats, *In Trace Element Metabolism in Man and Animals, Vol. 3*, M. Kirchgessner (ed), Freising-Weihenstephan, West Germany, pp. 185–187.

Grant, C. A., 1961. Morphological and etiological studies of dietetic microangiopathy in pigs ("mulberry heart"), *Acta Vet. Scand.* 2(Suppl. 3):1–107.

Grice, H. C., Munro, I. C., Wiberg, G. S., and Heggtveit, H. A., 1969. The pathology of experimentally induced cobalt cardiomyopathies. A comparison with beer drinkers cardiomyopathy, *Clin. Toxicol.* 2:273–287.

Groth, D. H., Vignati, L., Lowry, L, Mackay, G., and Stokinger, H. E., 1973. Mutual antagonistic and synergistic effects of inorganic selenium and mercury salts in chronic experiments, in *Trace Substances in Environmental Health*. Vol. 6, D. D. Hemphill, (ed.), University of Missouri Press, Columbia, Missouri, pp. 187–189.

Gunn, S. A., Gould, T. C., and Anderson, W. A. D., 1962. Interference with fecal excretion of Zn-65 by cadmium, *Proc. Soc. Exp. Biol. Med.* 111:559–562.

Gunn, S. A., Gould, T. C., and Anderson, W. A. D., 1963. The selective injurious response of testicular and epididymal vessel's cadmium and its prevention by zinc, *Am. J. Path.* 42:685–702.

Gunn, S. A., Gould, T. C., and Anderson, W. A. D., 1968. Specificity in protection against lethality and testicular toxicity from cadmium, *Proc. Soc. Exp. Biol. Med.* 128:591–595.

Halverson, A. W., and Palmer, I. S., 1975. The effect of substances which protect against selenium toxicity on selenium utilization by rats, *Proc. S. Dak. Acad. Sci.* 54:148–156.

Harada, H., Ito, L., Ebato, K., Takeuchi, M., Amemiya, T., Yamanobe, H., Suzuku, S., and Totani, T., 1975. Effect of selenium on the toxicity of methylmercury (II). Methylmercury and total mercury concentration of organs in rats administered methylmercury, selenium, and vitamin E, *Ann. Rep. Tokyo Metr. Res. Lab.* P. H. 26:123–128.

Harris, S. B., Wilson, J. B., and Printz, R. H., 1972. Embryotoxicity of methylmercuric chloride in golden hamsters, *Teratology* 6:139–142.

Hill, C. H., 1975. Interrelationships of selenium with other trace elements, *Fed. Proc. Fed. Am. Soc. Exp. Biol.* 34:2096–2100.

Hill, C. H., and Matrone, G., 1970. Chemical parameters in the study of *in vivo* and *in vitro* interactions of transition elements, *Fed. Proc. Fed. Am. Soc. Exp. Biol.* 29:1474–1481.

Hoekstra, W. G., 1975. Biochemical function of selenium and its relation to vitamin E, *Fed. Proc. Fed. Am. Soc. Exp. Biol.* 34:2083–2089.

Hollo, Z. M., and Slatarov, S., 1960. The prevention of thallium death by selenate, *Naturwissenschaften* 47:87.

Holmburg, R. E., and Ferm, V. R., 1969. Interrelationships of selenium, cadmium, arsenic, in mammalian teratogenesis, *Arch. Environ. Health* 18:873–877.

Hove, E. L., 1955. Anti-vitamin E stress factors as related to lipid peroxides, *Am. J. Clin. Nutr.* 3:328–336.

Hsieh, H. S., and Ganther, H. E., 1975. Acid-volatile selenium formation catalyzed by glutathione reductase, *Biochemistry* 14:1632–1636.

Hsu, F. S., Krook, L., Pond, W. G., and Duncan, J. R., 1975. Interactions of dietary calcium with toxic levels of lead and zinc in pigs, *J. Nutr.* 105:112–118.

Imura, N., and Naganuma, A., 1978. Interaction of inorganic mercury and selenite in rabbit blood after intravenous administration, *J. Pharm. Dyn.* 1:67–73.

Jenkins, K. J., and Kidiroglou, M., 1972. Comparative metabolism of ^{75}Se-selenite, ^{75}Se-selenate and ^{75}Se-selenomethionine in bovine erythrocytes, *Can. J. Physiol. Pharmacol.* 50:927–935.

Jensen, L. S., 1974. Interactions of silver and copper with selenium in chicks, *Fed. Proc. Fed. Am. Soc. Exp. Biol.* 33:694.

Jensen, L. S., 1975. Modification of a selenium toxicity in chicks by dietary silver and copper, *J. Nutr.* 105:769–775.

Jensen, L. S., Peterson, R. P., and Falen, L., 1974. Inducement of enlarged hearts and muscular dystrophy in turkey poults with dietary silver, *Poult. Sci.* 53:57–64.

Kagi, J. H. R., and Vallee, B. L., 1961. Metallothionein: a cadmium- and zinc-binding protein from equine renal cortex. II. Physicochemical properties, *J. Biol. Chem.* 236:2435–2442.

Kamstra, L. D., and Bonhorst, C. W., 1953. Effect of arsenic on the expiration of volatile selenium compounds by rats, *Proc. S. Dak. Acad. Sci.* 32:72–74.

Kari, T., and Kauranen, 1978. Mercury and selenium content of seals from fresh and brackish waters in Finland, *Bull. Environ. Contam. Toxicol.* 19:273–280.

Kemmerer, A. R., Elvehjem, C. A., and Hart, E. B., 1931. Studies on the relation of manganese to the nutrition of the mouse. *J. Biol. Chem.* 92:623–630.

Khera, K. S., 1973. Teratogenic effects of methylmercury in the cat: Note on the use of this species as a model for teratogenicity studies, *Teratology* 8:293–304.

Klug, H. L., Lampson, G. P., and Moxon, A. L., 1950a. The distribution of selenium and arsenic in the body tissues of rats fed selenium, arsenic, or selenium plus arsenic, *Proc. S. Dak. Acad. Sci.* 29:57–65.

Klug, H. L., Moxon, A. L., and Petersen, D. F., and Van Potter, R., 1950b. The *in vivo* inhibition of succinic dehydrogenase by selenium and its release by arsenic, *Arch. Biochem. Biophys.* 28:253–259.

Koeman, J. H., van de Ven, W. S. M., de Goedij, J. J. M., Tjioe, P. S., and van Haaften, J. L., 1975. Mercury and selenium in marine mammals and birds, *Sci. Total. Envir.* 3:279–287.

Krista, I., Carlson, C. W., and Olson, O. E., 1961. Effect of arsenic on selenium deposition in chicken eggs, *Poult. Sci.* 40:1365–1367.

Latshaw, J. D., 1975. Natural and selenite selenium in the hen and egg, *J. Nutr.* 105:32–37.

Leach, R. M., Muenster, A., and Wein, E. M., 1969. Studies on the role of manganese in bone formation. II. Effect upon chondroitin sulfate synthesis in chick epiphyseal cartilage, *Arch. Biochem.* 133:22–28.

Lee, H. J., and Jones, G. B., 1976. Interactions of selenium, cadmium and copper in sheep, *Aust. J. Agric. Res.* 27:447–452.

Lee, M., Chan, K. K. S., Sairenji, E., and Niikuni, T., 1979. Effect of sodium selenite on methylmercury-induced cleft palate in the mouse, *Environ. Res.* 19:39–48.

Levander, O. A., 1977. Metabolic interrelationships between arsenic and selenium, *Environ. Hlth. Perspect.* 19:159–164.

Levander, O. A., and Argrett, L. C., 1969. Effects of arsenic, mercury, thallium, and lead on selenium metabolism in rats, *Toxicol. Appl. Pharmacol.* 14:308–314.

Levander, O. A., and Baumann, C. A., 1966a. Selenium metabolism, V. Studies on the distribution of selenium in rats given arsenic, *Toxicol. Appl. Pharmacol.* 9:98–105.

Levander, O. A., and Baumann, C. A., 1966b. Selenium metabolism. VI. Effect of arsenic on the excretion of selenium in the bile, *Toxicol. Appl. Pharmacol.* 9:106–115.

Levander, O. A., Morris, V. C. and Higgs, D. J., 1973a. Acceleration of thiol-induced swelling of rat liver mitochondria by selenium, *Biochemistry* 12:4586–4590.

Levander, O. A., Morris, V. C., and Higgs, D. J., 1973b. Selenium as a catalyst for the reduction of cytochrome c by glutathione, *Biochemistry* 12:4591–4595.

Levander, O. A., Morris, V. C., Higgs, D. J., and Ferretti, R. J., 1975. Lead poisoning in vitamin E-deficient rats, *J. Nutr.* 105:1481–1485.

Levander, O. A., Morris, V. C., and Ferretti, R. J., 1977b. Filterability of erythrocytes from vitamin E-deficient lead-poisoned rats, *J. Nutr.* 107:363–372.

Levander, O. A., Ferretti, R. J., and Morris, V. C., 1977a. Osmotic and peroxidative fragilities of erythrocytes from vitamin E-deficient lead-poisoned rats, *J. Nutr.* 107:373–377.

Levander, O. E., Morris, V. C., and Ferretti, R. J., 1977c. Comparative effects of selenium and vitamin E lead-poisoned rats, *J. Nutr.* 107:378–382.

Lilas, R., and Fischbein, A., 1976. Chelation therapy in workers exposed to lead, *J. Am. Med. Assoc.* 235:2823–2824.

Lindquist, R. R., 1968. Studies on the pathogenesis of hepatolenticular degeneration. III. The effect of copper on rat liver lysosomes, *Am. J. Path.* 53:903–927.

Lynch, F. P., Smith, D. F., Fisher, M., Pike, T. L., and Weinland, B. T., 1976. Physiological responses of calves to cadmium and lead, *J. Anim. Sci.* 42:410–421.

Lyons, M., and Insko, W., 1937. Chondrodystrophy in the chick embryo produced by manganese deficiency in the diet of the hen, *Ky. Agr. Exp. Sta. Lexington Bull.* 371:61–75.

Magat, W., and Sell, J. L., 1979. Distribution of mercury and selenium in egg components and egg-white proteins, *Proc. Soc. Exp. Biol. Med.* 161:458–463.

Mason, K. E., and Young, J. O., 1967. Effectiveness of selenium and zinc in protecting against cadmium-induced vascular lesions in the testis and epididymis of the rat, *Acta Path. Microbiol. Scand.* 63:513–521.

Mason, K. E., Brown, J. A., Young, J. O., and Nesbit, R. I., 1964. Cadmium-induced injury of the rat testis, *Anat. Rec.* 149:135–148.

Matrone, G., 1974. Chemical parameters in trace-element antagonisms, in *Trace Element Metabolism In Animals*, Vol. 2, W. G. Hoekstra, J. W. Suttie, H. E. Ganther, and W. Mertz (eds.), University Park Press, Baltimore, pp. 91–102.

McConnell, K. P., 1942. Respiratory excretion of selenium studied with the radioactive isotope, *J. Biol. Chem.* 145:55–60.

McConnell, K. P., and Portman, O. W., 1952a. Excretion of dimethyl selenide by the rat, *J. Biol. Chem.* 195:277–282.

McConnell, K. P., and Portman, O. W., 1952b. Toxicity of dimethyl selenide in rat and mouse, *Proc. Soc. Exp. Biol. Med.* 79:230–231.

McGuire, S. O., Fehrs, M. S., Miller, W. J., Gentry, R. P., Neathery, M. W., Blackmon, D. M., Heinmiller, S. R., and Lassiter, J. W., 1981. Metabolism of selenium and copper in calves fed low and high copper and selenium diets, *Fed. Proc. Fed. Am. Soc. Exp. Biol.* 40:868.

Means, J. R., Carlson, G. P., and Schnell, R. C., 1979. Studies on the mechanism of cadmium-induced inhibition of the hepatic microsomal mono-oxgenase of the male rat, *Toxicol. Appl. Pharmacol.* 48:293–304.

Megel, H., and Karlog, O., 1980. Studies on the interaction and distribution of selenite, mercuric, methoxyethyl mercuric and methyl mercuric chloride in rats. II. Analyses of the soluble proteins and the precipitates of liver and kidney homogenates, *Acta Pharmacol. Toxicol.* 46:25–31.

Merali, Z., and Singhal, R. L., 1975. Protective effect of selenium on certain hepatotoxic and pancreotoxic manifestations of sub-acute cadmium administration, *J. Pharmacol. Exp. Ther.* 195:58–66.

Merali, Z., and Singhal, R. L., 1976. Prevention by zinc of cadmium-induced alterations in pancreatic and hepatic functions, *Br. J. Pharmacol.* 57:573–579.

Metz, E. N., 1969. Mechanism of hemolysis by excess copper, *Clin. Res.* 17:32.

Moxon, A. L., and Wilson, W. O., 1944. Selenium-arsenic antagonism in poultry, *Poult. Sci.* 23:149–151.

Moxon, A. L., and Rhian, M. A., Anderson, H. D., and Olson, O. E., 1944. Growth of steers on seleniferous range, *J. Anim. Sci.* 3:299–309.

Muth, O. H., Whanger, P. D., Weswig, P. H., and Oldfield, J. E., 1971. Occurrence of myopathy in lambs of ewes fed added arsenic in a selenium-deficient ration, *Am. J. Vet. Res.* 32:1621–1623.

Naganuma, A., and Imura, N., 1980. Bis(methylmercuric) selenide as a reaction product from methylmercury and selenite in rabbit blood, *Res. Commun. Chem. Pathol. Pharmacol.* 27:163–173.

Naganuma, A., Kojima, Y., and Imura, N., 1980. Interaction of methylmercury and selenium in mouse: Formation and decomposition of bis(methylmercuric) selenide, *Res. Commun. Chem. Pathol. Pharmacol.* 30:301–316.

Nobunaga, T., Satoh, H., and Suzuki, T., 1979. Effects of sodium selenite on methylmercury embryotoxicity and teratogenicity in mice, *Toxicology Appl. Pharmacol.* 47:79–88.

Nogawawa, K., Ishizaki, A., Fukushima, M., Shibata, I., and Hagina, N., 1975. Studies on the women with acquired Fanconi syndrome observed in Ichi river basin polluted by cadmium, *Environ. Res.* 10:280–307.

Obermeyer, B. D., Palmer, I. S., Olson, O. E., and Halverson, A. W., 1971. Toxicity of trimethylselenonium chloride in the rat with and without arsenite, *Toxicol. Appl. Pharmacol.* 20:135–146.

Olson, O. E., 1960. Selenium and the organic arsenicals, *Feed Age* 10:49–50.

Olson, O. E., Schulte, B. M., Whitehead, E. I., and Halverson, A. W., 1963. Selenium toxicity. Effect of arsenic on selenium metabolism in rats, *J. Agr. Food Chem.* 11:531–534.

Orent, E. R., and McCollum, E. V., 1931. Effects of deprivation of manganese in the rat, *J. Biol. Chem.* 92:651–678.

Palmer, I. S., and Bonhorst, C. W., 1957. Modification of selenite metabolism by arsenite, *J. Agr. Food Chem.* 5:928–930.

Palmer, I. S., Arnold, R. L., and Carlson, C. W., 1973. Toxicity of various selenium derivatives to chick embryos, *Poult. Sci.*, 52:1841–1846.

Parizek, J., 1957. The destructive effect of cadmium ion on testicular tissue and its prevention by zinc, *J. Endocrinol.* 15:56–63.

Parizek, J., 1964. Vascular changes at sites of oestrogen biosynthesis by parenteral injection of cadmium salts; the destruction of placenta by cadmium salts, *J. Reprod. Fertil.* 7:263–265.

Parizek, J., and Ostadalova, I., 1967. The protective effect of small amounts of selenite in sublimate intoxification, *Experientia* 23:142–143.

Parizek, J., Ostadalova, I., Benes, I., and Babicky, A., 1968a. Pregnancy and trace elements: the protective effects of compounds of an essential trace element—selenium—against the peculiar toxic effects of cadmium during pregnancy, *J. Reprod. Fertil.* 16:507–509.

Parizek, J., Ostadalova, I., Benes, I., and Pitha, J., 1968b. The effect of a subcutaneous injection of cadmium salts on the ovaries of adult rats in persistant oestrus, *J. Reprod. Fert.* 17:559–562, 1968.

Parizek, J., Benes, I., Kalouskova, J., Babicky, A., and Lener, J., 1969a. Metabolic interrelationships of trace elements. Effects of zinc salts on the survival of rats intoxicated with cadmium, *Physiol. Bohemoslov.* 18:89–93.

Parizek, J., Babicky, A., Ostadalova, I., Kalouskova, J., and Pavlik, L. 1969. The effect of selenium compounds on the cross-placental passage of [203]Hg, in *Radiation Biology of the Fetal and Juvenile Mammal*, M. R. Sikov and D. D. Mahlum (ed.), USAEC, Oak Ridge, Tennessee, pp. 137–170.

Parizek, J., Ostadalova, I., Kalouskova, J., Babicky, A., and Benes, J., 1971. The detoxifying effects of selenium. Interrelations between compounds of selenium and certain metals, in *Newer Trace Elements in Nutrition*, W. Mertz and W. E. Cornatzer (eds.), Marcel Dekker, New York, pp. 85–122.

Petering, H. G., 1974. The effect of cadmium and lead on copper and zinc metabolism, in *Trace Element Metabolism in Animals*, Vol. 2, W. G. Hoekstra, J. W. Suttie, H. E. Ganther, and W. E. Mertz (eds.), University Park Press, Baltimore, pp. 311–325.

Petering, H. G., Johnson, M. A., and Stemmer, K. L., 1971. Studies on zinc metabolism in the rat. 1. Dose-response effects of cadmium, *Arch. Envir. Hlth.* 23:93–101.

Piotrowski, J. K., Bem, E. M., and Werner, A., 1977. Cadmium and mercury binding to metallothionein as influenced by selenium, *Biochem, Pharmacol.* 26:2191–2192.

Potter, S. D., and Matrone, G., 1977. A tissue culture model for mercury–selenium interactions, *Toxicol. Appl. Pharmacol.* 40:201–215.

Powell, G. W., Miller, W. J., Morton, J. D., and Clifton, C. M., 1964. Influence of dietary cadmium level and supplemental zinc on cadmium toxicity in the bovine, *J. Nutr.* 84:205–214.

Rafter, G. W., 1980. Copper inhibition of glutathione reductase and its reversal by thiol reagents, *Fed. Proc. Fed. Am. Soc. Exp.* 39:1772.

Ramstoeck, E. R., Hoekstra, W. G., and Ganther, H. E., 1980. Trialkyllead metabolism and lipid peroxidation *in vivo* in vitamin E- and selenium-deficient rats, as measured by ethane production, *Toxicol. Appl. Pharmacol.* 54:251–257.

Rastogi, S. C., Clausen, J., and Srivastava, K. C., 1976. Selenium and lead: Mutual detoxifying effects, *Toxicology* 6:377–388.

Reddy, K. A., Omaye, S. T., Hasegawa, G. K., and Cross, C. E., 1978. Enhanced lung toxicity of intratracheally instilled cadmium chloride in selenium-deficient rats, *Toxicol. Appl. Pharm.* 43:249–257.

Rhian, M., and Moxon, A. L., 1943. Chronic selenium poisoning in dogs and its prevention by arsenic, *J. Pharmacol. Exp. Therap.* 78:249–264.

Rice, D. P., Murthy, L., Menden, E., and Petering, H. G. 1973. The impact of low level cadmium feeding on blood chemicals in male Sprague-Dawley rats, in *Trace Substances in Environmental Health*, Vol. VII, D. Hemphill (ed), University of Missouri Press, Columbia, Missouri, pp. 305–311.

Richards, M. P., and Cousins, R. J., 1976. Metallothionein and its relationshp to the metabolism of dietary zinc in rats, *J. Nutr.* 106:1591–1599.

Richards, M. P., and Cousins, R. J., 1977. Isolation of an intestinal metallothionein induced by parenteral zinc, *Biochem. Biophys. Res. Commun.* 75:286–294.

Roberts, K. R., Miller, W. J., Stake, P. E., Gentry, R. P., and Neathery, M. W., 1973. High dietary cadmium and zinc absorption and metabolism in calves fed for comparable nitrogen balance, *Proc. Soc. Exp. Biol. Med.* 144:907–909.

Rusiecki, W., and Brezezinski, J., 1966. Influence of sodium selenate on acute thallium poisonings, *Acta Pol. Pharmacol.* 23:69–74.

Sandholm, M., 1973a. The initial fate of a trace amount of intravenously administered selenite, *Acta Pharmacol. Toxicol.* 33:1–5.

Schnell, R. C., 1978. Cadmium-induced alteration of drug action, *Fed. Proc. Fed. Am. Soc. Exp. Biol.* 37:28–34.

Schroeder, H. A., 1965. Cadmium as a factor in hypertension, *J. Chronic Dis.* 18:647–656.

Schroeder, H. A., Balassa, J. J., and Vinton, W. M. Jr., 1964. Chromium, lead, cadmium, nickel and titanium in mice: Effect on mortality, tumors and tissue levels, *J. Nutr.* 83:239–250.

Schultz, J., and Lewis, H. B., 1940. The excretion of volatile selenium compounds after the administration of sodium selenite to white rats, *J. Biol. Chem.* 133:199–207.

Schwarz, K., and Foltz, C. M., 1957. Selenium as a integral part of factor 3 against necrotic liver degeneration, *J. Am. Chem. Soc.* 79:3292–3293.

Scott, M. L., Olson, G., Krook, L., and Brown, W. R., 1967. Selenium-responsive myopathies of myocardium and of smooth muscle in the young poult, *J. Nutr.* 91: 573–583.

Scrutton, M. C., Utter, M. F., and Mildvan, A. S., 1966. Pyruvate carboxylase, VI. The presence of tightly bound manganese, *J. Biol. Chem.* 241:3480–3487.

Seidell, A., 1952. *Solubilities, Inorganic and Metal-Organic Compounds*, Vol. l, 4th ed., American Chemical Society, Washington, D.C.

Sell, J. L., Guenter, W., and Sifri, M., 1974. Distribution of mercury among components of eggs following the administration of methylmercuric chloride to chickens, *J. Agric. Food Chem.* 22:248–251.

Shaver, S. L., and Mason, K. E., 1951. Impaired tolerance to silver in vitamin E deficient rats, *Anat. Rec.* 109:382–388.

Shull, L. R., and Cheeke, P. R., 1973. Antiselenium activity of tri-*o*-cresyl phosphate in rats and Japanese quail, *J. Nutr.* 103:560–568.

Sivertsen, T., Karlsen, J. T., Norheim, G., and Froslie, A., 1978. Concentration of selenium in liver in relation to copper level in normal and copper-poisoned sheep, *Acta Vet. Scand.* 19:472–474.

Skilleter, D. N., 1975. Decrease in mitochondrial substrate uptake caused by trialkyltin and trialkyllead compounds in chloride media and its relevance to inhibition of oxidative phosphorylation, *Biochem. J.* 146:465–471.

Snaith, S. M., and Levvy, G. A., 1969. Purification and properties of α-D-mannosidase from rat epididymis, *Biochem. J.* 114:25–33.

Snell, K., Ashby, S. L., and Barton, S. J., 1977. Distribution of perinatal carbohydrate metabolism in rats exposed to methylmercury in utero, *Toxicology* 8:277–283.

Srivastava, S. K., and Beuter, E., 1969. The transport of oxidized glutathione from the erythrocytes of various species in the presence of chromate, *Biochem. J.* 114:833–837.

Stillings, B. R., Lagally, H., Bauersfeld, P., and Soares, J., 1974. Effect of cystine, selenium and fish protein on the toxicity and metabolism of methylmercury in rats, *Toxicol. Appl. Pharmacol.* 30:243–254.

Stonard, M. D., and Webb, M., 1976. Influence of dietary cadmium on the distribution of the essential metals copper, zinc and iron in tissues of the rat, *Chem. Biol. Interact.* 15:349–363.

Stone, C. L., and Soares, J. H., Jr., 1976. The effect of dietary selenium on lead toxicity in the Japanese quail, *Poult. Sci.* 55:341–349.

Stowe, H. D., 1980. Effects of copper pretreatment upon the toxicity of selenium in ponies, *Am. J. Vet. Res.* 41:1925–1928.

Su, M. Q., and Okita, G. T., 1976. Embryocidal and teratogenic effects of methylmercury in mice. *Toxicol. Appl. Pharmacol.* 38:207–216.

Supplee, W. C., 1963. Antagonistic relationship between dietary cadmium and zinc, *Science* 139:119–120.

Szymanska, J. A., Mogilnicka, E. M., and Kaszper, B. W., 1977. Binding of bismuth in the kidneys of the rat: The role of metallothionein-like proteins, *Biochem. Pharmacol.* 26:257–258.

Szymanska, J. A., Zychowicz, M., Zelazowski, A. J., and Piotrowski, J. K., 1978. Effect of selenium on the organ distribution and binding of bismuth in rat tissues, *Arch. Toxicol.* 40:131–141.

Tamura, Y., Maki, T., Yamada, H., Shimamura, Y., Ochiai, S., Nishigaki, S., and Kimura, Y., 1975. Studies on the behaviour of accumulation of trace elements in various tissues of tuna, *Ann. Rep. Tokyo Metro. Res. Lab. P.H.* 26:200–204.

Thomson, G. G., and Lawson, B. M., 1970. Copper and selenium interaction in sheep, *N.Z. Vet. J.* 18:79–82.

Ukita, T., Takeda, Y., Sato, Y., and Takahashi, T., 1967. Distribution of [203]Hg Labeled mercury compounds in adult and pregnant mice determined by whole-body autoradiography, *Radioisotopes* 16:439.

Underwood, E. J., 1971. Manganese, in *Trace Elements in Human and Animal Nutrition*, 3rd ed., E. J. Underwood (ed.), Academic Press, New York, pp. 177–207.

Underwood, E. J., 1977. *Trace Elements in Human and Animal Nutrition*, 4th ed., Academic Press, New York.

Van Vleet, J. F., Boon, G. D., and Ferrans, V. J., 1981a. Induction of lesions of selenium-vitamin E deficiency in ducklings fed silver, copper, cobalt, tellurium, cadmium or zinc: Protection by selenium or vitamin E supplements, *Am. J. Vet. Res.* 42:1206–1217.

Van Vleet, J. F., Boon, G. D., and Ferrans, V. J., 1981b. Induction of lesions of selenium-vitamin E deficiency in weanling swine fed silver, cobalt, tellurium, zinc, cadmium and vanadium, *Am. J. Vet. Res.* 42:789–799.

Vigliani, E.C., 1969, The biopathology of cadmium, *Am. Ind. Hyg. Assoc. J.* 30:329–340.

Wada, O., Yamaguchi, N., Ono, T., Nagashashi, M., and Morimura, T., 1976. Inhibitory effect of mercury on kidney glutathione peroxidase and its prevention by selenium, *Environ. Res.* 12:75–80.

Wagner, P. A., Hoekstra, W. G., and Ganther, H. E., 1975. Alleviation of silver toxicity by selenite in the rat in relation to tissue glutathione peroxidase. *Proc. Soc. Exp. Biol. Med.* 148:1106–1110.

Wahlstrom, R. C., Kamstra, L. D., and Olson, O. E., 1955. The effect of arsanilic acid and 3-nitro-4-hydroxyphenylarsonic acid on selenium poisoning in the pig, *J. Anim. Sci.* 14:105–110.

Waites, G. M. H., and Setchell, B. P., 1966. Changes in blood flow and vascular permeability of the testis, epididymis and accessory reproductive organs of the rat after administration of cadmium chloride, *J. Endocr.* 34:329–342.

Walker, G. W. R., and Bradley, A. M., 1969. Interacting effects of sodium monohydrogenarsenate and selenocystine on crossing over in *Drosophilia melanogaster*, *Can. J. Genet. Cytol.* 11:677–688.

Washko, P. W., and Cousins, R. J., 1977. Role of dietary calcium and calcium-binding protein in cadmium toxicity in rats, *J. Nutr.* 107:920–928.

Webb, M., 1975. Cadmium, *Br. Med. Bull.* 31:246–250.

Webster, W. S., 1978. Cadmium-induced fetal growth retardation in the mouse, *Arch. Environ. Health*, 33:36–42.

Weser, U., and Koolman, J., 1970. Reactivity of some transition metals on nuclear protein biosynthesis in rat liver, *Experientia* 26:246–247.

Whanger, P. D., 1976. Selenium versus metal toxicity in animals, in *Industrial Health Foundation Symposium on Selenium—Tellurium in the Environment*. Industrial Health Foundation, Pittsburgh, pp. 234–240.

Whanger, P. D., Weswig, P. H., Schmitz, J. A., and Oldfield, J. E., 1976. Effects of selenium, cadmium, mercury, tellurium, arsenic, silver, and cobalt on white muscle disease in lambs and effect of dietary forms of arsenic on its accumulation in tissues, *Nutr. Repts. Int.* 14:63–72.

Whanger, P. D., and Weswig, P. H., 1978. Influence of 19 elements on development of liver necrosis in selenium and vitamin E deficient rats, *Nutr. Rep. Int.* 18:421–428.

Wiberg, J. S., and Neuman, W. F., 1957. The binding of bivalent metals by deoxyribonucleic and ribonucleic acids, *Arch. Biochem.* 72:66–83.

Wilgus, H. S., Norris, L. C., and Heuser, G. F., 1936. The role of certain inorganic elements in the cause and prevention of perosis, *Science* 84:252–253.

Wilson, R. H., de Eds, F., and Cox, A. J., 1941. Effects of continued cadmium feeding, *J. Pharmacol. Exp. Ther.* 71:222–235.

Windell, C. C., and Tata, J. R., 1966. Studies on the stimulation by ammonium sulfate of the DNA-dependent RNA polymerase of isolated rat liver nuclei, *Biochim. Biophys. Acta* 123:478–492.

Witting, L. A., and Horwitt, M. K., 1964. Effects of dietary selenium, methionine, fat level and tocopherol on rat growth, *J. Nutr.* 84:351–360.

Environmental Occurrence of Selenium

6.1 Geochemistry of Selenium

Selenium is very widely distributed in the earth's crust, but is rarely present in any material in concentrations exceeding 500 $\mu g/g$. The amount of selenium in any soil is dependent upon a number of factors: first, the presence or absence of selenium in the parent material of the soil; second, the possible removal of selenium by leaching during soil formation; and third, processes subsequent to soil formation added selenium or removed it. Figure 6-1 shows a simplified geochemical cycle of selenium, from molten rocks to igneous rocks or to the atmosphere, which can be converted to sediments, or to the hydrosphere, which can be taken up by plants and soils. From plants and soils selenium can enter the food chain of humans and animals.

The geochemistry of selenium in a crystallizing magma parallels that of sulfur. Sulfides of iron, cobalt, and nickel tend to separate from the magma during the early stages of magma formation. Selenium separates at the same time and is associated with, and sometimes a part of, these minerals. Goldschmidt (1954) ranks selenium as the ninth most abundant element in these sulfides. The ore deposits at Sudbury, which is north of Lake Huron in Canada, have sulfides of this type and are the largest nickel and selenium source in North America. As magma crystallization continues, there is residual fluid phase in which the volatile components, including sulfur and selenium, are concentrated. The residual fluid phase eventually becomes a supercritical aqueous solution rather than a melt. When such a solution moves through fractures or by solutions of adjacent rocks, volatiles are able to separate from the parent magma. The volatiles can also form sulfide ore bodies. In a system which is relatively closed, sulfur and selenium are depleted in the silicate rock formed from a magma. However, in a more open system, for example a volcanic one, large

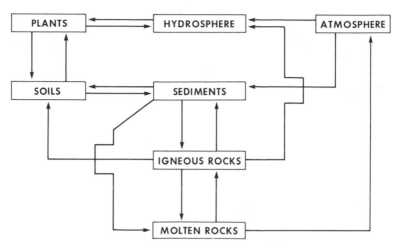

Figure 6-1. Geochemical cycle of selenium. *Symposium Selenium in Biomedicine,* O. H. Muth (ed.), DVM; see H. W. Lakin, p. 28. Permission AVI Publishing Co., Newport, Connecticut.

quantities of sulfur as well as smaller quantities of selenium escape as volatile gases.

It is difficult to estimate the abundance of these elements in the earth's crust because of the separation of such a large part of the sulfur and selenium from the magma during crystallization. As the earth's crust is composed of igneous rocks, the mean sulfur and selenium content of igneous rocks is the accepted value for their crustal abundance. Goldschmidt (1954) has estimated that igneous rocks contain an average of 520 μg/g of sulfur and about 0.09 μg/g of selenium. Frost (1967) has estimated that sea water, the earth's crust, and animal and plant matter contain in μg/g about 0.004, 0.09, 1–20, and 0.02–4000, respectively. This would indicate that animals and plants have the ability to concentrate selenium from the earth's crust.

Selenium mainly occurs in the earth's crust associated with sulfide minerals or as selenides of silver, copper, lead, mercury and nickel or other metals. Some of the highest selenium contents of sulfides are associated with uranium ores from the sandstone-type deposits in the Western United States. The following maximum values have been reported for selenium in these ores: pyrite, 5%; marcasite, 0.65%; and chalcocite, 5%. The chalcopyrite–pentlandite–pyrrhotite deposits are related to the basic and ultrabasic rocks of the Precambrian age and contain the largest amounts of selenium (Coleman and Delevaux, 1957). Goldschmidt (1954) has suggested that selenium content is higher in high-temperature hydrothermal sulfides than in low-temperature sulfides. Takimoto *et al.* (1958) have studied the distribution of selenium in sulfur minerals. They concluded that the selenium content of sulfide materials is highest in those of earliest formation which may be related to the highest temperature of formation and not affected by the kind of sulfide mineral that occurs in close paragenetic relation.

Selenium can be easily oxidized from Se^0 (elemental selenium) to Se^{4+} (SeO_3^{2-}) and to Se^{6+} (SeO_4^{2-}). The selenites (SeO_3^{2-}) are stable in alkaline to mildly acid conditions and should occur in nature. An insoluble precipitate is formed with the composition of basic ferric selenite $[Fe_2(OH)_4SeO_3]$ when extremely dilute solutions of selenites react with ferric chloride. Lakin and Trites (1958) have described the oxidation of Se^{2-} to SeO_3^{2-} and the subsequent reduction is found in a uranium mine in Baggs area, Carbon County, Wyoming. Seleniferous pyrite is oxidized to give ferric sulfate, sulfuric acid, and hydroselenous acid. A partial neutralization of the sulfuric acid results in the precipitation of ferric hydroxide, which contains any associated selenium. Both the selenious acid and the ferric sulfate will continue downward until they enter the reducing conditions of the water table. There the sulfate ions are reduced to S^{2-} ions, the ferric ions to Fe^{2-}, and the selenate ions to Se^0 and some to Se^{2-}, the latter entering the immediate vicinity of sandstone-type uranium deposits which have concentrations of as much as 1000 µg/g selenium.

Because sedimentary rocks cover more than three quarters of the earth's surface, they are the major parent material of agricultural soils. The concentrations of selenium in sedimentary rocks range from 0.08 to 1 µg/g and are much higher than the estimates for the earth's crust. The sandstone content of selenium is quite variable and ranges between 0.05 and 1 µg/g. Shales average about 0.6 µg/g selenium. Shales are the principal sources of selenium-toxic soils of the Great Plains, the Rocky Mountain foothills of the United States, Australia, and Ireland. Carbonate rock ranges from 0.0 to 2.0 µg/g but some carbonaceous limestones contain as much as 30 µg/g. Selenium is often present in phosphorites in relatively high concentration. Coal has abundant amounts of selenium, ranging from 0.1 to 4 µg/g. When the seleniferous coal or oil is burned, selenium is introduced into the atmosphere from which selenium is redistributed to the earth's surface in rain and snow. Dust from air conditioner filters has been found to contain from 0.05 to 10 ppm Se. Selenium from these sources is probably in the form of insoluble oxides or in the elemental form and may not be of immediate value to plants and animals.

6.2 Soil Selenium

Of the rocks exposed at the weathering subsurface that may serve as parent materials shales represent about 40% of the parent rocks and are the most abundant. The sandstones, limestones, and igneous rocks are about equally abundant at about 20% each. Of these the soils most likely to be uniformly deficient in selenium are those derived from igneous rocks.

Slater *et al.* (1937) have measured the selenium contents of 11 soil profiles and found that they ranged from 0.02 to 2.5 μg/g. The most ferruginous horizons of the soils are also the most seleniferous. In acid ferruginous soils selenium is bound as a basic ferric selenite or strongly absorbed on ferric oxide. Vegetation grown on these soils do not produce toxicity as the selenium is only slightly available to plants. Lateritic soils of the continental United States that have been analyzed also contain 0.5–2.4 μg/g of selenium in the iron-rich horizons, but do not produce toxic vegetation. In South Dakota, Wyoming, Nebraska, Kansas and Colorado many soils have a high selenium content and produce toxic

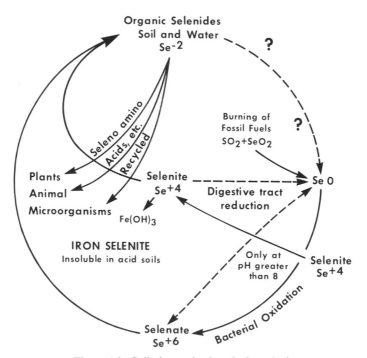

Figure 6-2. Soil-plant animal cycle for selenium.

seleniferous plants. The soil selenium of these states is not bound to the ferruginous horizons of the soil and would be available for plants. As a general rule, more alkaline soil contains less selenium. At pH values greater than 6.5, selenium might be more easily oxidized to the more soluble selenate. The reaction $2H_2SeO_3 + O_2 \rightarrow H_2SeO_4$, takes place more easily in an alkaline environment. These soils occur in either low-rainfall or poorly drained areas. If the soils are irrigated, selenium may be moved in or out of these soils by leaching.

Acid soils are mainly composed of decaying organic matter. The reduction of selenium to selenites or even to selenides probably takes place in the horizons rich in acid organic matter. In such soils selenium moves from the soil to vegetation and back to the soil if the vegetation dies and decomposes where it stands. If the vegetation is removed by the harvesting of crops, for example, selenium depletion will occur unless selenium is replaced by fertilizers, ground or surface water selenium, or seleniferous dusts.

Certainly the natural source of selenium for agriculture and most other biological uses is the soil, from which the element is accumulated to varying amounts by plants which are ultimately consumed by animals. A fairly straightforward chain of reactions takes place, allowing the diagrammatic presentation of a selenium cycle in nature (Figure 6-2).

6.3 Uptake and Concentration of Trace Elements in the Roots, Stem, and Leaves of Plants

Johnson *et al.* (1967) have reported that selenium is taken up by plants either as selenate, selenite, or organic selenium. According to Leggett and Epstein (1956), selenate is taken up through the same binding sites in the plant roots as sulfate. They found that the two ions were taken up by the same active absorption process in competition with each other, whereas selenite is taken up through other sites. In duckweed, Butler and Peterson (1967) found that selenite was absorbed and assimilated about three times more easily than selenate. Gissel-Nielsen (1976) has studied the uptake and translocation of selenite by means of ^{75}Se in water culture experiments with barley and maize. After 10 min the ^{75}Se activity was found in the green part of the plant. This corresponds to a translocation speed from the roots of 1–2 cm/min. After 30 min, the experiment showed that more than 15% of the selenium absorbed by the roots was translocated as selenite, and greater than 80% was present in the amino acid fraction, probably as selenomethionine. The selenium concentration in the xylem sap ranged from 1 to 5 μg Se/ml. After fractionation, of the selenium in the leaves, the results showed that 25%–30% of the selenium was water soluble, and 10% of the total was present as selenite. If the protein was hydrolyzed with pronase, the water-soluble fraction of ^{75}Se increased to 96%, of which 73%–76% were in the amino acid fraction. The predominant selenium containing compound was selenomethionine.

Corn grown in culture solutions containing 5 ppm of selenite or organic selenium accumulated 200 and 1000 ppm of selenium, respectively (Ganje, 1966). When the selenium level was increased to 10 ppm, corn accumulated 300 ppm from the selenite form and more than 1500 ppm from the organic selenium form. Ganje (1966) has reported that the organic forms of selenium and selenate were the most available forms for corn grown in culture solutions. In general, resistance to selenium toxicity among plant species varies so widely that a general toxicity level cannot be reliably estimated.

Selenium uptake by plants also depends on sulfur (see Chapter 4 on S–Se interactions), calcium, phosphorous, and nitrogen concentration. In calcium-rich soils the mobility of selenium increases because of its oxidation and greater solubility. Selenium is also more available to plant consumption in calcium-rich soils than in podzolic soils where the contents are increased in vegetation. Several selenium poisonings have been observed in cattle that have consumed seleniferous plants in calcareous areas such as Ireland (Walsh *et al.*, 1951). Quite early in the investigation of white muscle disease in Oregon, it was observed that the incidence of the disease was sometimes increased following the application of gypsum, $CaSO_4 \cdot 2H_2O$, to the soils on which the animals' feed was grown (Muth, 1955). The uptake of selenium by winter wheat was decreased from 110 to 2 ppm by the application of gypsum or of elemental sulfur (Hurd-Karrer and Kennedy, 1936).

Davies (1966) has made a comparison of the content of selenium in plants fertilized with sodium selenite alone, or mixed with superphosphate on a peat soil base. They found that the selenium was lower where the superphosphate was included (0.57 vs. 0.43 ppm, respectively). The lower content of selenium was attributed to preferential stimulation of the plant growth by the superphosphate, rather than the competitive absorption between sulfate and selenite by the plants. Singh and Singh (1977) have conducted experiments on wheat to ascertain the effect of phosphorous application on the detoxification of selenium. The dry matter at 50 days' maturity decreased significantly with an increase in selenium in the growth medium from 0 to 10 ppm. The application of phosphorous at 50 ppm increased dry matter yield twofold at 50 days. Maturity and grain yield increased about threefold. The increased selenium concentration, even with 50 ppm phosphorous, showed a similar trend. If phosphorous was further increased to 100 ppm, phosphorous tended to depress the dry matter and grain yield, but when selenium was increased up to 5 ppm the dry matter was increased and grain yield was increased up to 2.5 ppm.

Sharma *et al.* (1981) found that with up to 5 ppm of selenium application, there was a significant increase in concentration and uptake of phosphorous. However, the uptake of sulfur decreased to deficiency level showing an antagonistic relationship. Singh and Malhotra (1976) have noted that phosphorous applied in conjunction with selenium has increased sulfur concentration significantly in *berseem*, demonstrating a synergistic relationship between sulfur and phosphorous. Gissel-Nielson (1974) have also reported such complicated interactions. He observed that phosphorous decreased selenium in barley at high concentrations of nitrogen and sulfur, but increased selenium when nitrogen level was low and the sulfur level high.

Gissel-Nielsen (1979) has studied the influence of different levels of nitrogen on the uptake, transport and distribution of selenium on different fractions of *Zea mays* plants which were grown in nutrient solutions at three levels of nitrogen. After eight weeks, ^{75}Se-labeled selenite was added 4 hr before the tops were severed and the xylem sap was collected. Increasing the nitrogen levels caused an increase in the selenium concentration of the roots, but there was a decrease in the leaves to a greater extent than could be explained by the higher yield of dry matter. The increased selenium content of the roots resulted from a much greater content of selenite (from 7% to 35% of the total selenium), whereas the relative content of selenium-containing amino acids decreased to some extent. Similarly in the leaves, nitrogen increased the fraction of water-soluble selenium, but here the amino acids were increased from 10% to 36%, while the content of selenite was only 2% of the total selenium.

Nitrogen also plays a role in decreasing the selenium content in crops. Singh (1975) has studied the effect of nitrogen fertilizers on the selenium content of fodders and observed that the application of 200 ppm of urea nitrogen was effective in reducing the selenium content of sorghum. Ammonium sulfate (200 ppm) application even reduced the dry matter yield to a greater extent. However, ammonium sulfate was less effective in reducing selenium content in comparison to urea. The exact mechanism of the role of nitrogen in decreasing selenium content in plants is not known, but it is possible that the decreased selenium may be due to a dilution effect due to the increase of yield. The role of sulfur, phosphorous, and nitrogen in decreasing selenium uptake and also increasing amino acid synthesis is complicated and more studies are needed in this regard.

6.4 Forage Selenium

Kubota *et al.* (1967) have prepared a map showing the regional distribution of selenium concentrations in crops of the United States (Figure 6-3). In addition to the Great Plains and Rocky Mountain foothill areas which produce high to very toxic forages, they were able to identify large areas of apparent adequacy as well as where selenium toxicity is only occasionally and locally a problem. There were three major areas of the United States that were of interest because of the incidence of selenium-responsive diseases. Parts of the Pacific Northwest, including portions of Washington, Oregon, and Northern California, grow forages which are very low in selenium (less than 0.05 ppm) and in these areas selenium-responsive diseases are often observed. The Pacific Northwest is an area of recent volcanic deposits and the soils formed on them or transported from them are inherently very low in selenium. In addition, because of low pH and high sesquioxides, the low native selenium is plant unavailable.

In the Northeast portion of the United States, acid soils derived from very old sedimentary rocks that predate the major Cretaceous period of selenization of North America, produce crops which have only slightly more selenium than those of the Pacific Northwest. Selenium-responsive

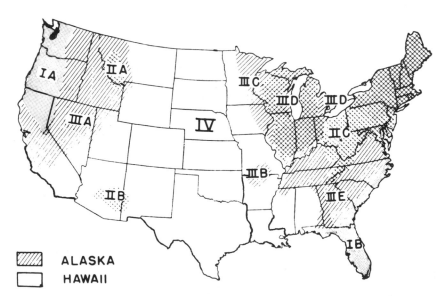

Figure 6-3. Selenium in forages and grains of the United States. Median concentration of selenium (ppm). IA, 0.03; IB, 0.02; IIA, IIB, IIC, IIIB, 0.05; IIIA, IIIC, 0.09; IIID, 0.10; IIIE, 0.06; and IV, 0.26. Reprinted with permission from *Ag. Food Chem.* 15:448-453, 1967, by American Chemical Society.

diseases have been reported in this area, but the incidence is low because of the intermingling of feeds from adequate areas.

There is a third low-selenium area which is located along the South Atlantic seaboard where a combination of soil parent material with low total selenium and low availability to plants appears to account for the low selenium in the crops.

6.5 Selenium in Water

Schutz and Turekian (1965) estimate an average value of 0.09 μg/liter for selenium in the major oceans. In acid waters having a pH of 2.4 to 3.0 and in weakly alkaline waters having a pH of 7.4 to 8.0, higher concentrations of selenium have been observed. In general, water contains less than 1 μg/liter. The selenium concentration that was greatest was the 400 μg/liter reported by Scott and Voegeli (1961) in a sample from a tributary of the Animas River in Colorado. According to the standards for drinking waters of the United States Public Health Service, amounts of selenium in potable waters exceeding 0.01 ppm may be potentially dangerous and constitute grounds for rejection of the water supply (Code of Federal Regulations, 1967).

6.6 Selenium in Food

The selenium content of food can vary widely depending on where the plants are grown. Morris and Levander (1970) have determined by fluorometry the selenium content of a wide variety of foods representing a cross section of the American diet. Most fruits and vegetables contained less than 0.01 μg/g of selenium (Table 6-1). Exceptions were garlic, mushroom, and radish containing 0.25, 0.13, and 0.04 μg/g, respectively. Grain products vary widely in their selenium content with cornflakes containing as low as 0.025 μg/g and barley cereal as high as 0.66 μg/g (Table 6-2). Whole wheat flour and whole wheat bread contained two to four times more selenium than did white flour and white bread. Of the milk products assayed, table cream was the lowest in selenium and skim milk the highest (0.005 vs. 0.05 μg/g) (Table 6-3). Dried skim milk powder samples ranged from 0.095 to 0.24 μg/g. Egg yolk contains about three times as much selenium as does egg white, and brown sugar contains about four times as much selenium as white sugar.

Meat samples ranged from about 0.01 μg/g for chicken muscle to as much as 1.9 μg/g for pork kidney with most values between 0.2 and 0.5

Table 6-1. Selenium Content of
Vegetables and Fruits[a]

Product	μg Se/g
Vegetables	
Carrots	
Fresh	0.022
Canned	0.013
Cabbage, fresh	0.022
Cauliflower, fresh	0.006
Corn	
Fresh	0.004
Canned	0.004
Garlic, fresh	0.250
Green pepper, fresh	0.007
Green beans	
Fresh	0.006
Canned	0.009
Lettuce, fresh	0.008
Mushrooms	
Fresh	0.130
Canned	0.109
Onions, white, fresh	0.015
Potatoes	
Sweet, fresh	0.007
White, fresh	0.004
White, canned	0.009
Radishes, fresh	0.040
Tomatoes	
Fresh	0.005
Canned	0.010
Turnips, fresh	0.007
Fruits	
Apples, fresh, peeled	0.004
Applesauce, canned	0.002
Bananas, fresh, peeled	0.010
Oranges, fresh, peeled	0.013
Peaches	
Fresh, peeled	0.004
Canned	0.003
Pears	
Fresh, peeled	0.006
Canned	<0.002
Pineapples	
Fresh	0.006
Canned	0.010

[a] Permission *J. Nutr.* 100:1385–1386, 1970.

Table 6-2. Selenium Content of Grains and
Cereal Products[a]

Product	μg Se/g
Barley cereal	0.659
Bread	
White	0.277
Whole wheat	0.665
Corn flakes	0.026
Flour	
White	0.192
Whole wheat	0.636
Noodles, egg	0.623
Oat breakfast cereal, prepared[b]	0.428
Oats, quick	0.110
Rice	
Polished	0.318
Brown	0.387
Rice breakfast cereal, puffed, prepared	0.028
Wheat cereal[c]	0.241
Wheat breakfast cereal, prepared[d]	0.105

[a] Permission *J. Nutr.* 100:1385–1386, 1970.
[b] Cheerios, General Mills, Minneapolis, Minnesota.
[c] Wheatena, Standard Milling Co., Kansas City, Missouri.
[d] Wheaties, General Mills, Minneapolis, Minnesota.

μg/g (Table 6-4). The content of seafood was generally higher, ranging from 0.4 to 0.7 μg/g. The amounts of selenium in baby food reflected the amounts found in the food from which it was derived, except that amounts of selenium were proportionately lower in strained baby food (Table 6-5).

Arthur (1972) has determined by fluorometric chemical analysis the selenium content of a variety of foods consumed by Canadians. The results of these analyses were similar to those reported by Morris and Levander (1970). The average values for corn and rice products were 0.07 and 0.11 μg Se/g, respectively. Foods prepared from grains from a low-selenium province in Ontario (C) contained less Se than those from grains of the high-selenium prairie provinces (W). Average values (μg Se/g) were: oat 0.15C, 0.45W; wheat 0.08C, 0.56W, 1.27 durum variety; and bran preparations 0.07C, 0.87W.

Ferretti and Levander (1972) have studied by a fluorometric method the effect of milling and processing on the selenium content of grains and cereal products. Wheat flour and farina contained 14% and 29% less selenium, respectively, than the original raw wheat. Corn meal, flour, and grits contained 10%–28% less selenium than did the raw corn. Milling

Table 6-3. Selenium Content of Dairy
Products, Eggs, Saccharin, and
Sugar[a]

Product	μg Se/g
Cheese	
American, processed	0.090
Cottage	0.052
Swiss	0.100
Cream, table	0.005
Cream substitute	0.034
Milk	
Evaporated, canned	0.012
Skim	0.045
Skim, powdered, dried #1	0.096
Skim, powdered, dried #2	0.240
Whole homogenized	0.012
Egg	
Yolk #1	0.184
Yolk #2	0.181
White #1	0.057
White #2	0.045
Saccharin	0.005
Sugar	
Brown	0.011
White	0.003

[a] Permission *J. Nutr.* 100:1385–1386, 1970.

of oats and rice caused little change in the selenium level of these grain fractions which were destined for consumer use. There was little or no decline in the selenium concentration during the manufacture of nonsugared corn or wheat breakfast cereals. However, there was a significant decrease in the concentration of selenium in sugared corn and sugared wheat breakfast cereals which was apparently due to dilution of the selenium by the sugar. The authors also observed that the decrease in selenium concentration in consumer products due to milling or processing of grains appeared to be less than the decreases reported with several other nutritionally essential trace elements. For example, Czerniejewski *et al.* (1964) have reported decreases of over 80% in the concentration of cobalt, manganese, and magnesium when wheat was processed into flour. They also observed decreases of over 65% in the concentrations of copper, iron, and zinc.

Lorenz (1978) has found that losses of 50% of the selenium content of the original wheat occurred during wheat processing. The decrease in selenium content from wheat to flour depended on the patent per cent, indicating that selenium, as most other mineral elements in wheat, is not

Table 6-4. Selenium Content of Meats
and Seafoods[a]

Product	μg Se/g
Meats	
Beef	
Round steak	0.340
Ground	0.199
Liver	0.431
Kidney	1.55
Pork	
Chop	0.239
Kidney	1.89
Lamb	
Chop	0.178
Kidney	1.42
Chicken	
Breast	0.116
Leg	0.136
Skin	0.150
Seafoods	
Lobster tail	0.637
Shrimp, shelled, deveined	0.588
Cod, fillet	0.427
Flounder, fillet	0.335
Oysters	0.652

[a] Permission *J. Nutr.* 100:1385–1386, 1970.

Table 6-5. Selenium Content of Strained Baby
Foods[a]

Product	μg Se/g
Beef	0.116
Chicken	0.107
Lamb	0.130
Liver	0.258
Pork	0.124
Carrots	0.002
Green beans	0.005
Peaches	0.003
Pears	0.003
Oatmeal cereal with applesauce and bananas	0.030
Rice cereal with applesauce and bananas	0.021
Vanilla custard pudding	0.016

[a] Permission *J. Nutr.* 100:1385–1386, 1970.

evenly distributed in the wheat kernal. The selenium content of the hard wheat flours decreased as both the total flour extraction and the flour extraction of the patent decreased.

Higgs *et al.* (1972) have studied the effect of cooking on the selenium content of a variety of foodstuffs typically found in the American diet. Little or no loss of selenium occurred as a result of broiling meats, baking seafoods, frying eggs, or boiling cereals. If the cereals were dry heated, 7%–23% of the selenium was lost. Boiling two vegetables that contain relatively high amounts of selenium, asparagus and mushrooms, led to 29% and 44% losses of selenium, respectively. Higgs *et al.* (1972) concluded that, in general, most ordinary cooking techniques probably did not result in major losses of selenium from most foods.

6.7 Intakes and Recommended Daily Allowance in Humans

The average intakes of selenium per day in the United States and other countries have been estimated by several investigators: average intake per day in the Northeastern United States has been estimated at 62 µg/day (Schroeder *et al.*, 1970); Canada, 110–220 (Thompson *et al.*, 1975); New Zealand 25 (Robinson, 1976); The Netherlands, 110; France, 166; Italy, 141 (Cresta, 1976); Japan, 208 (Yasumoto *et al.*, 1976) or 100–200 (Noda *et al.*, 1980); and Britain 60 (Thorn *et al.*, 1978). Watkinson (1974) has estimated the intakes of three countries: New Zealand, 56; United States, 132; and Canada, 150.7. Swedish hospital diets provide from 23–210 µg Se/day (Wester, 1974). In addition, Schrauzer, *et al.* (1977) has estimated the intake of an American couple at 100 µg/day. Welsh *et al.* (1981) have estimated the diets of Maryland residents to contain about 81 µg/day. Levander *et al.* (1981) have reported that the gastrointestinal tract is about 80% efficient in absorbing food and that six North American men needed a dietary selenium intake of about 70 µg/day to replace losses and maintain body stores.

The Food and Nutrition Board (1980) of the National Academy of Science has recommended a daily intake of 50–200 µg/day for adults. Recommendations for other age groups were: infants 0–0.5, 10–40 µg; infants 0.5–1.0, 20–60 µg; children 1–3, 20–80 µg; age 4–6, 30–120 µg; adolescents 7–11, 50–200 µg.

6.8 Regulations in Regard to Animal Diets

Responsibility for regulations of essential nutrient additions to animal diets has been assumed by the Food and Drug Administration (FDA) through the interpretation of the Federal Food, Drug and Cosmetic Act. These substances are regulated as food additives or through qualification for the generally recognized as safe (GRAS) list. Even though essential nutrients, such as selenium, are not optional additives to animal diets that are demonstrably deficient, supplementation with unapproved nutrients places one in violation of the FDA interpretation of law. Approval was granted for selenium supplements to swine and certain poultry diets in 1974 (FDA, 1974). It was believed at that time that the inability to supplement deficient poultry and swine diets had caused annual losses of over $82 million. The amounts approved were up to 0.1 ppm in a complete feed of growing chickens up to 16 weeks of age. Selenium addition to complete turkey feed was approved up to 0.2 and 0.1 ppm was approved for complete swine feed. A further amendment was published on January 26, 1979 (FDA, 1979), which included beef, dairy cattle, and all ages and sexes of sheep. Annual losses to the beef cattle, dairy cattle, and sheep industry due to selenium deficiency was estimated at 545 million dollars in 1975. In sheep, additions of selenium to complete feed were not to exceed a level of 0.1 ppm. A feed supplement for limit feeding was not to exceed 0.23 mg/head/day. Approval was also given for up to 30 ppm in a salt-mineral mixture for free choice feeding at a rate not to exceed an intake of 0.23 mg/head/day. The complete feed for beef cattle was not to exceed a level of 0.1 ppm. A feed supplement for limit feeding was not to exceed an intake of 1 mg/head/day. Up to 20 ppm in a salt–mineral mixture for free choice feeding was also approved, but there was a limit of 1 mg/head/day.

References

Arthur, D., 1972. Selenium content of Canadian foods, *Can. Inst. Food Sci. Technol. J.* 5:165–169.

Butler, G. W., and Peterson, C. J., 1967. Uptake and metabolism of inorganic forms of selenium-75 by *Spirodela oligorrhiza*, *Aust. J. Biol. Sci.* 20:77–86.

Code of Federal Regulations: Title 42. Public Health, U.S. Government Printing Office, Washington, D.C. 1967.

Coleman, R. G., and Delevaux, M. H., 1957. Occurrence of selenium in sulfides from some sedimentary rocks of the United States, *Econ. Geol.* 52:499–527.

Cresta, M., Allegrini, M., and Casadei, E., 1976. Nutritional considerations on trace elements in the diet, *Food. Nutr. (FAO)* 2:8–18.

Czerniejewski, C. P., Shank, C. W., Bechtel, W. G., and Bradley, W. B., 1964. The minerals of wheat, flour and bread, *Cereal Chem.* 41:67–72.

Davies, E. B., 1966a. Uptake of native and applied selenium by pasture species. I. Uptake of selenium by browntop, ryegrass, cocks foot and white clover from Atiamuri sand, *New Z. J. Agric. Res.* 9:317–327.

FDA, 1974. Food additives: Selenium in animal feed, *Fed. Regist.* 39:1355 (Tuesday, January 8).

FDA, 1979. Food additives permitted in fed and drinking water of animals: Selenium, *Fed. Regist.* 44:5392 (Friday, January 26).

Ferretti, R. J., and Levander, O. A., 1974. Effect of milling and processing on the selenium content of grains and cereal products, *Ag. Food Chem.* 22:1049–1051.

Food and Nutrition Board. *Recommended Dietary Allowances*, ninth revised ed., Washington, D.C.: National Academy of Sciences, National Research Council, 1980.

Frost, D. V., 1967. Significance of the symposium, in *Symposium on Selenium and Biomedicine*, O. H. Muth (ed.), Avi Publishing Co., Westport, Connecticut, pp. 7–26.

Ganje, T. J., 1966. Selenium, in *Diagnostic Criteria for Plants and Soils*, H. L. Chapman (ed.), Riverside, University of California Division of Agricultural Science, pp. 394–404.

Gissel-Nielson, G., 1974. Effect of fertilization on uptake of selenium by plants, in *Plant Analysis and Fertilizer Problems*, Vol. 1., *Proceedings of the International Colloquium on Plant Analysis and Fertilizer Problems*, Hanover, Federal Republic of Germany, pp. 111–116.

Gissel-Nielsen, G., 1976, Selenium in soils and plants, *Proceedings of the Symposium on Selenium-Tellurium in the Environment, Industrial Health Foundation*, Pittsburgh, Pennsylvania, pp. 10–25.

Gissel-Nielsen, G., 1979. Uptake and translocation of selenium-75, in *Zea Mayes, Symposium on Isotopes and Radiation in Research on Soil-Plant Relationships*, International Atomic Energy Agency, Vienna, Austria, pp. 427–436.

Goldschmidt, V. M., 1954. *Geochemistry*, Oxford Press, London.

Higgs, D. J., Morris, V. C., and Levander, O. A., 1972. Effect of cooking on selenium content of foods, *J. Agr. Food Chem.* 20:678–680.

Hurd-Karrer, A. M., 1936. Inhibiting effect of sulphur in selenized soil on toxicity of wheat to rats, *J. Agric. Res.* 52:933–942.

Johnson, C. M., Asher, C. J., and Broyer, T. C., 1967. Distribution of selenium in plants, in *Selenium in Biomedicine*, O. H. Muth (ed.), Avi Publishing Co., Westport, Connecticut, pp. 57–75.

Kubota, J., Allaway, W. H., Carter, D. L., Cary, E. E., and Lazar, V. A., 1967. Selenium in crops in the United States in relation to selenium responsive diseases of animals, *J. Agric. Fd. Chem.* 15:448–453.

Lakin, H. W., and Trites, A. R., 1958. The behavior of selenium in zone of oxidation, Int. Geol. Cong., Mexico, *Symposium de Exploracion Geoquimica*, 1:113–124.

Leggett, J. E., and Epstein, E., 1956. Kinetics of sulfate absorption by barley roots, *Plant Physiol.* 31:222–226.

Levander, O. A., Sutherland, B., Morris, B. A., and King, J. C., 1981. Selenium balance in young men during selenium depletion and repletion, *Am. J. Clin. Nutr.* 34:2662–2669.

Lorenz, K., 1978. Selenium in wheats and commercial wheat flours, *Cereal Chem.* 55:287–294.

Morris, Y. C., and Levander, O. A., 1970. Selenium content of foods, *J. Nutr.* 100:1383–1388.

Muth, O. H., 1955. White muscle disease (myopathy) in lambs and calves. I. Occurrence and nature of the disease under Oregon conditions, *J. Am. Vet. Assoc.* 126:355–361.

Nada, K., Hirai, S., and Danbara, H., 1980. Selenium contents of Japanese foodstuffs by neutron activation analysis, *Eiyo to Shokuryo* 33:93–99.

Robinson, M. F., 1976. The moonstone: more about selenium, *J. Hum. Nutr.* 30:79–91.

Schrauzer, G. N., White, D. A., and Schneider, C. J., 1977. Selenium in human nutrition: dietary intakes and effects of supplementation, *Bioinorg. Chem.* 8:303–318.

Schroeder, H. A., Frost, D. V., and Balassa, J. J., 1970. Essential trace elements in man: selenium, *J. Chron. Dis.* 23:227–243.

Schutz, D. F., and Turekian, K. K., 1965. The investigation of the geographical and vertical distribution of several trace elements in sea water using neutron activation analysis, *Geochim. Cosmochim. Acta* 29:259–313.

Scott, R. C., and Voegeli, P. T., 1961. Radiochemical analyses of ground and surface water in Colorado, 1954–1961, *Water Conservation Board Basic—Data Report. No. 7.*

Sharma, S., Singh, R., and Bhattacharyya, A. K., 1981. Perspective of selenium research in soil-plant-animal system in India, *Fert. News* 26:19–28.

Singh, M., 1975. Effect of N-carriers, soil type and genetic variation on growth and accumulation of selenium, nitrogen, phosphorous and sulfur in sorghum and cowpea, *Forage Res.* 1:68–74.

Singh, M., and Malhotra, P, K., 1976. Selenium availability in Berseem, *Plant Soil* 44:261–266.

Singh, M., and Singh, N., 1977. Effect of S and Se and S containing amino acids and quality of oil in Raya in normal and sodic soil, *Ind. J. Pl. Physiol.* 20:56–62.

Slater, C. S., Holmes, R. S., and Byers, H. G., 1937. Trace element in the soils from the erosion experiment stations, with supplementary data on other soils *U.S. Dept. Agric. Tech. Bull.*, 552.

Takimoto, K., Minoto, T., and Hiroko, S., 1958. On the distribution of selenium in some sulfide minerals showing the relations of intimate paragenesis, *J. Japan Assoc. Miner. Petrologists Econ. Miner.* 42:161–170.

Thompson, J. N., Erdody P., and Smith D. C., 1975. Selenium content of food consumed by Canadians, *J. Nutr.* 105:274–277.

Thorn, J., Robertson, J., and Buss, D. H., 1978. Trace nutrients. Selenium in British food, *Brit. J. Nutr.* 39:391–396.

Walsh, T., Fleming, G., and O'Connor, R., 1951. Selenium toxicity associated with an Irish soil series, *Nature* 168:881

Watkinson, J. H., 1974. The selenium status of New Zealanders, *N.Z. Med. J.* 80:202.

Welsh, S. O., Holden, J. M., Wolf, W. R. and Levander, O. A., 1981. Selenium in self-selected diets of Maryland residents, *J. Am. Diet. Assoc.* 79:277–285.

Wester, P.O., 1974. Trace element balances in relation to variations in calcium intake, *Atherosclerosis* 20:207–215.

Yasumoto, K., Iwami, K., Yoshida, M., and Mitsuda, H., 1976. Selenium content of foods and its average daily intake in Japan, *Eiyo To Shokuryo* 29:511–515.

Toxicity of Selenium 7

7.1 Introduction

Selenium is present at least in trace amounts in all soils and in all natural feeds and its occurrence is not restricted to any specific area of the earth. Some soils contain an excess of selenium in forms which are available to plants. In this case plants absorb the selenium in amounts that make them toxic to animals. In some other cases the selenium is in a form which is not available to plants. The soils and plants with large available amounts of selenium have been referred to as "seleniferous" or "indicator plants," which implies excessive levels of the element. Rosenfeld and Beath (1964) have divided the indicator plants into two categories. According to their classification the first category was called primary indicators and included over 20 species of *Astragalus* (1000 ppm Se) *Machaeranthera*, Oonoposis (800 ppm Se), *Haplopappus*, Zylorhiza (120 ppm Se), and Stanleya (700 ppm Se) which appear to require selenium for their growth. High levels of selenium normally accumulate in these species, sometimes as much as several thousand parts per million.

The second category is referred to as secondary indicators because they do not appear to require selenium for their growth. They will accumulate the element when they grow on soils of high available selenium content. They belong to a number of genera which include *Aster*, (72 ppm Se), *Atriplex*, (50 ppm Se), Gutierrezia (60 ppm Se), Castilleja, Comandra, and Grindelia (38 ppm Se).

Finally, there is a third group of plants which includes the grasses and grains. This group does not normally accumulate selenium in excess of about 50 ppm under field conditions. When livestock, usually cattle, eat toxic plants, three types of selenium poisoning have been described (Rosenfeld and Beath, 1964): (1) acute; (2) chronic, blind staggers; (3) chronic, alkali disease.

7.1.1 Acute Toxicity

Acute poisoning comes about when high selenium (10,000 ppm Se) content plants are consumed in large quantities. Usually animals will avoid the indicator plants because they do not have a good taste. However, under poor grazing conditions brought about by bad weather or overgrazing of an area, cattle or sheep may sometimes be forced to eat these toxic indicator plants. The symptoms which are severe include abnormal movement and posture, anorexia, watery diarrhea, labored breathing, bloat, elevated temperature and pulse rate, prostration, and often death from respiratory failure. The pathological changes include hepatic necrosis, hemorrhage, edema, nephritis, and congestion and ulceration of numerous tissues. Because the highly seleniferous plants are not palatable, deaths from the acute type are quite rare. In individual herds a few cases of large losses have been observed.

An incident of acute poisoning has occurred on a farm where white muscle disease had been diagnosed in calves that had not receive supplementary selenium (Shortridge *et al.*, 1971). Selenium had been previously administered to 557 steer and heifer Aberdeen Angus calves when they were three months old. They were weaned when six months old and were in excellent health. Their weights were estimated to be between 150 and 200 kg with most of the calves near the heavier weight. After a vaccination all of the calves received an injection of 5 ml of sodium selenite solution subcutaneously behind the scapula. The final injection solution should have contained 2.4 mg Se/ml, but an error was made in the preparation and analysis of the solution showed that it contained 20 mg Se/ml, which was 8.4 times the correct amount. Each calf therefore received 20 mg Se/ml, which was approximately 8.4 times the correct amount. Therefore, each calf received 100 mg selenium (about 0.5 mg/kg live weight) instead of 12 mg selenium as was intended.

Within 2 hr of receiving the injection, some calves were severely depressed, salivating, and showed respiratory distress. The first deaths occurred 2 hr after treatment. Eighteen died within 24 hr, 38 within 48 hr, and a total of 75 within 72 hr. After five weeks, 376 (67%) of the 557 calves died. Two hundred and four (80%) of the 254 heifers died and 172 (56%) of 303 steers died. The most obvious clinical sign in calves that died later in the course of the episode was dyspnea.

Four of the animals that died within four days of receiving the injection were necropsied. Gross lesions in all these animals were similar. At the injection site there was an area of approximately 15 cm diameter containing serous exudate and petechiae. The oral, ocular, and vaginal mucosae were pale. An excess of clear yellow fluid was in the thoracic cavity. The lungs were dark red and edematous. The interlobular septa were distended by serous fluid. Ecchymoses were present in the coronary grooves and toward the apex of the heart, and the epicardial surface of the heart was pale. When the surface of the myocardium was cut it was pale yellow with red streaks. The liver was swollen and congested and there were subcapsular hemorrhages on the kidneys.

Within 24 hr the only tissues examined histologically from one of the animals that died were the liver and kidney. The liver showed multiple necrosis which was centrilobular in distribution. Most of the liver cells contained large fat vacuoles. The kidney showed congestion, especially in the area of the corticomedullary junction. In addition, there were degenerative changes which had progressed to the stage of necrosis in some of the epithelial cells of the convoluted tubules.

Histological examinations were also made from the liver, kidney, lung, and myocardium from two calves that died between 48 and 74 hr after dosing. In both calves there was dilation of the hepatic sinusoids, hydropic degeneration of most hepatocytes with foamy vacuolation of the cytoplasm, and swollen vesicular nuclei. In the kidney congestion in the are of the corticomedullary junction was the only abnormality recognized in the kidney. The lungs showed congestion with marked intra-alveolar fibrinous exudate, intra-alveolar hemorrhage, and edema of the interlobular septa. Degenerative changes in the walls of many arterioles were characterized by rounding of the normally elongated smooth muscle nuclei, pyknosis and karyorrhexis of some of these nuclei, and vacuolation or hyalinization of some muscle fiber cytoplasm. A poor affinity for eosin was exhibited by the myocardium of the left ventricle. Some isolated muscle fibers had strongly eosinophilic cytoplasm and pyknotic nuclei. There were also some scattered localized areas of myocardial necrosis. Moderate numbers of plump active fibroblasts were contained in these necrotic areas. Some arterioles showed evidence of degenerative changes similar to those seen in the lung arterioles.

Tissues were also examined from one animal that died four weeks after being injected. Although the liver showed marked vacuolation of centrilobular cells, no abnormalities were seen in the kidneys. The lungs were congested and there was marked exudation of proteinaceous and fibrinous exudate into the alveoli. There were several areas of alveolar collapse and large numbers of septal macrophages were present. The myocardium showed no recent changes, but the cytoplasm of muscle fibers contained lipofuscin granules.

7.1.2 Blind Staggers

Blind staggers is a more advanced type of chronic selenium poisoning which results from the consumption of primary indicator plants (100–10,000 ppm Se) in smaller amounts over a longer period of time. In Wyoming, animals are found from time to time to be acutely poisoned by selenium. In the early stages of blind staggers, cattle wander, stumble, have impaired vision, and also lose their appetite for food and water. Later the front legs become weak, and if extreme enough, there is finally respiratory failure and paralysis preceding death. Other symptoms include stumbling over obstacles, weak forelegs, and paralysis of the tongue and throat. Lesions include hepatic necrosis, nephritis, hyperemia, and ulceration of the upper intestinal tract. Sometimes there are delayed symptoms which do not appear until the animals are in the feed lot being fattened for market. This type of poisoning has a similar pathology in sheep, but it is not as readily diagnosed, because the signs and lesions in sheep are less distinct.

7.1.3 Alkali Disease

The first report of alkali disease in livestock was made by Madison (1860), an army surgeon stationed at Fort Randall in the territory of Nebraska. He described a fatal disease of horses which had grazed in a certain area near the fort. The horses lost the long hair from the mane and tail, and their feet became sore. In 1891 after farmers settled around Fort Randall, the farmers experienced difficulty with the same disease in their livestock. Experiments by Franke (1934) led to the discovery of selenium as the cause of alkali disease.

Chronic poisoning of the alkali disease type results from the continuous ingestion of feeds that contain more than 5 ppm but usually less than 40 ppm of selenium. It is found primarily in South Dakota. In horses, cattle, and swine the most common symptoms of the poisoning have been described as follows: rough hair coat, lack of vitality, loss of appetite, malformation and sloughing off of hoofs, lameness due to joint erosion of the long bones, inflammation with swelling along the coronary band, and emaciation. In advanced cases liver cirrhosis, atrophy of the heart, and anemia occur. A "bobtail" appearance can occur in horses if they lose the long hair from the mane and tail. In horses other lesions include alopecia and elongated and weak, cracked hoofs. On occasion hogs lose their body hair and cattle lose the long hair from their tails. Sheep seem to be more tolerant and get a milder form of the disease. They do not have similar symptoms to cattle and horses. However, sheep may suffer a loss of appetite and show a reduced rate of weight gain. Inorganic selenium has been found to affect rats (Franke and Potter, 1935), dogs (Rhian and Moxon, 1943), swine (Wahlstrom *et al.*, 1956), poultry (Moxon, 1937), and the embryos of chicken eggs (Franke *et al.*, 1935) in a similar way as do seleniferous grains, causing the typical "alkali disease" symptoms.

In growing chicks, depressed rates of weight gain, rough feathers, and characteristics of nervousness have been observed. In chickens and turkeys, egg production is delayed and as a result of deformed embryos a lower percentage of hatchability is observed. The most common deformities are missing or short upper beaks, deformed feet and legs, edema of the head and neck, missing eyes, and a wiry down. In sheep and pigs reduced reproducibility is also observed. From the economic standpoint, reduced reproductive performance may be the most significant effect of excessive selenium intake.

7.1.4 Toxicity in Rabbits

The 24-hr LD_{50} for sodium selenite given orally to 72 rabbits was found to be about 8.6 mg/kg live weight (Berschneider *et al.*, 1976a). The oral LD_{50} is four times that of the intravenous value. If the rabbits were given 15 mg Na_2SeO_3/kg live weight by mouth, all of them were dead within 21 hr.

If the rabbits were given a dietary supplement of 30 ppm for 50 days, they thrived and grew faster than the unsupplemented controls. However, when the supplementation continued beyond 50 days, the rabbits declined clinically in a precipitous fashion. Findings after necropsy included catarrhal enteritis, toxic liver dystrophy, scattered pulpous splenic outgrowths, and interstitial nephritis. Another group of rabbits that was given 60 ppm of selenium supplementation in the diet developed faster than unsupplemented controls for about 60 days. At that time they went into a decline, just as was observed in the rabbits on the 30-ppm-Se supplement. The decline was evidenced by generally poor health and overall malnutrition. Necropsy findings were almost identical with those after the 30-ppm-Se supplement.

In rabbits, fetal and maternal damage results from the administration of two types of sodium selenite, Uroselevit Proinj, and Uroselevit Premix (Bernschneider *et al.*, 1976b). The general physical condition of the mothers was adversely influenced by a high dose of selenium. The maternal health was excellent while the rabbits received the 2.5% Uroselevit Premix. Variations from the controls in weight, length, and maturation as well as pre- and postimplantation losses were observed in fetuses from mothers dosed with selenium. Fetal mortality rose from control levels of 3%–5% to 70%–80% in selenium-dosed mothers. Some surviving fetuses were below control weights.

7.1.5 Toxicity in Hamsters

Hadjimarkos (1970) has studied the effects of chronic selenium toxicity in groups of eight weanling hamsters by adding 6,9, or 12 ppm of selenite–selenium to their drinking water. The weights after one year and mortality after one year were compared to a control group. After one year all of the hamsters were alive. The groups given 9 and 12 ppm showed about a 10% weight loss. The results demonstrated that hamsters were far more resistant to selenium toxicity compared to other animals.

7.1.6 Toxicity in Sheep

Morrow (1968) has studied the effect of an accidental acute toxicosis in 20 young 4.5-kg lambs who were orally administered 10 mg of sodium selenite. Seven died within 10–16 hr. Diarrhea developed in eight others, but they recovered. Five others were unaffected. Several clinical signs were exhibited in the experimentally poisoned ewes. These were depression, ataxia, dyspnea, frequent urination, elevated temperature, and increased pulse and respiratory rates. The terminal signs included cyanotic mucous membranes, dilated pupils, tympany, and recumbency. Gross and microscopic pathologic changes were numerous. They included serumlike fluid in the thoracic cavity. The heart had moderate bilateral dilation and the lungs were severely hyperemic and edematous. The bronchi and distal half of the trachea were filled with frothy edema. Moderate numbers of small submucosal hemorrhages occurred in the trachea and bronchi. The bronchial and mediastinal lymph nodes were hyperemic and edematous. On microscopic examination the lungs showed severe edema, moderate hyperemia, and hemorrhages, and the aveoli were lined with red blood cells. The kidneys were hyperemic, especially in the medulla. Microscopically in the kidneys there was severe acute necrotizing nephrosis which primarily affected the proximal convoluted tubules. Frequent and extensive hemorrhages were observed in the cortex. In a limited number of glomeruli there was edema. The medulla had severe hypermia but no inflammation. The bladders were distended with urine and contained petechial hemorrhages. Other organs such as the thymus and the brain were moderately hyperemic.

Glenn *et al.* (1964) have described the pathogenesis of selenium toxicosis in 30 ewes fed subtoxic to toxic levels of sodium selenate for one to five months. Seventeen of the 30 died of selenium toxicosis.

The most severe and consistent pathologic changes were found in the myocardium and lungs. Myocardial alterations were focal to diffuse degeneration, necrosis, and early replacement fibrosis. Irregular-shaped, pale focal areas with indistinct margins were distributed throughout the thick muscular walls of the left ventricle and the interventricular septum. In seven ewes, the pericardial fluid was increased to as much as 40 cc and was serous or serofibrinous. The earliest changes seen in the myocardial cells were slight swelling and loss of striations. The entire intracellular mass gradually became more acidophilic and distinctly granular, which indicates hydropic degeneration. Many of the intercalated disks were homogeneous, distinctly acidophilic, and readily visible. The nuclei were more densely basophilic and pyknotic; karyorrhexis and lysis occurred later. Edema of the myocardium was a common finding. Ten of the 17 ewes that died had excessive qualities of serous to serofibrinous pleural fluid, ranging from a few to as much as 500 cc. Microscopic examination revealed pulmonary congestion and edema in all 17 ewes. These pulmonary changes, consisting of edema as well as interstitial hemorrhages, were typical of the degenerative processes which characterize passive congestion of the lungs resulting from left ventricular insufficiency. Atrophy of lymphoid centers in the spleen and lymph nodes was common. Degenerative changes were occasionally found in other tissues such as the liver, kidneys, and the gastrointestinal tract.

7.1.7 Toxicity in Rats

The most important clinical symptoms of subacute selenium poisoning in the white rat have been described by Franke (1934) and may be summarized as follows: marked restriction of food intake, decreased growth, marked progressive anemia, and definite pathological changes especially in the liver. Rats fed a diet producing subacute toxicity restrict their food consumption to as little as 25% of that of rats fed a similar diet without selenium. A characteristic hunched posture can be observed in most of the rats. In a few cases the hind legs became paralyzed. The fur becomes rough and is stained a dark yellow around the genitals. Usually gross pathology is not evident in rats that die in a short time. The most marked symptom is vein dilation in the visceral region. The reproductive organs are undeveloped and may show degeneration. The lungs and liver usually have a congested appearance and the intestines show evidence of hemorrhage. The animals are jaundiced and very emaciated.

Chronic poisoning is observed in rats fed a ration with a lower content of selenium. They may show slight increases in weight and appear to be in fair condition, but their reproduction may be impaired (Franke and Potter, 1936). In many rats a definite breakdown of vital organs takes place. They develop an anemia which is progressive and die with hemoglobin levels as low as 2.0 g/100 cc blood. The most outstanding pathology is in the liver, which is atrophied, necrotic, cirrhotic, and hemorrhagic in varying degrees. If the liver happens to be in active regeneration there is usually atrophy of the right lateral and/or the caudate lobes. The left lateral and central lobes are most susceptible to atrophy. Sometimes the liver appears shriveled and mottled. Ascites and pleural edema are common and in many cases the heart and the spleen are enlarged. In some instances death results from internal hemorrhage from the liver. Microscopically one can observe liver hyperplasia, increased cystic sinusoids, and focal myelosis (Fitzhugh *et al.*, 1944). After a few months there is more of a chronic type of liver damage without intrahepatic hemorrhage or necrosis, with less of the cystic sinusoids and myeloid foci previously seen, but with increasing portal fibrosis, distortion of normal architecture, focal capsular retraction, hepatic cell atrophy, and focal cell hyperplasia. These findings seem to resemble portal cirrhosis. Sometimes nodularity was so pronounced that it resembled miniature bunches of grapes. Ascites was occasionally present.

Palmer and Olson (1974) have studied the effect of several levels of sodium selenite or sodium selenate in the drinking water of 21-day-old Sprague–Dawley rats receiving a corn- or rye-based diet to determine the relative toxicities of the two selenium forms. Levels of 2 or 3 ppm of either form of selenium produced a small decrease in weight in four to six weeks as compared to the control. When rats were exposed to water containing 6 or 9 ppm of selenium as either selenite or selenate, considerable mortality occurred. Even though more rats receiving selenite died than did rats receiving selenate, the overall toxicities of the two forms of selenium were similar.

7.1.8 Effect of Diet on Toxicity

Early in the investigation of "alkali disease," investigators tried to find a dietary factor (or factors) that might prevent or reduce the toxicity of seleniferous rations. Changing the calcium and phosphorous content

(2.8%–11.2% tricalcium phosphate) and the Ca:P ratio of the ration (1:2 to 1:6) had no beneficial effect. In addition, vitamins A and D had no close relationship to selenium poisoning. This was shown by feeding a ration containing 4% cod liver oil. Yeast as a source of vitamin B complex (up to 0.8g per rat per day as a supplement) had no beneficial effect. Improved growth was observed when the rats were fed the combined supplements of orange juice, dry yeast, and cod liver oil, but the toxicity of selenium was not changed. Moxon and Rhian (1943) have observed that the daily administration of crystalline vitamin B (given orally by injection) to rats fed a diet containing 12 ppm of selenium greatly increased the severity of the symptoms. If up to 5% cystine was added to the diet there were no alleviating effects on selenium poisoning and the 5% level seemed to be toxic itself. Schneider (1936) also fed cystine to rats at concentrations of 0.2, 0.4, and 0.6% of the diet and noted no beneficial results on selenium poisoning. If sulfur was fed at 0.5% of the diet, no beneficial effect was observed on selenium poisoning. Sulfur itself was slightly toxic to rats and chickens (Moxon, 1937).

The effect of the protein content of diets containing toxic amounts of selenium has been studied (Moxon, 1937; Smith and Stohlman, 1940). From feeding diets of high, normal, and low protein content (with equal caloric value), Moxon (1937) concluded that selenium poisoning was less severe on the diet with a high protein content. This finding has been substantiated by Smith and Stohlman (1940), who have postulated that the protein content may have some relationship to previously observed species differences in susceptibility to selenium poisoning. In contrast to an earlier finding of Smith and Stohlman (1940) and Klug et al. (1953) that neither cystine nor methionine were themselves of value in reducing the toxicity of selenium, Schultz and Lewis (1940) found that addition of methionine to a methionine-deficient diet did have beneficial effects. Smith and Stohlman (1940) have suggested that the protein effect noted may be due more to the methionine content of the diet. The proteins which have definite value in the alleviation or prevention of chronic selenium poisoning are casein, lactalbumin, ovalbumin, gelatin, and proteins which are derived from wheat, dried brewer's yeast, desiccated liver, and zein.

Moxon and Rhian (1943) have observed that of the commercial proteins fed to rats crude casein and linseed meal were the only ones that prevented the characteristic liver lesions. Excellent growth response was derived from casein diets and linseed meal. They gave the best growth of the vegetable proteins used but neither was as good as casein or dried liver. Dried whole beef liver, purified casin, and whole milk powder all gave good growth but failed to prevent or reduce the pathological lesions of selenium poisoning. In addition, meat scraps and corn gluten meal gave poor growth as well as no protection against selenium poisoning. Halverson *et al.* (1955) have tested linseed oil and purified casein against the toxicity of selenium in rats. They found linseed oil to be the more effective of the two proteins tested against toxicity. Jensen and Chang (1967) have observed that 20% linseed oil in the diet alleviated the toxic effect of 10 ppm of selenium in growing single-comb white leghorn cockerel chickens. Growth rates were only slightly reduced with 20 ppm selenium. Including 20% soybean meal failed to modify the toxicity. Levels of 5% and 10% linseed meal were less effective in counteracting selenosis than was 20%. Fractionation studies showed that the protective factor in linseed meal was extracted by methanol and ethanol, but the factor was not destroyed by autoclaving. The results indicated that linseed meal contains a heat-stable, organic polar factor.

Palmer *et al.* (1980) have isolated two new cyanogenic glycosides, linustatin and neolinustatin, from linseed oil meal. Each of the compounds were fed to rats in a corn-based diet at levels of 0.1% and 0.2%. At the 0.2% level, both substances gave significant protection against growth depression caused by 9 ppm sodium selenite. Linustatin and neolinustatin were found to be present in linseed oil meal at levels of 0.17% and 0.19%, respectively. Linamarin, which resembles the structures of linustatin and neolinustatin, was fed at the level of 0.2% and gave significant protection against growth depression and liver damage. A related cyanogenic glycoside, amygdalin, appeared to give small but nonsignificant protective responses.

The mechanism by which linseed meal counteracts a selenium toxicity in animals is not known. The factor in linseed oil could be interfering with the absorption of selenium or facilitating its removal from the tissues. Levander *et al.* (1970) have indicated that selenium accumulates in the liver and kidneys of rats fed linseed oil. These results suggest that a factor in linseed oil may be complexing with selenium to form a less harmful compound in the tissues.

In experiments with dogs, dried beef liver (18% of the ration) gave excellent growth and prevented all symptoms of selenium poisoning on a ration containing 10 ppm selenium. Crude casein gave poor growth and yielded no protection against selenium poisoning. Linseed meal also gave good protection in dogs, and it was the only protein to protect both rats and dogs. Smith and Stohlman (1940) have reported good protection by the proteins of wheat and corn for the rat, but Moxon and Rhian (1943) have shown no particular protective action for these proteins with dogs. The most severe cases of poisoning in dogs seem to have occurred on rations containing large amounts of these proteins.

There is no agreement concerning the effect of fat content of the diet on selenium poisoning. Moxon (1937) has reported that in a ration containing 37.5 ppm of selenium, the percentage of fat in the ration has very little effect on the toxicity of selenium. Smith (1939) has observed some beneficial effect from a ration low in protein and with a high fat content, but suggested that the result may come from the protein-sparing action of the fat.

7.1.9 Biochemical Lesions

Moxon and Franke (1935) have found that traces of selenium will greatly reduce the fermentation of glucose by yeast. Labes and Krebs (1935) have shown that the oxygen consumption of a muscle powder suspension is inhibited by selenite. Wright (1938) has reported that both sodium selenite and selenate inhibit the oxygen consumption *in vitro* of liver, kidneys, brain, muscle, and tumor slices. At first in livers from healthy rats there was an increase of oxygen consumption, which was followed by a decrease. Selenite-poisoned tissues were not able to oxidize glucose, succinic, lactic, pyruvic, or citric acids, but were rapidly able to oxidize *p*-phenylenediamine.

Potter and Elvehjem (1936) have measured the oxygen in the absence and presence of sodium selenite. Uptake of living yeast used the following substrates: glucose, mannose, fructose, lactic acid, pyruvic acid, acetic acid, alcohol, and succinic acid in the absence and presence of sodium selenite. About 80% inhibition of the oxygen uptake on the sugars and less than 10% inhibition on lactate and pyruvate were found with sodium selenite. Succinate was the most poorly oxidized of the substrates. Comparative experiments also showed that selenium does not cause as complete inhibition of the oxygen uptake of yeast as cyanide. The oxygen uptake in yeast is almost completely inhibited by cyanide. This suggests that selenium does not inactivate the cytochromes or indophenol oxidase but rather that it attacks certain dehydrogenating systems.

Collett (1924) has demonstrated that selenium inhibited succinic dehydrogenase of minced muscle. Klug *et al.* (1950) have studied the effect of dietary casein levels on rat liver succinic dehydrogenase activity and the *in vivo* inhibition of the enzyme by selenite together with the release of the inhibition by arsenite. Selenium lowers the liver succinic dehydrogenase levels of rats fed the casein diet. The inclusion of arsenic in the selenium diet restores the succinic dehydrogenase levels to normal after a relatively short period of depression. The authors suggest that the toxic action of selenium in the animal body is due, in part, to the inactivation of succinic dehydrogenase. Klug *et al.* (1953) added selenium compounds to a manometric assay for succinic dehydrogenase. Seleno-cystine was the most potent inhibitor and selenomethionine was the least toxic for this system. Sodium selenite was about intermediate in toxicity. Other materials listed in order of decreasing toxicity were selenohomo-cystine, diselenodipropionic acid, selenocystathionine, colloidal selen-ium, sodium selenite, methyl isoselenourea sulfate, selenomethionine, and oxidized glutathione. Red and black elemental selenium were non-toxic. When injected in rats the toxicity of a group of compounds was in the following order: methyl isoselenourea sulfate, selenohomocys-tine, selenocystine, selenomethionine, selenium tetraglutathione, and *bb'*-diselenodipropionic acid. The authors conclude that there was little correlation between toxicity to animals and the inhibition of succinic dehydrogenase. There of course could be other factors involved in the conversion *in vivo* of selenium to the toxic form. Certainly inhibition of succinic dehydrogenase seems to be an important factor in the toxicity of selenium.

Wright (1940) has reported that selenium in the form of selenite, selenate, and diselenodiacetic acid all exert an inhibiting action on rat liver urease. Urease is dependent on the presence of sulfhydryl groups for its activity. Because selenium catalyzes the oxidation of sulfhydryl groups (Voegtlin *et al.*, 1931), it is likely that the inhibition of urease activity by selenium is a result of a loss of sulfhydryl groups by oxidation. The mechanism of the inhibition of urease seems to be similar to the inhibition of succinic dehydrogenase, which also appears to be dependent on sulfhydryl groups for its activity (Hopkins and Morgan, 1938). Arginase from rat liver was not inhibited by selenium (Wright, 1940). Selenium has been observed to inhibit fatty acid synthetase in chick liver (Donaldson, 1977).

McConnell and Portman (1952) have observed that inorganic selen-ium is metabolized to the relatively nontoxic dimethyl selenide. The pro-duction of dimethyl selenide has been considered to be a detoxification mechanism. Sternberg *et al.* (1968) have suggested that the toxicity of selenite might involve depletion of S-adenosylmethionine, because of the initial rapid phase of methylation following selenite injection. This hy-pothesis appears reasonable in view of the central role of S-adenosyl-methionine in methyl, aminopropyl, and carboxyaminopropyl group transfers, as well as in transsulfuration from methionine to cysteine. The enzyme responsible for S-adenosyl-methionine synthesis is methionine adenosyltransferase (ATP: L-methionine S-adenosyltransferase, EC 2.5.1.6). Hoffman (1977) has given mice toxic doses of selenite. The toxic doses were found to rapidly decrease mouse liver S-adenosylmethionine and increase S-adenosylhomocysteine, indicative of an increased rate of trans-methylation. S-adenosylmethionine levels remained depressed beyond the time when dimethyl selenide synthesis ceased, suggesting that selenite inactivated methionine adenosyltransferase. The effect of graded doses of selenite on the conversion of the methionine analog ethionine to S-adenosylethionine *in vivo* suggested that selenite inactivated methionine adenosyltransferase. *In vitro* studies also indicated inactivation of this enzyme by selenite. Liver homogenates from selenite-injected mice were found to have less than 50% of the methionine adenosyltransferase activity of saline-injected controls. The exact mechanism of the inactivation of methionine adenosyltransferase remains to be established, but possibly involves a reaction of selenite with sulfhydryl groups.

The findings of Hoffman (1977) suggest a biochemical link between the hepatotoxic effects of selenite and of methionine deficiency because both involve a shortage of active methyl groups required for normal metabolism. Dini *et al.* (1981) have observed that ip administration of 1 mg/kg selenium as sodium selenite in guinea pigs produced a selective myocardial mitochondria damage. The damage was prevented by ip administration of methionine 10 mg/kg or pyruvate at 7 mg/kg ip for three days.

7.1.10 LD$_{50}$ of Various Selenium Compounds

The type of selenium compound and the route of administration can be expected to cause real differences in measurements of toxicity. Methods include intravenous, intraperitoneal, or subcutaneous injection, administration in the food or water, application to the skin, and subjection to vapors.

Different species are affected by selenium in different ways, and some species are more resistant than others. Therefore, humans may not respond to toxic doses as livestock or experimental animals do, and extrapolation of toxicity from animals to human beings should be done very cautiously.

Young animals seem more susceptible to the poisoning than do older ones, and embryos—especially the embryos in eggs—seem to be the most vulnerable.

7.1.10.1 Rats and Rabbits

The toxicity of various selenium compounds in rats or rabbits and the mode of administration are summarized in Table 7-1.

Table 7-1. Acute Toxicity of Some Selenium Compounds[a]

Compound	Experimental animal	Mode of administration	Toxicity[b]
Sodium selenite	Rat	Intraperitoneal injection	MLD[c] 3.25–3.5 mg Se/kg body wt
	Rat	Intravenous injection	MLD[d] 3 mg Se/kg body wt
	Rabbit	Intravenous injection	MLD[d] 1.5 mg Se/kg body wt
	Rat	Injection	MLD 3–5.7 mg Se/kg body wt
	Rabbit	Injection	MLD 0.9–1.5 mg Se/kg body wt
Sodium selenate	Rat	Intraperitoneal injection	MLD[c] 5.5–5.75 mg Se/kg body wt
	Rat	Intravenous injection	MLD[d] 3 mg Se/kg body wt
	Rabbit	Intravenous injection	MLD[d] 2–2.5 mg Se/kg body wt
Selenium oxychloride	Rabbit	Application to skin	83 mg of compound caused death in 5 hr; 4 mg caused death in 24 hr
Hydrogen selenide	Rat	In air	All animals exposed to 0.02 mg/liter of air for 60 min died within 25 days
D,L-selenocystine	Rat	Intraperitoneal injection	MLD[b,c] 4 mg Se/kg body wt
D,L-selenomethionine	Rat	Intraperitoneal injection	MLD[b] 4.25 mg Se/kg body wt
Diselenodipropionic acid	Rat	Intraperitoneal injection	LD$_{50}$ 25–30 mg Se/kg body wt
Dimethyl selenide	Rat	Intraperitoneal injection	LD$_{50}$ 1600 mg/kg body wt
Trimethylselenonium chloride	Rat	Intraperitoneal injection	LD$_{50}$ 49.4 mg Se/kg body wt

[a] Reproduced from *Selenium*: Medical and Biologic Effects of Environmental Pollutants, National Academy Press, New York, 1979.
[b] MLD, minimum lethal dose; LD$_{50}$, dose causing death in one half of test animals.
[c] Smallest amount that would kill 75% of the rats in less than two days.
[d] Smallest amount that would kill 40%–50% of the animals.

7.1.10.2 Horses

Miller and Williams (1940) reported that the MLD of selenium as sodium selenite for equines when administered orally in a single dose was 2.2 mg/kg. In another experiment they gave two horses feed containing the salt at the rate of 11 ppm Se for two months; then, as the feed was not relished, they drenched the horses with the same amount for four months. For the next four months the dose was increased to 22 ppm followed by 44 ppm for 15 months. The horses died in 17 and 17.5 months, respectively, after the treatment.

7.1.10.3 Cattle

Miller and Williams (1940) reported the MLD of sodium selenite in cattle to be 9.0 mg Se/kg. Maag *et al.* (1960) found steers unable to tolerate 1.1 mg Se/kg body weight when administered as sodium selenite three times over an extended period.

7.1.10.4 Dogs

Heinrich and MacCanon (1957) have reported that dogs can develop a tolerance for Na_2SeO_3 under suitable conditions, but that the 24-hr LDH_{50} of selenium for dogs was between 0.875 and 1.0 mg/kg when sodium selenite was given intravenously. Anderson and Moxon (1942) have studied the blood changes in dogs given subcutaneous sodium selenite and reported the MLD of selenium to be between 0.59 and 0.92 mg/kg when the dogs were under "barbital depression."

7.1.10.5 Swine

Miller and Williams (1940) found the MLD of selenium to be 15 mg/kg when given to swine orally as a single dose. Orstadius (1960) found that 2.0 and 1.2 mg of Se/kg, when given subcutaneously in single doses, were lethal in four hours and five days, respectively. Miller and Schoening (1938) reported that selenium as sodium selenite was toxic for swine when fed at the rate of 11.3 ppm, even though one lived 98 days at that level.

7.1.10.6 Poultry

Moxon (1937) has studied the effect of selenium as selenite in the rations of poultry. Evidence of toxicity was found when hens were fed 26 ppm; pullets, 6.5 ppm; and growing chicks, 8 ppm. All of these trials resulted in suppression of feed intake and loss of weight.

7.1.10.7 Sheep

Tucker (1960) has found that ewes tolerated 0.27 mg Se/kg/day when sodium selenate was administered orally over a period of two consecutive pregnancies without apparent harm to either ewes or their lambs. When the dose was doubled, intoxication resulted.

Three sheep were given 2.2 mg/kg of the salts, sodium selenite (1.0 mg Se/kg), sodium selenate (0.76 mg Se/kg), and Se dioxide (1.56 mg Se/ kg) by subcutaneous injection, and all died with 18 hr (Muth and Binns, 1964). Three sheep were also given the same doses orally, but only the one with sodium selenate died in 72 hr while the two dosed with sodium selenite and selenium dioxide survived. Elemented selenium in the form of selenium black was found to be nontoxic in two sheep.

In another trial, Muth and Binns (1964) have found that selenite given by subcutaneous injection killed in a matter of hours from the first doses of 20–30 mg, and in five days when 10 mg was given daily. If the selenite was given orally, 30 mg/day of selenite killed in six days, 20 mg/day in 11 days, and 10 mg/day failed to kill in 21 days. The selenate and selenium dioxide given at 30 mg/day failed in 8 and 11 days, respectively.

Skerman (1962) has given oral weekly doses of 0.1, 1.0, and 5.0 mg Se as selenium dioxide in aqueous solution over a period of 54 weeks to 29 kg lambs who were kept on perennial ryegrass–white clover pastures. A periopherolobular fatty metamorphosis was observed in the livers of treated lambs, but no other changes were noted.

Kuttler *et al.* (1961) have used sodium selenite suspended in peanut oil containing 2% beeswax which was injected subcutaneously at levels of 0.20, 0.40, 0.80, and 1.50 mg Se/kg body weight. One animal died out of two receiving 0.80 mg Se/kg. The two animals receiving 0.40 Se/kg were anorectic for 24 hr, but no other toxic effects were observed.

Rosenfeld and Beath (1946) have given sheep 10 mg Se/day orally as sodium selenite for 21 days, then increased the dose to 20 mg Se/day until signs of toxicity developed.

Muth and Binns (1964) outlined feeding trials wherein organic and inorganic Se was fed to sheep at two levels. The organic material was the seleniferous *Astragalus prussii* which contained 45.4 ppm of selenium. At the 0.33 mg/kg day level, death occurred in five and six days in those receiving the plant material, whereas those receiving 0.33 mg Se/kg/day of sodium selenite, or an aqueous extract of *Atriplex canesc* containing an equal amount of Se, were normal.

A marked difference in species susceptibility to selenium poisoning appears to exist. When parenteral administration is compared in the sheep and rat, the ratio of the MLD in these two species may be as much as

1:6, the rat being more tolerent. It also appears that among ruminants, cattle may be more tolerant than sheep when selenium is given orally to both species.

7.2 Industrial Medical Aspects

Selenium can occur in three forms: (1) red amorphous powder, (2) red crystals, and (3) a gray form of the substance. Selenium is mainly produced as a by-product that occurs in the sludges and sediments that result from the refining of copper. Selenium is also found at times in the flue dust that results from the manufacture of sulfuric acid.

7.2.1 Occupational Hazards

The greatest amount of selenium that is produced in the United States goes into the manufacture and fabrication of rectifiers. Some selenium is used as a coloring agent to produce ruby glass and it also is used as a coloring agent in paints and dyes. Selenium also acts as a decolorizing agent in the manufacture of green glass. Selenium is used in the vulcanization of rubber. Other known uses include the following: insecticides, manufacture of electrodes, photocells, selenium cells, semiconductor fusion mixtures, toning baths in photography, dehydrogenation of organic compounds, photostatic work, X-ray xerography, radioactive scanning of the pancreas in medical diagnosis, and as a chemical catalysts in the Kjeldahl test. Selenium has also been used in the form of an alloy with stainless steel, and cast steel.

Selenium hexafluoride, SeF_6, is a gaseous compound of selenium and is used for insulating various types of electrical equipment. A variety of chemical compounds such as $SeOCl_2$, selenium oxychloride, finds use as solvents, platicizers, and agents for various chemical processes. Some selenium compounds also are used as flameproofing agents for clothing materials and for wire cable coverings.

In most ores that are sulfide in nature selenium is a contaminant. These ores include copper, gold, nickel, and silver. During a chemical process when the selenium is removed from these ores, it is possible for the chemical workers to be exposed to the selenium. Workers may also be exposed to selenium hydride, H_2Se, or hydrogen selenide, which could be generated by the reaction of acids or water with metal selenides or the reaction of hydrogen with selenium compounds that are soluble.

Selenium exposure may occur in the following list of occupations.

This list is not meant to be all inclusive, but it does represent most kinds of occupations in which selenium may pose an industrial threat: arc light electrode makers, copper smelter workers, electric rectifer makers, glass makers, organic chemical synthesizers, pesticide makers, makers of photographic chemicals, pigment makers, plastic workers, pyrite roaster workers, rubber makers, semiconductor makers, sulfuric acid makers, and textile workers.

7.2.2 Permissible Limits for Selenium Exposure

The federal standards are: selenium compounds (as Se), 0.2 mg/m^2, selenium hexafluoride, 0.05 ppm, 0.4 mg/m^3; and hydrogen selenide, 0.05 ppm, 0.2 mg/m^3 (Wilber, 1980). In studies on 200–300 selenium workers over a period of 17 years, Glover (1967) found an average urinary level of 84 mg/liter. The highest figure was 490 mg of selenium per liter for a selenium worker who had inhaled selenium dust. From these studies Glover suggested that in public, rural, and industrial health situations, the selenium urinary level should be below 100 mg/liter.

7.2.3 Toxicity in Humans

Selenium enters the body when it is inhaled in the form of a dust or a vapor. Liquid solutions of selenium compounds can also pass through the skin. Finally, selenium compounds may be swallowed and exert their poisonous action by being absorbed from the digestive tract. Elemental selenium and some of its natural compounds are not particularly irritating to the body and are not well absorbed by the body. Certain chemical compounds of selenium such as selenium oxychloride ($SeOCl_2$) and selenium chloride Se_2Cl_2, however, are strong vesicants. $SeOCl_2$ can blister and severely burn the skin and thus destroy it. Certain compounds such as selenium dioxide and selenium oxychloride are strong irritants which are able to attack the upper parts of the respiratory tract and the eyes; these vesicants can also irritate the mucous lining of the stomach.

Certain selenium compounds may also cause dermatitis or inflammation of the skin if it is exposed to these compounds. An allergy to selenium dioxide may develop in some individuals; the general appearance is in the form of a rash of the urticarial type. The allergy to selenium dioxide may manifest itself by causing a pink discoloration of the eyelids and palpebral conjunctivitis. Another mode of entry for selenium dioxide is to penetrate under a bed of finger nails. This penetration under the nails can produce excruciatingly painful nail irritation and a painful condition called paronychia. Certain selenium compounds can be absorbed through the unbroken skin. This results in stomach upsets or toxicity. In some cases selenium sulfide, which is found in some shampoos, can penetrate through scalp wounds and cause generalized toxic reactions.

Hydrogen selenide poisoning is also known to occur, and the general effects are much like the effects seen after exposure to various other irritating gases used or released by industrial processes. The most common sign of hydrogen selenide poisoning is irritation of the mucous membranes of the nose, eyes, and the upper respiratory tract. These responses are followed by a sensation of tightness in the chest. When the victim is removed from the area of exposure, these signs and symptoms disappear. If there is severe enough exposure, or if the individual is especially susceptible to hydrogen selenide, pulmonary edema may suddenly develop.

Hydrogen selenide is considered 15 times more dangerous than hydrogen sulfide. Even though hydrogen sulfide claims its annual toll of deaths every year, hydrogen selenide has never caused a death or an illness lasting more than ten days in a human being (Glover, 1976). There are two reasons for this lack of severe toxicity: first is the fact that in industry hydrogen sulfide is used by the tanker load, whereas hydrogen selenide is never used in quantity; the second reason is that hydrogen selenide is very easily oxidized back to the red selenium on the surface of the mucous membranes of the nose and probably in the alveoli of the lungs. It is likely that this quick breakdown of the hydrogen selenide to the harmless red elemental selenium has prevented death.

There have been no toxic incidents reported in the use of organic selenium compounds except where hydrogen selenide has suddenly been evolved during their synthesis. In addition, organic selenium compounds seem to be less toxic than the inorganic forms.

From the diagnostic point of view, the first and probably the most characteristic indication of selenium exposure and absorption by the human body is the garlic odor imparted through the breath of the victim. Excretion in the breath of small amounts of dimethyl selenide seems to be a cause of this odor. The odor disappears completely in somewhere between seven to ten days after the victim is removed from selenium exposure. Garlic breath is not an absolutely certain guide to the absorption of selenium. An early more subtle sign which many victims experience but never complain about is a metallic taste of the mouth. There are also other fairly generalized effects of selenium absorption. They include pallor, lassitude, irritability, vague complaints of indigestion, and giddiness. Vital organs do not ordinarily seem to be harmed by the absorption of selenium in humans. However, results of animal experimentation suggest that the possibility of liver and kidney damage in human beings exposed to toxic levels of selenium compounds should not completely be discounted. Damage to the liver and other toxic effects is well recognized in livestock that grazed on plants containing high levels of selenium.

Smith and Westfall (1937) have attempted to correlate the symptomology and selenium intake of 50 families from highly seleniferous areas of South Dakota and Nebraska. The majority of the subjects had lived on seleniferous farms from 10 to 40 years, and no one had a residence of less than three years. The per cent distribution of observed symptoms were: gastrointestinal disturbances, 20.7; bad teeth, 18.0; icteroid discoloration of the skin, 18.7; history of recurrent jaundice, 3.3; vitiligo, 1.3; pigmentation of the skin 2.0; sallow and paled color in younger individuals, 11.3; dermatitis, 3.3; rheumatoid arthritis, 2.0; pathological nails, 2.0; and cardiorenal disease, 1.3. None of the above symptoms can be regarded as specifically due to selenium. The high incidence of gastrointestinal disturbances may be of some importance in the light of observations made on lower animals. The incidence of skin discoloration may be related to the occurrence of bilirubinemia which has been observed in experimental animals which have been subjected to chronic selenium poisoning (Fimiani, 1951).

References

Anderson, H. D., and Moxon, A. L., 1942. Changes in the blood picture of the dog following subcutaneous injections of sodium selenite, *J. Pharmacol. Exp. Therap.* 76:343–354.

Bernschneider, F., Hess, M., Neuffer, K., and Willer, S., 1976a. LD$_{50}$ and Selenkonzentration in den Organen von Kaninchen nach oraler Applikation von Natriumselent and Prüfung der Toxizität von Uroselevit-Prämix, *Arch. Exp. Veterinaermed.* 30:525–531.

Berschneider, F., Hess, M., Neuffer, K., and Willer, S., 1976b. LD$_{50}$ und Senenkonzentration in den Organen von Kaninchen nach oraler Applikation von Natriumselenit und Prüfund der Toxizität von Uroselevit pro inj., *Arch. Exp. Veterinaermed.* 30:627–632.

Collett, M. E., 1924. The specificity of the intracellular hydrogenases in frog's muscle, *J. Biol. Chem.* 58:793–797.

Dini, G., Franconi, F., and Martini, F., 1981. Mitochondrial alterations induced by selenium in guinea pig myocardium, *Exp. Molec. Pathol.* 34:226–235.

Donaldson, W. E., 1977. Selenium inhibition of avian fatty acid synthetase complex, *Chem. Biol. Interact.* 17:313–320.

Fimiani, R., 1951. Glutationemia nell intossicazione cronica sperimentale da selenio, *Folia Med.* (Naples) 34:260–263.

Fitzhugh, O. G., Nelson, A. A., and Bliss, C. I., 1944. The chronic oral toxicity of selenium, *J. Pharmacol. Exp. Therap.* 80:289–299.

Franke, K. W., 1934. A toxicant occurring naturally in certain samples of plant foodstuffs. I. Results obtained in preliminary feeding trials, *J. Nutr.* 8:596–608.

Franke, K. W., Moxon, A. L., Poley, W. E., and Tully, W. C., 1935. Monstrosities produced by the injection of selenium salts into hens' eggs, *Anat. Rec.* 65:15–22.

Franke, K. W., and Potter, V. R., 1935. A new toxicant occurring naturally in certain samples of plant foodstuffs. IX. Toxic effects of orally ingested selenium, *J. Nutr.* 10:213–221.

Franke, K. W., and Potter, V. R., 1936. A new toxicant occurring naturally in certain samples of plant foodstuffs. XIV. The effect of selenium containing foodstuffs on growth and reproduction of rats at various ages, *J. Nutr.* 12:205–214.

Glenn, M. W., Jensen, R., and Griner, L. A., 1964a. Sodium selenate toxicosis: pathology and pathogenesis of sodium selenate toxicosis in sheep, *Am. J. Vet. Res.* 25:1486–1494.

Glover, J. R., 1967. Selenium in human urine: a tentative maximum allowable concentration for industrial and rural populations, *Ann. Occupat. Hyg.* 10:3–14.

Glover, J. R., 1976. Environmental health aspects of selenium and tellurium, *Proceedings of the Symposium on Selenium–Tellurium Environments*, Industrial Health Foundation, Pittsburgh, Pennsylvania, pp. 279–292.

Hadjimarkos, D. M., 1970. Toxic effects of dietary selenium in hamsters, *Nutr. Rep. Int.* 1:175–179.

Halverson, A. W., Hendrick, M., and Olson, O. E., 1955. Observations on the protective effect of linseed oil meal and some extracts against chronic selenium poisoning in rats, *J. Nutr.* 56:51–60.

Heinrich, H., and MacCanon, D. M., 1957. Toxicity of intravenous sodium selenite in dogs, *Proc. S. Dak. Acad. Sci.* 36:173–177.

Hoffman, J. L., 1977. Selenite toxicity, depletion of liver S-adenosylmethionine, and inactivation of methionine adenosyltransferase, *Arch. Biochem. Biophys.* 179:136–140.

Hopkins, F. G., and Morgan, E. J., 1938. The influence of thiol-groups on the activity of dehydrogenases, *Biochem. J.* 32:611–620.

Jensen, L. S., and Chang, C. H., 1976. Fractionation studies on a factor in linseed meal protecting against selenosis in chicks, *Poult. Sci.* 55:594–599.

Klug, H. L., Moxon, A. L., Petersen, D. F., 1950. The *in vivo* inhibition of succinic dehydrogenase by selenium and its release by arsenic, *Arch. Biochem. Biophys.* 28:253–259.

Klug, H. L., Moxon, A. L., Petersen, D. F., and Painter, E. P., 1953. Inhibition of rat liver succinic dehydrogenase by selenium compounds, *J. Pharm. Exp. Ther.* 108:437–441.

Kuttler, K. L., Marble, D. W., and Blincoe, C., 1961. Serum and tissue residues following selenium injections in sheep, *Am. J. Vet. Res.* 22:422–428.

Labes, R., and Krebs, H., 1935. Die verschiedene Angriffsart von Tellurit, Selenit, Arsenit und anderen Giften auf die Dehydridase—und Oxydaseatmung des Muskelgewebes, *Fermentforsch.* 14:430–442.

Levander, D. A., Young, M. L., and Meeks, S. A., 1970. Studies on the binding of selenium by liver homogenates from rats fed diets containing either casein or casein plus linseed meal, *Toxicol. Appl. Pharmacol.* 16:79–87.

Maag, D. D., Osborn, J. S., and Clopton, J. R., 1960. The effect of sodium selenite on cattle, *Am. J. Vet. Res.* 21:1049–1053.

Madison, T. C., 1860. Sanitary report—Fort Randall. Statistical report on the sickness and mortality in the Army of the United States, 36th U.S. Congress, 1st Session, *Senate Ex. Doc.* 52:37.

McConnell, K. P., and Portman, O. W., 1952. Excretion of dimethyl selenide by the rat, *J. Biol. Chem.* 195:277–282.

Miller, W. T., and Williams, K. T., 1940. Minimum lethal dose of selenium, as sodium selenite, to horses, mules, cattle and swine, *J. Agr. Chem.* 60:163–174.

Morrow, D. A., 1968. Acute selenite toxicosis in lambs, *J. Am. Vet. Med. Assoc.* 152:1625–1629.

Moxon, A. L., and Franke, K. W., 1935. Effect of certain salts on enzyme activity, Effect of sodium selenate, selenite, selenide, tellurite, sulfate, sulfite, sulfide, arsenite, and vanadate on the rate of carbon dioxide production during yeast fermentation, *Ind. Eng. Chem.* 27:77–81.

Moxon, A. L., 1937. Alkali disease or selenium poisoning, *S. Dak. Agric. Exp. Sta. Bull.* No. 311:1–91.

Moxon, A. L., and Rhian, M., 1943. Selenium poisoning, *Physiol. Rev.* 23:305–337.

Muth, O. H., and Binns, W., 1964. Selenium toxicity in domestic animals, *Ann. N.Y. Acad. Sci.* 111:583–590.

Orstadius, K., 1960. Toxicity of a single subcutaneous dose of sodium selenite in pigs, *Nature* 188:1117.

Palmer, I. S., and Olson, O. E., 1974. Relative toxicities of selenite and selenate in the drinking water of rats, *J. Nutr.* 104:306–314.

Palmer, I. S., Olson, O. E., Halverson, A. W., Miller, R., and Smith, C., 1980. Isolation of factors in linseed oil meal protective against chronic selenosis in rats, *J. Nutr.* 110:145–150.

Potter, V. R., and Elvehjem, C. A., 1936. XXIX. The effect of selenium on cellular metabolism. The rate of oxygen uptake by living yeast in the presence of sodium selenate, *Biochem. J.* 30:189–196.

Rhian, M., and Moxon, A. L., 1943. Chronic selenium poisoning in dogs and its prevention by arsenic, *J. Pharmacol. Exp. Therap.* 78:249–264.

Rosenfeld, I., and Beath, O. A., 1946. The influence of protein diets on selenium poisoning, *Am J. Vet. Res.* 7:52–56.

Rosenfeld, I., and Beath, O. A., 1964. *Selenium*, Academic Press, New York.

Schneider, H. A., 1936. Selenium in nutrition, *Science* 83:32–34.

Schultz, J., and Lewis, H. B., 1940. The excretion of volatile selenium compounds after the administration of sodium selenite to white rats, *J. Biol. Chem.* 133:199–207.

Shortridge, E. H., O'Hara, P. J., and Marshall, P. M., 1971. Acute selenium poisoning in cattle, *N.Z. Vet. J.* 19:47–50.

Skerman, K. D., 1962. Observations on selenium deficiency of lambs in Victoria, *Proc. Aust. Soc. Anim. Prod.* 4:22–27.

Smith, M. I., and Westfall, B. B., 1937. Further studies on the selenium problem in relation to public health, *U.S. Pub. Health Rep.* 52:1375–1384.

Smith, M. I., 1939. The influence of diet on the chronic toxicity of selenium, *Pub. Health Repts.* 54:1441–1453.

Smith, M. I., and Stohlman, E. F., 1940. Further observations on the influence of dietary protein on the toxicity of selenium, *J. Pharmacol. Exp. Therap.* 70:270–278.

Sternberg, J., Brodeur, J., Imbach, A., and Mercier, A., 1968. Metabolic studies with seleniated compounds. 3. Lung excretion of selenium 75 and liver function, *Int. J. Appl. Radiat. Isotop.* 19:669–684.

Tucker, J. O., 1960. Preliminary report of selenium toxicity in sheep, *Proc. Am. Coll. Vet. Toxicol.* 7:41–45.

Voegtlin, C., Johnson, J. M., and Rosenthal, S. M., 1931. The oxidation catalysis of crystalline glutathione with particular reference to copper, *J. Biol. Chem.* 93:435–453.

Wahlstrom, R. C., Kamstra, L. D., and Olson, O. E., 1956. Preventing selenium poisoning in growing and fattening pigs, *S. Dak. Agric. Exp. Sta. Bull.* 456:1–15.

Wilber, C. G., 1980. Toxicology of selenium: A review, *Clin. Toxi.* 17:171–230.

Wright, C. I., 1938. Effect of sodium selenite and selenate on the oxygen consumption of mammalian tissues, *Pub. Health Rep.* 53:1825–1836.

Wright, C. I., 1940. Effect of selenium on urease and arginase, *J. Pharm. Exp. Ther.* 68:220–230.

Selenium in Health and Disease

8.1 Selenium and Cancer

One of the more exciting effects of selenium on health is selenium's anticarcinogenic effect against experimentally induced cancer in several animal systems.

8.1.1 Skin Cancer

In an effort to modify experimental carcinogenesis during its development and also to check if free radical damage due to peroxidation might be involved during carcinogenesis, sodium selenide as well as other antioxidants were applied concomitantly along with croton oil to female Swiss albino mice, after the mice had been initiated with 7,12-dimethylbenzanthracene (DMBA). In the initial experiment sodium selenide greatly reduced the number of the mice with skin tumors ($P < 0.05$) (Shamberger and Rudolph, 1966). In addition, the total number of tumors was reduced not only by sodium selenide, but also by vitamin E. Riley (1968) has also observed a reduction in DMBA–croton oil carcinogenesis by sodium selenide.

The tumor promotion experiments were expanded and repeated. In five of six nondietary tumor-promotion experiments, sodium selenide significantly reduced the number of mice with tumors. In two of these six experiments, vitamin E also significantly reduced the number of mice with tumors (Shamberger, 1970). In the same experiments ascorbic acid did not significantly reduce the number of animals with tumors (Shamberger, 1972). However, in all of the experiments including the one with ascorbic acid the total number of tumors was reduced. Because all of these effects were with the DMBA–croton oil or the DMBA–croton resin system, there was a possibility that the effects observed were unique to these experimental tumor systems. Both sodium selenide and vitamin E significantly reduced the number of mice with tumors induced by DMBA–phenol. In addition the total number of tumors was reduced. In this experiment vitamin E was somewhat more effective than sodium selenide.

In the same experiments the complete carcinogen 3-methylcholan-threne (MCA) was also applied concomitantly with sodium selenide. After 19 weeks of daily application of 0.01% MCA in acetone, 87% of the control mice had tumors. In the sodium selenide group 68% of the mice had papillomas. The probabilities of the numbers of mice with papillomas at 19 weeks and cancers at 30 weeks being significantly different from the control group was <0.10. However, the total numbers of papillomas and cancers were greatly reduced when compared to those in the controls. Even though the numbers of animals with tumors were not always sin-gificantly reduced in the selenium–vitamin E experiments, the consistent feature of all of these experiments was that the total number of tumors was reduced when sodium selenide or vitamin E was applied.

The effect of selenium-deficient diets and selenium-adequate diets on DMBA–croton and benzopyrene carcinogenesis were then studied (Shamberger, 1970). Dietary sodium selenite markedly reduced papillo-mas induced by the tumor promotor, DMBA–croton oil. In this experi-ment 14/35 mice fed 1.0 ppm sodium selenite had papillomas at 20 weeks. Of the mice fed selenium-deficient diets, 26/36 had papillomas ($P < 0.01$). At 27 weeks there was also a reduction in the number of animals with cancers in the group supplemented with 1.0 ppm sodium selenite.

A similar tumor pattern also was observed with benzopyrene. At 22 weeks, 31/36 mice fed selenium-deficient diets had papillomas, while 16/36 mice with diets containing 1.0 ppm sodium selenite had papillomas ($P < 0.001$). At 27 weeks the numbers of cancers and the size of the cancers were decreased in the mice fed 1.0 ppm of sodium selenite.

Shamberger (1972) has observed that mice initiated with DMBA and treated with 0.04% croton oil had more tumors and a greater percentage of mice with tumors than a group of mice treated with DMBA–0.004% croton resin. Because croton oil has a much greater content of polyun-saturated fatty acids than croton resin, the increased tumorigenicity of croton oil could be due to its greater content of polyunsaturated fatty acids. In a second experiment addition of corn oil to the croton resin increased the number of tumors. Skin treated with DMBA and croton oil had greater peroxidation (malonaldehyde increase) and more tumors than skin treated with DMBA and croton resin. In addition, peroxidation in two experiments increased after DMBA initiation, but returned to normal at about day 45. Antioxidants applied on days 2–21 in another experiment, at the same time as the peroxidation increase, decreased the tumor incidence. The regression of tumors observed at 16–18 weeks may be linked to an increase of lysosomal enzyme activity observed before regression.

Because of the structural similarity of malonaldehyde to the known carcinogens β-propiolactone and glycidaldehyde and the increase of malonaldehyde levels or peroxidation, malonaldehyde was tested as an initiator and a tumor promotor. One application of malonaldehyde, β-propiolactone, glycidaldehyde, or DMBA (initiation) was applied to each group of mice (Shamberger *et al.*, 1974). Each group was treated with croton oil for 30 weeks. After 30 weeks the groups initiated with malonaldehyde, β-propiolactone, glycidaldehyde, or DMBA had an incidence of 52%, 44%, 40%, and 95% tumors, respectively. Applications of DMBA, benzopyrene, and 3-methylcholanthrene to mouse skin increased malonaldehyde levels.

8.1.2 Liver Cancer

Clayton and Bauman (1949) have reported that the inclusion of 5 ppm of selenium in a purified diet reduced the incidence of liver tumors in rats induced by 3-methyl-4-dimethylaminoazobenzene (DAB). Selenium as sodium selenite was fed during a four-week period of interruption between two four-week periods during which 0.064% DAB was fed. In two separate experiments there was a reduction of about 50% in the incidence of liver tumors when selenium was fed during the intermediate period. The incidence of tumors in the groups fed selenium and in the control groups were 22% (2/9) vs. 40% (4/10) in one series and 31% (4/13) vs. 62% (8/13) in the second series. The animals on the selenium diet gained weight at a slower rate than control animals during the intermediate period. The slower gains were due in part to a decrease in food intake. Five ppm of selenium in the diet is considered by selenium researchers to be toxic with three ppm being borderline toxic. In general, cancer researchers avoid toxic levels of a test substance or a carcinogen, because the effect against carcinogenicity may be due to toxicity.

Harr *et al.* (1973) have reported an experiment in which they fed rats diets containing selenium in varying concentrations and a known carcinogen. Eighty female OSU-known rats were used. These rats were divided into four groups with 20 rats per group. The rats, which were supplemented with vitamin E, were selenium deficient and were fed the carcinogen 2-acetylaminofluorene (AAF) along with varying amounts of selenite. The results indicate a dramatic decrease in the rates of induced hepatic and mammary cancers following increased additions of selenium.

The rats fed 0, 0.1, 0.5, and 2.5 ppm of selenium had a combined incidence of 60%, 60%, 10%, and 0% of hepatic or mammary cancer. In a preliminary experiment (Harr *et al.*, 1972) reported that selenium supplementation reduced the number of tumors in rats given 0.03% of dietary 2-acetylaminofluorene. The rats were given 4 ppm Se (Na_2SeO_3) in water for 14 weeks. A final incidence of tumors of 9/13 rats was observed in the group given the basal diet alone, while 4/14 rats developed hepatomas when given the selenium supplement. In the control animals fed the carcinogen-free basal diet but maintained on the selenium supplement for the entire period, no tumor or liver changes were observed.

The effects of selenium on 3-methyl-4-dimethylaminoazobenzene(DAB)-induced hepatocarcinogenesis was determined (Griffin and Jacobs, 1977). Three groups of male Sprague–Dawley rats were fed 0.05% of DAB: one of group served as control; one group received 6 ppm Se (Na_2SeO_3) in the drinking water and one group received 6 ppm of selenium which added to the diet in the form of high-selenium yeast. The azo compound was incorporated into the diet for eight weeks and then removed and the two types of selenium supplements were continued for an additional four weeks. At this time evaluation revealed a 92% tumor incidence (11/12 animals) in the controls. Selenium in the drinking water reduced the incidence to 46% (7/15) and the dietary supplementation reduced tumor incidence to 64% (9/14).

In another study (Daoud and Griffin, 1980) seven groups of Sprague–Dawley male rats were fed diets containing 0.05% DAB for nine weeks. One group served as control. A second group received 4 ppm of Se as sodium selenite in the drinking water for nine weeks. A third group had only 4 ppm of Se in the drinking water for the initial three weeks. Group four had only 4 ppm of Se in the drinking water for the final four weeks. Group 5 had 2 ppm of Se in the drinking water for nine weeks. Group 6 received 1% sorbic acid which was added to the diet for the entire period, and finally the seventh and last group received 0.5% butylated hydroxytoluene which was added to the diet for the entire period. The tumor incidence by group was: 1, 12/15; 2, 2/14; 3, 7/13; 4, 5/13; 5, 6/15; 6, 15/15; and 7, 0/15. The experiment confirmed the findings of the inhibition of selenium upon azo dye carcinogenesis (Griffin and Jacobs, 1977). It was also of interest that in group 4 there was a reduction of hepatocarcinogenesis when the selenium was given during the later stages of the study.

Balanski and Hadsiolov (1979) have fed diethylnitrosamine (DEN) to rats and in one experimental group fed 1 ppm of sodium selenite. In a second experiment they also fed 5 or 10 ppm of sodium selenite to rats along with the DEN. In rats fed 1 ppm of sodium selenite along with the DEN there were 15/20 of the rats with liver tumors. In this group there were 2.26 tumors per rat bearing tumors. In the control group 17/20 of the rats had tumors with an average of 2.9 tumors per rat with tumors. Even though the number of animals with tumors was not significantly different when the control was compared to the group fed 1 ppm sodium selenite, there were 50 tumors in the control and only 34 tumors in the group fed 1 ppm sodium selenite. There was also a reduction in the number of tumors per tumor-bearing animal in a second experiment. When the animals were fed 5 ppm or 10 ppm of sodium selenite. The control group had 68 tumors, while the group fed 5 ppm had 40 and the group fed 10 ppm had a total of 15 tumors. The tumor-reducing effect in this second experiment may have been due to toxicity. Over 3 ppm of sodium selenite in the diet is considered toxic. Dzhioev (1978) has also tested diethyni-trosamine carcinogenesis in rats. Selenium treatment reduced the percentage of rats with tumors from 100% to 27%. In another experiment selenium reduced the percentage of mice with DEN-induced lung tumors from 76% to 54% (Dzhioev, 1978).

Marshall et al. (1979) have studied the effects of the addition of 4 ppm of Se as sodium selenite to the drinking water of male albino rats fed diets containing 0.03% AAF for 14 weeks. There was about a 50% reduction in tumor incidence with 9/13 of the control rats having tumors and in the group with selenium in the drinking water 4/14 had tumors. Selenium in the drinking water also provided protection against hepatic damage. An in vitro assay system utilizing microsomes from selenium supplemented or nonsupplemented 3-methylcholanthrene(MC)-induced rats was used to determine the effect of oral selenium intake on the metabolism of AAF. Oral selenium administration led to a decrease in N-hydroxylation and an increase in ring hydroxylation. When selenium was added to the microsomal assay system, 3-OH-AAF formation increased and N-OH-AAF formation decreased. These changes shifted the balance of metabolism toward the detoxification pathways and away from the carcinogenic pathways.

In vivo, MC has been demonstrated to reduce the carcinogenicity of AAF in rats (Lotlikar et al., 1967). In the experiment of Marshall et al. (1979) MC-induced microsomes were used. The in vitro reduction in the formation of N-OH-AAF seems to parallel the in vivo finding of Lotliker et al. (1967). Lotliker et al. (1978), on the other hand, has reported that MC pretreatment leads to increases in both ring and N-hydroxylation of AAF. Therefore the mechanism of inhibition of AAF carcinogenesis by MC may be a result of lower liver levels of N-OH-AAF due to increased biliary excretion of N-OH-AAF rather than decreased formation of N–OH AAF. In addition, dietary selenium had no effect upon the MC-induced mixed-function oxidase activity, which metabolizes AAF. This result with rat liver microsomes agrees with the previous observation of Rasco

et al. (1977) that selenium does not affect the induction of aryl hydrocarbon hydroxylase activity in cultured human lymphocytes.

Wortzman *et al.* (1980) have fed male weanling Charles River CD
rats a Torula yeast-based selenium-deficient diet or the same diet supplemented with selenium (0.5 ppm) as sodium selenite. The effect of
dietary selenium on the interaction between AAF and rat liver DNA was
studied *in vivo*. There was no difference between selenium-deficient and
selenium-supplemented rats with respect to the total amount of AAF
covalently bound to liver DNA *in vivo* at 1, 4, 16, 24, 96, or 168 hr
following a single intraperitoneal injection of labeled AAF. However,
alkaline sucrose gradient analysis revealed the production of DNA single-
strand breaks in the livers of selenium-deficient rats at 4 hours after
intraperitoneal injection of AAF (10 mg/kg). These lesions were apparently repaired 24 hr after injection of the carcinogen. Under the same
experimental conditions, AAF failed to produce evidence of DNA damage
in the livers of selenium-supplemented rats. The dose of 20 mg/kg AAF
resulted in extensive degradation of hepatic DNA in both groups of rats.
This damage was repaired at 48 and 72 hr after administration and was
not affected by the selenium status of the animals. Neither the selenium-
supplemented nor the selenium-deficient animals exhibited any significant
alteration in lipid peroxidation which was measured by determining the
hepatic malondialdehyde content of rats after AAF injection. In contrast,
the administration of carbon tetrachloride resulted in a statistically significant increase in the hepatic malondialdehyde concentration of selenium-deficient animals and a smaller increase in the selenium-supplemented rats.

8.1.3 Colon Cancer

Jacobs *et al.* (1977a) have studied the effect of selenium on colon
carcinogenesis. Sprague–Dawley rats were injected weekly with either
1,2-dimethylhydrazine (DMH) or methylazoxymethanol acetate (MAM).
Selenium was added to the drinking water at the level of 4 ppm (sodium
selenite). After 17 weeks the extent of colon carcinogenesis was evaluated. Again the total number of tumors were decreased in the selenium
treated groups. In the DMH control there were 39 tumors and in the
selenium treated, only 11 tumors. In the rats treated with MAM there
were 73 tumors in the control group and only 42 in the selenium-treated
animals. The incidence of MAM-induced tumors was 93% (14/15) with
the selenium additive and 100% (14/14) without the selenium supplement.

However, selenium reduced the incidence of DMH-induced colon tumors by more than 50%. The control group had 13/15 with tumors, whereas the selenium-treated group had only 6/15 with tumors.

Jacobs et al. (1981) have also studied the effect of adding a 4-ppm selenium supplement to the drinking water before, during, and after 20 weekly injections of 20 mg/kg of 1,2-dimethylhydrazine (DMH). The incidence of colon tumors in groups provided selenium before DMH, before and during DMH, and only during DMH treatment were reduced to 39% (11/28), 43% (13/30), and 36% (10/28), respectively. The incidence in the DMH only control was 63% (19/30). At ten-week intervals throughout the study, selected blood and tissue components were analyzed. Several hematological changes were correlated with DMH treatment; serum glutamic oxalacetic transaminase increased two-fold, serum alkaline phosphatase increased 24%, serum protein content decreased 14%, the white blood count increased two- to three-fold, and hemoglobin decreased 67%. It is likely that the enzyme and blood changes are the changes reflecting host tumor interaction rather than early carcinogenic reactions.

Soullier et al. (1981) have studied the effects of selenium supplementation on azoxymethane-induced intestinal cancer in Sprague–Dawley rats given eight weekly injections of azoxymethane (8 mg/kg body wt), and fed a 30% beef fat diet. There were two groups: one group served as controls and the other group was selenium-supplemented by receiving 8 ppm H_2SeO_3 in the drinking water. The average number of intestinal tumors was 6.5 in the control group and 3.1 in the selenium-supplemented group. A significant reduction was observed in tumor incidence in the proximal half of the colon of the selenium-treated rats. An increased concentration of tissue selenium was observed in the proximal half of the colon of these rats. In the supplemented rats blood selenium levels increased rapidly the first nine weeks of the experiment, followed by a plateau significantly higher than that for nonselenium controls. A significant increase in the liver and the intestinal selenium levels were also seen in the supplemented groups. These increases in the selenium levels of the intestinal and the proximal part of the colon seem to parallel the anticarcinogenic action of selenium. The lack of a protective effect in the distal part of the colon may reflect the decrease of selenium in that part of the colon. This segment-specific reduction in tumors correlates well with the observed preferential uptake of supplemental selenium.

Harbach and Swenberg (1981) have studied the effects of selenium on DMH metabolism, DNA alkylation, and the rate of cell turnover of the colon tissue and the effects of selenium pretreatment (4 ppm in the drinking water, for two, four, or six weeks) on DMH metabolism. These effects were monitored in male Sprague–Dawley rats by measuring expired $^{14}CO_2$ and azo (^{14}C) methane over a 12-hr period after a subcutaneous injection of (^{14}C) DMH (20 mg/kg body wt). When the selenium-pretreated rats were compared to control rats, which received only (^{14}C) DMH, selenium pretreatment caused an increase in exhaled azomethane (31%–69%) and a corresponding decrease in $^{14}CO_2$ (4%–33%) as the duration of the treatment increased from two to six weeks.

When the extent of DNA alkylation was measured as N-7 and O^6-methylguanine formation after selenium pretreatment, a reduction of 20%–27% was observed in the liver and alkylation was increased 40%–43% in the colon. In addition, metabolic incorporation of (^{14}C) from (^{14}C) DMH into adenine and guanine (presumably via C, pathways) was reduced 69%–72% in colon DNA of selenium-treated rats and (^3H) thymidine incorporation was reduced 61%–65%. These decreases may have been due to decreased cell turnover. A similar response with regard to DNA alkylation was not observed in the liver. The observations of Harbach and Swenberg (1981) suggest that selenium decreases hepatic DMH metabolism, and that this decrease may be compensated by an increase in extrahepatic metabolism and alkylation. Although colon alkylation is increased by selenium pretreatment, fewer tumors result. The decrease in the number of tumors may be due to a decrease in DNA synthesis in the colon. On the other hand other mechanisms as yet undetermined may be important in the inhibition of DMH carcinogenesis by selenium. Banner et al. (1981) have observed that methylazoxymethanol (MAM) acutely inhibits the synthesis of RNA and DNA. Selenium had no effect on the inhibition of RNA and DNA by MAM both in the liver and the colon. MAM does not require microsomal activation. The observation suggested that the tumor-preventative effects of selenium are probably due to a mechanism other than with carcinogen activation and interaction.

8.1.4 Breast Cancer

Schrauzer and Ishmael (1974) have given 2 ppm of selenium in the form of SeO_2 in the drinking water for 15 months to 30 virgin C_3H strong, female mice. This type of mouse is especially susceptible to spontaneous mammary tumors. The exposure of the mice lowered the incidence of spontaneous mammary tumors from 82% (in the untreated controls) to 10% in the selenium treated mice. In the same experiment another group of mice were given arsenic in the drinking water, at levels of 10 ppm. The tumor incidence in this group was reduced to 27%, but there was a significant enhancement of the growth rate of spontaneous or transplanted mammary tumors. In contrast, the mammary tumors in the control group grew about twice as fast as those in the selenium group.

In a subsequent study the spontaneous tumor incidence was observed in the C_3H/St mice that were given 2, 5, and 15 ppm of selenium and arsenic at the 10- and 80-ppm level (Schrauzer *et al.*, 1976). Another group was given a Se/Zn supplement at the ratio of 5/200 ppm. The percentage tumor incidence in the 2-, 5-, and 15-ppm selenium groups were 10%, 36%, and 33%, respectively. The 10- and 80-ppm arsenic-treated groups had an incidence of 27% and 40%, respectively. The Se/Zn group had an incidence of 94%, which was a mammary tumor incidence close to the 82% observed in the nontreated control group. The result in the Se/Zn group might be predicted because of the known antagonistic interactions between selenium and zinc. Similar known antagonistic reactions are known for arsenic and selenium, but arsenic and selenium were not tested together in this experiment. Schrauzer *et al.* (1978) using a similar experimental design jointly administered 2 ppm of arsenic as arsenite along with 2 ppm of selenium as selenite in the drinking water of inbred female C_3H/St mice. Arsenic and selenium were also tested alone. The tumor incidence was 41% in the control group, 36% in the arsenic group, 17% in the selenium group, and 62% in the As–Se-treated group. The increase in carcinogenesis with As–Se treatment seems to be consistent with the known interactions between As and Se.

Concentrations of arsenic and selenium in lung, liver, and kidney tissue from dead smelter workers and from a control group have been measured with the aid of neutron activation analysis (Wester *et al.*, 1981). A sevenfold increase of arsenic was found in lung tissue from the exposed workers compared with the control group. The median value of arsenic in lung tissue from workers who died from lung cancer was not higher than corresponding amounts from workers dead from other malignancies or from cardiovascular or other diseases. As the time of retirement increased, the arsenic content decreased in the liver, but not in the lung tissue. The workers dead from malignancies had a higher As/Se ratio (0.61 ± 0.42) than workers from other diseases (0.34 ± 0.26). Controls had a value of 0.06 ± 0.05. This result seems to parallel the report of Schrauzer *et al.* (1978). However, accumulation of antimony, cadmium, lead, and lanthanum were observed in lung tissue from the exposed workers. This finding indicates a possible multifactorial cause behind the excess mortality from lung cancer in smelter workers.

Shamberger and Bratush (1980) have measured the cadmium, selenium, and zinc levels of the kidney cortex from 123 patients who died from different diseases. Selenium was significantly lower in patients with cancer metastases. Cancer patients also showed a trend toward a lower selenium. In contrast, zinc was significantly greater in kidneys from patients with cancer or patients with cancer metastases. The Zn/Se ratio had a greater significant difference from controls than did selenium or zinc alone.

Nonsmokers and cigar smokers had a significant lower kidney cadmium content than cigarette smokers. Nonsmoking cancer patients with metastases had a greater kidney cadmium content than nonsmokers with other diseases. The Cd/Se ratios were greater in cancer patients and cancer patients with metastases than the controls. There was a positive significant correlation between zinc and cadmium in 123 patients.

Schrauzer *et al.* (1981) has also observed that exposing mice to high lead:selenium ratios leads to a low tumor incidence. The 19 months of exposure to 25 ppm of lead causes little if any toxicity and produces an overall tumor incidence of about 20%. Exposure to only 5 ppm lead under identical conditions markedly shortens the tumor latency period and produces high tumor incidence, a pattern similar to that of arsenic when it was jointly administered with selenium.

Schrauzer *et al.* (1978) have reported that inbred female C_3H/St mice exhibit the normal incidence of spontaneous mammary adenocarcinoma of 80%–100% if they are maintained on a standard commerical laboratory diet containing 0.15 ppm of selenium with meat and dried skimmed milk as major protein sources. If animals of the same strain are kept on a diet containing 0.45 ppm of selenium with fishmeal as the main source of protein, the tumor incidence decreased to 42%. The tumor incidence decreases further to 25%, 19%, and 10% if the animals in addition receive 0.1, 0.5, and 1.0 ppm of selenium in the drinking water. Selenium-supplemented groups of animals also remained tumor free for longer periods than the unsupplemented controls.

Schrauzer *et al.* (1980) have infected four groups of female inbred C_3H/St mice with the Bittner milk particle. Since selenium is present in most foods in the organic form rather than the inorganic form and the drinking water is not a normal source of the element, they decided to investigate its effects upon addition to the diet in the proper nutritional form. One group was maintained on a Torula yeast diet supplemented with 1 ppm of selenium (organically bound in yeast) and had a 27% incidence. The mice that were switched from the 1-ppm selenium diet to a diet containing only 0.15 ppm selenium after reaching the age of 13.8 months rapidly develop mammary tumors during their remaining lifespan with the overall tumor incidence reaching 69%. This is not statistically different from the 77% incidence of tumors observed in animals maintained on the 0.15-ppm selenium diet over their entire postweaning life span. In contrast, animals changed from the 0.15-ppm selenium diet to that containing 1.0 ppm selenium at the age of 13.8 months develop mammary tumors with a total incidence of only 46%, which is significantly lower ($P < 0.05$) than in the 0.15-ppm selenium control group. Their study demonstrated that dietary selenium prevents and retards tumor development as long as it is supplied in adequate amounts.

Thompson and Becci (1980) have tested the effect of two types of diets with or without selenium supplementation on N-methyl-N-nitrosourea(MNU)-induced mammary carcinogenesis in noninbred female Sprague–Dawley rats. Mammary cancer was induced by a single intravenous injection of MNU. Twenty-three rats fed Purina rodent chow type 5002 with 5 ppm sodium selenite had a tumor incidence of 85% and 2.30 cancer per rat, whereas 25 controls fed the same diet but with no added selenium had a tumor incidence of 95% and 3.95 cancers/rat. Twenty-five rats fed a Torula yeast diet with 1 ppm of added sodium selenite showed an incidence of 68% and 1.49 cancers/rat. The control group with 25 rats was fed 0.01 ppm sodium selenite and had an incidence of 68% and 2.11 cancers/rat. Selenium feeding prolonged the latency of mammary tumor appearance. The results of Thompson and Becci (1980) suggested that dietary selenium had an effect on the postinitiation stage of mammary carcinogenesis.

Thompson et al. (1981a) has used a similar experimental design with either 4 ppm sodium selenite, 300 ppm retinyl acetate, or 4 ppm sodium selenite and 300 ppm retinyl acetate added to a Purina 5001 laboratory chow. The effect on the postinitiation stage of mammary carcinogenesis was a reduction in the number of cancers/rat in all three groups. The number of cancers/rat were 3.68 in the control, 2.84 in the selenium-treated group, 2.44 in the retinoic acid treated group, and 1.72 in the combined retinoic-acid- and selenium-treated group.

Thompson and Tagliaferro (1980) have studied the effect of selenium on 7,12-dimethylbenzanthracene(DMBA)-induced mammary tumorigenesis. One group of female Sprague–Dawley rats was fed a diet which was unsupplemented and another group was fed a diet supplemented with 5 mg selenium as sodium selenite per kilogram. At 35 days of age the selenium-supplemented rats were randomized into one of two groups and fed the selenium-supplemented diet from 35 to 63 days of age. On day 64, one selenium-treated group was changed to an unsupplemented diet. Rats were maintained on these diets until the study was terminated. At 59 days of age, all rats were administered 20 mg DMBA dissolved in sesame oil via gastric intubation. Rats were palpated twice each week for the detection of mammary tumors. After 90 days post-DMBA, the rats fed unsupplemented diet (control group) had a 95% tumor incidence with an average of 2.9 tumors per rat and mean time to first tumor appearance (MTA) of 50 days. Feeding the selenium-supplemented diet from either 35 to 63 days of age or throughout the study reduced tumor incidence and the average number of tumors per rat and prolonged the MTA (70% and 45%); 1.5 and 0.9 tumors per rat; and 61 and 71 days, respectively.

Thompson *et al.* (1981b) repeated and modified their experiment to test either low (7.5 mg) or high (15.0 mg) of DMBA. At 28 days of age the rats were randomized into four groups of 45 rats each and were fed a diet containing per kilogram either 0.05, 0.15, 1.05, and 2.06 μg selenium. At 50 days of age, 20 rats in each group were given 15 mg DMBA; the remaining rats were given 7.5 mg DMBA intragastrically. All rats were changed to a diet containing 0.21 μg selenium per kg. Almost all of the rats got tumors but the number of tumors per rat was reduced by the selenium diets. The average number of tumors per rat across treatments was 9.1, 7.4, 7.1, and 5.0 and 5.8, 6.0, 4.5, and 3.8 at the high and low doses of DMBA, respectively. Antineoplastic activity was dependent on the amount of selenium ingested. The data suggested that selenium inhibits some aspect of the initiation stage of DMBA-induced mammary tumorigenesis which results in a significant reduction in the number of tumor occurrences.

Ip and Sinha (1981) have examined the effect of selenium depletion on mammary tumorigenesis following DMBA administration to female Sprague–Dawley rats that were fed different levels and types of fats. Four selenium-deficient basal diets were used containing either 1%, 5%, or 25% corn oil or 24% hydrogenated coconut oil. Only 0.1 ppm of selenium was considered a selenium-adequate diet. In animals that received an adequate supplement of selenium, an increase in fat intake was accompanied by an increased tumor incidence when corn oil was used in the diets. On the other hand, a high saturated fat diet was much less effective in this regard.

In a second experiment to test the effect of greater amounts of dietary selenium on DMBA-induced mammary carcinogenesis, Ip (1981a) fed rats 0.1, 0.5, 1.5, or 2.5 ppm (as sodium selenite) along with either a 5% or 25% corn oil diet and 5 mg DMBA. The total number of tumors were as follows (30 rats/group): 26, 23, 19, and 10 in the 5% corn oil group; and 65, 66, 41, and 21 in the 25% corn oil group. In a second experiment rats were given 10 mg of DMBA, and selenium was added to the diets at 0.1, 2.5, and 5.0 ppm. Tumor yields were found to be 71, 32, and 15, respectively, in the high-fat group. In these experiments there was a longer latency period of tumor appearance with selenium supplementation. High dietary selenium levels were able to protect against mammary carcinogenesis but the rats on a high corn oil diet still developed more tumors than those on a low corn oil diet at a comparable selenium supplementation.

Ip (1981b) has also studied the effect of selenium supplementation in the initiation and promotion phase of DMBA-induced mammary carcinogenesis in rats fed a high-fat diet. In this experiment control animals were fed 0.1 ppm of selenium (as sodium selenite), while the experimental groups were supplemented with 5 ppm of selenium for various periods of time: -2 to 24 weeks; -2 to $+2$, $+2$ to $+24$, $+2$ to $+12$; $+12$ to $+24$ and -2 to $+12$. The time of DMBA administration was taken as zero. Selenium again retarded tumorigenesis to various degrees. Ip (1981b) made the following conclusions: (1) Selenium can inhibit both the initiation and promotion phases of carcinogenesis; (2) a continuous intake of selenium is necessary to acheive maximal inhibit of tumorigenesis. Schrauzer (1980) made the same observation in his experiments; (3) the inhibitory effect of selenium in the early promotion phase is probably reversible; and (4) the efficacy of selenium is attenuated when it is given long after carcinogenic injury. Ip (1981b) also assessed the effectiveness of selenium in inhibiting the reappearance of mammary tumors that had regressed after ovariectomy. If the tumor-bearing animals were supplemented with 5 ppm of selenium immediately after endocrine ablation, the tumors reappeared at a slower rate compared to the controls. Ip's (1981b) results suggest that selenium may not only be effective in chemoprevention, but that selenium can also be used as an adjuvant chemotherapeutic agent.

Although Ip's (1981b) study shows that there is an association between the susceptibility of the mammary tissue to carcinogenesis and its lability to peroxidation as a result of increased fat intake similar to that observed with croton oil and croton resin on DMBA-initiated mouse skin (Shamberger, 1972), this does not necessarily mean that lipid peroxidation is the primary mechanism by which dietary fat promotes cancer development. Other factors that might be involved include elevated circulating prolactin (Ip et al., 1980), lymphocyte dysfunction (Kollmorgan et al., 1979), and the involvement of prostaglandins through the cellular immune system (Hillyard and Abraham, 1979). Pharmacological levels of selenium have been reported to potentiate the immune response of the host (Spallholz et al., 1975). It remains to be seen whether the immune response exerts an influence on the promotion of chemical carcinogenesis.

Some of these factors that might affect tumor growth have been examined. Ip (1981a) has found that selenium does not affect the levels of circulating prolactin and estrogens, hormones that are important for the development of the DMBA-induced mammary tumors (Bradley et al.,

1976). Ip (1981c) has fed rats diets containing 1%, 5%, and 25% of to-
copherol-stripped corn. The rats were fed either 0.1 ppm of selenium in
the form of sodium selenite added to the basal diet or just the basal diet
which contained less than 0.02 ppm of selenium. Mammary tumors were
induced by intragastric administration of 5 mg of DMBA. The incidence
of tumors and glutathione peroxidase levels in mammary and liver tissues
were determined as well as the thiobarbituric acid values per 100 mg of
mammary tissue. There was a marked increase of liver glutathione peroxidase in the groups given the 0.1-ppm selenium supplement. The groups
fed the basal diet became rapidly depleted in liver glutathione peroxidase
activity. There was a marked increase in the number of tumors per rat
in the selenium-depleted animals fed the 25% corn oil diet. There was
also a significant increase in the thiobarbituric acid values (malonaldehyde) in the selenium-depleted animals fed the 25% corn oil diet.

Ip's (1981c) results suggest that there is an association between the
susceptibility of the mammary gland to carcinogenesis and its lability to
peroxidation, and that both parameters are regulated by fat intake in
conjunction with the selenium status of the animal. Ip and Ip (1981) have
fed rats both retinyl acetate and selenium and markedly suppressed
DMBA-induced mammary cancer formation. The final tumor yield was
reduced to 8% of the control as compared with 51% and 36%, respectively,
for selenium and retinyl acetate. A continuous intake of both agents was
necessary to sustain the chemopreventive effect.

Young and Milner (1981) have studied the effect of 0.1, 2.5, and 5.0
ppm of selenium in a 30% torula yeast diet upon DMBA induced mammary
tumors. Tumor incidence eight weeks after intubation with DMBA was
53%, 14%, and 27% for the three diets. The total number of tumors at
autopsy was significantly decreased from 237 for controls to 92 and 113
for 2.5 and 5.0 ppm selenium, respectively.

Welsch et al. (1981) have divided 147 Sprague–Dawley rats into five
groups and treated them at 60 days of age with 5 mg of DMBA. Selenium,
as selenium dioxide (SeO_2), was administered in the drinking water to
four of the five groups (30 rats/group) at two doses (2 and 4 mg/liter) from
30 to 90 days of age (series 1) and from 90 to 150 days of age (series 2)
before the onset of palpable mammary tumors. One group of 27 rats
served as controls. The total number of carcinomas which developed in
each group were: controls, 60; series 1, 2-mg Se dose, 27, 4-mg Se dose,
29; series 2, 2-mg Se dose, 24; 4-mg Se dose, 32. Each dose level of Se
in both series significantly reduced the incidence of mammary carcinomas. In a second experiment 226 nulliparous and 99 multiparous GR mice
were treated daily with estrogen and progesterone for 13–16 weeks. SeO_2
was administered in the drinking water (2 mg/liter) to one half of these

mice. The total number of mammary carcinomas in the control nulliparous and multiparous mice were 119 and 90, respectively; in the selenium-treated group the nulliparous and multiparous mice had 113 and 81 tumors, respectively. The results of this experiment indicated that selenium did not affect mammary carcinoma incidence in hormone-treated nulliparous and multiparous GR mice.

Medina and Shepherd (1981) have studied the effect of SeO_2 on DMBA-induced mammary tumorigenesis in BALB/c, C3H/StWi, and BD2F mice. All mice were fed DMBA once a week for either two or six weeks (1 mg/mouse/week) when the mice were eight weeks old. SeO_2 was dissolved in water (SeO_2 = 6 mg/liter) and administered *ad libitum* starting when the mice were six weeks old. In one group of BD2F mice, a pituitary gland was grafted under the kidney capsule in order to stimulate lobulo-alveolar differentiation in the mammary glands. The results of this experiment in BD2F mice fed 2 or 6 mg DMBA, in C3H/StWi mice fed 6 mg DMBA, and in BALB/c mice fed 2 mg DMBA indicated that selenium supplementation markedly inhibited the mammary tumor incidence. The inhibition of tumor incidence ranged from 42% to 85%. In the BD2F, mice fed 2 mg DMBA and containing a pituitary isograft, the hormonal stimulation exerted by the isograft, partially counteracted the inhibition mediated by selenium. In the latter group, the mean latent period of tumor formation was delayed by the presence of selenium even though the tumor incidence was not significantly different between the selenium-treated and -untreated mice.

The effect of selenium on the induction of mammary preneoplastic dysplasias was examined in three groups of mice (Medina and Shephard, 1981). The term *dysplasia* denoted pathological lesions which are not neoplasms. Previous studies have indicated that both alveolar and ductal hyperplasias are precursor states of mammary neoplasms. Selenium supplementation reduced the incidence of mice with ductal hyperplasias from 61% to 15% in five-month-old BD2F mice and from 75% to 25% in eight-month-old BALB/c mice. In six-month-old, MMTV-positive BALB/cfC3H mice, selenium supplementation reduced the incidence of hyperplastic alveolar nodules from 45% to 10%.

In a third experiment, Medina and Shepherd (1981) studied the effect of selenium on the growth rate of first transplant generation mammary tumors in BALB/c, BD2F, and BALB/cfC3H mice. Of the 36 mammary tumors examined, selenium supplementation influenced the growth rate of only four tumors. The growth rates of three mammary tumors were inhibited and the growth rate of one tumor was enhanced. The results indicated that selenium did not alter the growth of established mammary tumors.

The results of Medina and Shepherd (1981) demonstrate that supplemental selenium inhibits both chemical- and viral-induced mouse mammary tumorigenesis, and secondly that the development of preneoplastic lesions, an early stage in mammary tumorigenesis, is very sensitive to selenium-mediated inhibition.

8.1.5 Tracheal Cancer

Thompson and Becci (1979) have studied the effect of graded levels of selenium on tracheal carcinomas induced by 1-methyl-1-nitrosourea (MNU) in male Syrian golden hamsters. No significant differences were observed among the groups in the incidence of either benign lesions or carcinomas.

8.1.6 Chemotherapeutic Effect of Selenium

Abdullaev *et al.* (1973) have observed that sodium selenite even in a single dose (1 mg/kg parenteral) retarded the growth of transplanted Ehrlich ascites, Guerin carcinoma, and sarcoma M-1 neoplasms in rats and mice. The selenium-containing salt was most effective when administered as soon as the tumor was palpable. When the chemotherapuetic effect was combined with X-ray therapy, the growth retardant effect of Na_2SeO_3 against Ehrlich ascites cells was enhanced. If the sarcoma M-1 or Guerin carcinoma suspensions are exposed to Na_2SeO_3, plus heat (40°C) before inoculation, impaired tumor growth was observed *in vivo*. The immune activity of the blood was increased by Na_2SeO_3 treatment.

Poirier and Milner (1979) have observed that sodium selenite, selenium dioxide, seleno-D,L-cystine, and seleno-D,L-methionine dramatically decreased the viability of Ehrlich ascites tumor cells (EATC) as measured by dye exclusion. Sodium selenate only marginally decreased EATC viability. Intraperitoneal injections of selenite in mice previously inoculated with EATC significantly inhibited tumor development. If the intraperitoneal injections of selenite were delayed five to seven days after the mice were inoculated with EATC, the effectiveness of this nutrient on the inhibition of EATC growth was reduced.

Greeder and Milner (1980) have investigated the effect of selenium dioxide, sodium selenite, sodium selenate, selenomethionine, and selenocystine on groups of ten mice which were treated with EATC. All selenium compounds were administered at a dose of μg/g of initial body weight. None of the treated mice had tumors, whereas all control mice developed Ehrlich ascites tumors. In a second experiment the dosages were lowered to 1.0 or 0.25 μg of the same five selenium compounds per gram of body weight and five mice per group were treated with EATC. Selenium dioxide, sodium selenite, sodium selenate, and selenocystine at 1μg/g all completely inhibited ascitic and solid tumor incidence. Selenomethionine was ineffective as all five mice developed ascites tumors. At the 0.25-μg/g dose all five of the selenomethionine-treated mice developed ascites tumors and two of five selenocystine-treated mice had ascites tumors. Some of the mice developed solid tumors.

Selenomethione may have been ineffective because selenomethionine is actively transported (Whanger *et al.*, 1976), whereas selenite and selenocystine are not. Greeder and Milner (1980) suggest that permeability may be a factor in the efficacy of the various selenium compounds in reducing tumor growth. Selenium also seemed to alter tumor growth selectively, since it had no significant effect on the growth of the host animals.

Selenopurine has been shown to retard the growth of L1210 leukemic cells (Mautner and Jaffe, 1958). The L1210 cell line is known to be susceptible to antifolics and to purine and pyrimidine analogs. Milner and Hsu (1981) have studied the effect of other forms of selenium on the L1210 cells both *in vitro* and *in vivo*. Selenium administration as sodium selenite was shown to be more effective in increasing the longevity of L1210-inoculated mice than was treatment with sodium selenate, selenocystine, or selenomethionine. Treatment at levels of 20, 30, or 40 μg/day of sodium selenite in mice inoculated with 10^2 cells resulted in 50%, 80%, and 90% cures, respectively. A larger number of cells in the inoculum decreased the cure rate. Longevity of L1210-inoculated mice was increased by about 30% when the drinking water was supplemented 3 ppm selenium as sodium selenite. The death of L1210 cells *in vitro* as indicated by trypan blue exclusion was dependent upon the form and concentration of selenium tested. If selenium at 1 μg/ml was incubated with L1210 cells for 1 hr prior to inoculation into mice, selenium significantly retarded the ability of the cells to propagate *in vivo*. Combined therapy with selenium (30 μg/day) and methotrexate resulted in a significantly longer life span of L1210-treated mice than resulted from either compound administered separately.

Medina and Oborn (1981) have examined the effect of selenium on the growth potential of normal, preneoplastic, and neoplastic mammary cells grown in primary monolayer cell cultures and on three established mammary cell lines. Selenium, present as Na_2SeO_3 in serum-free Dulbecco's modified Eagle's medium, inhibited all mammary cell cultures at $1 \times 10^{-5} M$. The growth of primary cell cultures of normal mammary cells and C4 preneoplastic cells and the established line YN-4 was inhibited by selenium at 5×10^{-7} M. This concentration of selenium did not stimulate the growth of D2 preneoplastic cells and tumors in primary cell cultures and established cell lines CL-S1 and WAZ-2t. The differential responses of cells from preneoplastic outgrowth lines C4 and D_2 as well as D2 primary tumors in $vitro$ correlated with the sensitivity when these same cell populations were subjected to selenium-mediated inhibition of growth and tumorigenesis in $vivo$.

Ip et $al.$ (1981) have studied the effect of dietary selenium deficiency and supplementation on the growth of the transplantable MT-W9B mammary tumor in female Wister–Furth rats. When the diet was supplemented with 2 ppm of selenium, tumor growth was inhibited and the final tumor weight was reduced by approximately 50% compared to the control rats receiving 0.1 ppm of selenium. The inhibitory response was not likely due to toxicity because no weight loss was induced in the rats. On the other hand, selenium deficiency (< 0.02 ppm) had no influence on the growth of the tumor.

Randleman (1980) has studied the effect of 1 ppm of selenium in the drinking water on the ovarian tumors which develop after normal ovaries are sewn into the spleen of albino laboratory rats (SAF/SD strain). Six of nine rats subjected to the ovary-to-spleen transfer with no selenium supplementation had tumors, a 66.6% incidence. In contrast, two of 29 rats, having the same ovary transfer but receiving selenium supplement, had tumors (6.8%). Histological examination showed the tumors to be adenocarcinoma of the ovary. The tumors of both of the selenium-fed animals were smaller and less advanced than those of the control rats. The tumors of the control rats were solid masses, averaging 1.5 cm in diameter. The two tumors of the selenium-supplemented rats were less than 0.5 cm, in both cases. Sodium selenite has also been observed to inhibit the vascularization induced by amelanotic tumor implants (A-Mel-4B32) in the Syrian hamster cheek pouch membrane (Jacobs et $al.$, 1980).

Exon et $al.$ (1980) have exposed mice to selenium-supplemented or selenium-deficient rations and then inoculated them with an oncogenic virus. Rauscher leukemia virus (RLV). Splenic lesions were not altered by dietary selenium supplementation or depletion. Under the experimental conditions utilized selenium did not affect the neoplasia induced by RLV in mice.

Watson-Williams (1920) has summarized the treatment of inoperable cancer with selenium. Watson-Williams has reviewed 90 previous cases treated with colloidal selenium and found considerable or appreciable improvement in 72 cases and no improvement in 18 cases. Eight cases were described as clinical cures. Watson-Williams then described the treatment of 20 additional cases with colloidal selenium-β. Selenium has five allotrophic forms of which two are red and three are black. The brick red selenium-β was first used medicinally for cancer in 1833 and has appeared in various powders and pastes employed by the medical profession for the relief of malignant ulcers. Selenium-β is a colloidal suspension of selenium which is a coral-red fluid. Selenium-β was stated to be electrically prepared. The particles were said to have a diameter of 5–20 μm and were isotonized by the addition of sodium chloride and sterilized. In the 20 patients that Watson-Williams treated there was apparent arrest of the malignant process in six of these; five more showed improvement; the condition of the remainder was inconclusive. Fourteen patients showed relief of their symptoms to varying degrees.

Prowse (1937) has described two cases of breast carcinoma with widespread metastases. Both cases received frequent mild doses of high-voltage X-rays and weekly intravenous injections of a compound colloid of selenium and sulfur. The two cases completely recovered. In contrast, Gillett and Wakely (1922) have described 100 cases treated with selenium injections. In general, little or no improvement was observed. Because there was no description of the type of selenium used and owing to the fact that many forms of selenium exist, there is a possibility that Gillett and Wakely used a different preparation than either Prowse (1937) or Watson-Williams (1920).

8.1.7 Epidemiological Relationship

There have been two major types of epidemiological relationship established in two different populations. Even though both relationships were inverse to selenium bioavailability and paralleled the animal results, this does not necessarily mean that there is a real relationship. A series of epidemiological studies (Shamberger and Willis, 1971; Shamberger et al., 1976) has related selenium bioavailability in various cities and states to the human cancer mortality in the cities and states of the United States. Statistically significant differences were found in the age-specific cancer death rates among states with high, medium, and low selenium levels. The death rates for several types of cancer showed larger difference in males than in females in the states with high selenium levels. The greater difference between males and females may have been due to sex difference or to the fact that males are heavier smokers and are more likely to be exposed to industrial pollution. In the states with high selenium forage levels, there was significantly lower mortality in both males and females from several types of cancer, particularly the environmental problem indicators, such as the gastrointestinal and urogenital types of cancer. A similar observation was observed in 17 paired large cities and 20 paired small cities (Shamberger and Willis, 1971). Alberta and Saskatchewan also have a higher selenium bioavailability and also had a lower human cancer death rate.

Schrauzer et al. (1977a) have correlated the age-corrected mortality from cancer at 17 major body sites with the apparent dietary selenium intakes estimated from food consumption data in 27 countries. There were also significant inverse correlations observed for cancers of the large intestine, rectum, prostate, breast, ovary, and lung, and for leukemia; weaker inverse associations were found for cancers of the pancreas, skin, and the bladder. In another study similar inverse correlations were found between cancer mortalities at the above sites and the selenium concentrations in whole blood which had been collected from healthy human donors in the United States and several other countries. Schrauzer et al. (1977a) postulated that the cancer mortalities in the United States as well as other Western industrialized nations would markedly decline if the dietary selenium intakes were increased to about twice the current average amount supplied by the U.S. diet.

The per capita intakes of zinc, cadmium, copper, and chromium were also estimated from food consumption data in 28 countries and these intakes were correlated with the age-corrected mortalities from cancers of the intestine, prostate, breast, skin, other organs, and leukemia (Schrauzer et al., 1977b). These correlations suggest that the anticarcinogenic effect of selenium may be counteracted by other trace elements. Manganese correlated inversely with cancer of the pancreas and arsenic intakes correlated inversely with male lung cancer mortalities. Zinc concentrations in whole blood which was collected from healthy donors in the United States correlated directly with regional mortalities from cancers of the intestine, breast, and other sites.

Jansson and Jacobs (1976) have reported that Seneca County in the Finger Lakes region of New York has an incidence of colorectal cancer about 1000 rankings removed from the surrounding counties. The mortality rate for white males in Seneca was 125/100,000/year, while the other 57 countries in New York had rates ranging from 145 to 216/100,000/year. The mortality rates for white females were 108/100,000/year in Seneca County and 118 to 160/100,000/year for the other counties.

Selenium might contribute to these low rates because the selenium level in the community water system in Seneca Falls, the largest community in the county, is about twice the average level in New York. Another more important factor may be the fact that the soil in Seneca is alkaline, while it is acid in most other parts of New York. In addition, there is a low level of rainfall around Seneca County. These two factors may facilitate the uptake of selenium in plants. Another factor which may be important is the phenomenon that Lake Seneca and Lake Cayuga are both deeper and lower, relative to sea level, than all the other Finger Lakes. Both lakes penetrate the horizontal salt stratum that underlies the region, resulting in concentrations of Na, K, and Cl ions which are about 20 times as high in Seneca and Cayuga Lakes as in the other Finger Lakes. With such differences between the levels of the macro elements, it is likely that there are also great differences in the amounts of different trace elements.

Bogden *et al.* (1981) have measured selenium, polonium-210, Alternaria spore counts, as well as the tar and nicotine contents of tobaccos from countries with high and low incidences of lung cancer. The tobacco concentrations of polonium-210 were similar in cigarettes from high- and low-incidence countries, as were levels of cigarette smoke tar and nicotine. However, tobaccos from low-incidence countries had significantly lower Alternaria spore counts. In addition, the mean selenium concentrations of tobaccos from the high-incidence countries (0.16 ± 0.05 µg/g) were significantly lower than those of tobaccos from the low-incidence countries (0.49 ± 0.22 µg/g). The differences in the selenium content of tobacco probably reflect soil levels of selenium. The major tobacco-growing regions in the Southeast of the United States have relatively low selenium concentrations in the soil. In contrast, Colombia and Mexico have been noted for high selenium levels in soil for several decades.

8.1.8 Selenium Blood Levels in Cancer Patients

Shamberger *et al.* (1973a) have compared selenium levels in the blood of 48 healthy individuals (27 men and 21 women) with selenium levels in the blood of 87 patients with gastrointestinal cancer, nine patients with hematologic cancers, and 39 patients with other types of cancers. No significant differences were observed in selenium levels between men and women or between persons below 50 years of age and those above 50 years of age. These results were in contrast to other findings that the selenium level in the blood begins decreasing at about 40–50 years of age (Dickson and Tomlinson, 1967).

Measured in µg/dl, the selenium levels were found to be 22.3 ± 0.6 for normal men and 23.6 ± 0.8 for normal women compared to 15.8 ± 0.4 for colon cancer patients, 15.3 ± 0.6 for stomach cancer patients, 13.2 ± 1.5 for patients with pancreatic cancer, $15.0 \pm$ for patients with liver cancer, and 20.7 ± 1.1 for patients with rectal cancer. However, three rectal cancer patients with liver metastases had very low selenium levels, 13.0 ± 1.1 μg/dl. With the exception of rectal cancer, Shamberger *et al.* (1973) found that patients with gastrointestinal cancer had the lowest blood selenium levels.

McConnell *et al.* (1975) have determined the levels of serum selenium in 110 patients with carcinoma. Thirty-six patients had primary neoplasms of the reticuloendothelial system, 28 patients had medical and surgical nonmalignant disorders, and 18 were healthy nonhospitalized persons. McConnell *et al.* (1975) found that the carcinoma patients had significantly lower selenium levels and that patients with gastrointestinal cancer had lower levels than those with primary lesions. They also noted, in agreement with Broghamer *et al.* (1976), that the lower the level of selenium among carcinoma patients, the greater the risk of metastases, multiple primaries, recurrence and early death.

McConnell *et al.* (1977) have also compared 30 women with breast cancer and 18 controls and found significantly lower serum selenium levels among the breast cancer patients ($P < 0.001$). In another study McConnell *et al.* (1980) have compared the serum from 35 women with breast cancer to the serum samples from 27 women known to be free of breast cancer. Again serum selenium was lower in the patients with breast cancer ($\bar{x} = 1.25$; SEM 0.04) than the mean level for the control group ($\bar{x} = 1.57$; SEM 0.08). Capel and Williams (1979) have measured the plasma selenium and the erythrocyte selenium and glutathione peroxidase in 15 women with breast cancer and compared these values to similar assays done on 14 postmenopausal controls and 11 premenopausal controls. In agreement with the study of McConnell *et al.* (1977) the plasma selenium level of the breast cancer patients was significantly lower than the controls ($P < 0.001$). Erythrocyte selenium and glutathione peroxidase levels were significantly higher than the controls. In addition, the levels of erythrocyte glutathione peroxidase were significantly higher ($P < 0.05$) in premenopausal women who were using steroidal–oral contraceptives and were more similar to the postmenopausal control levels.

Broghamer *et al.* (1978) have determined serum selenium levels by neutron activation analysis on 59 patients with a variety of reticuloendothelial tumors. The mean serum concentration for the control group of nonhospitalized healthy individuals was 1.48 ± 0.07 mg/g of dry serum, whereas the mean serum level of selenium for the 59 primary reticuloendothelial malignancies was 1.61 ± 0.16 mg/g of dry serum. Chemotherapy, particularly less than six weeks' initiation, produced elevations in the serum selenium levels. In contrast, Calautti *et al.* (1980) have also studied serum levels in malignant lymphoproliferative diseases. Selenium was determined by proton induced X-ray emission in 34 nonhospitalized healthy individuals and 38 patients with malignant lymphoproliferative disease (MLD). The mean serum levels of selenium in Hodgkin's disease and non-Hodgkin malignant lymphoma were not different from those of the control group. Lowered mean serum concentrations were observed in the group with chronic lymphocytic leukemia (5.2 ± 0.7 mg/dl) as compared to normal individuals (7.9 ± 0.3 mg/dl). The difference was significant ($P < 0.005$). A second selenium test was performed in 11 out of the 38 patients within eight weeks from the beginning of radiotherapy or chemotherapy. No significant changes in selenium levels were found.

Schrauzer *et al.* (1973) have observed that the methylene blue reduction times (MBRT) of human plasma samples were inversely related to the total plasma selenium content. The MBRT of human plasma was introduced in 1947 as a chemical test for malignancy, but was later found to be insufficiently specific or accurate for routine cancer diagnosis. From several of the previously mentioned blood selenium studies in cancer patients, it is apparent that selenium is lowered in many of the patients, but many others have normal levels. Therefore, the MBRT which relies on selenium levels would not be a useful cancer marker.

Robinson *et al.* (1979) have measured the selenium and the glutathione peroxidase activity in the whole blood, erythrocytes, and plasma in 66 noncancer surgical patients, 80 patients with cancer, and 104 healthy Otago residents from the South Island of New Zealand. Older residents over 60 years of age had lower blood selenium levels than the young and middle aged. Blood selenium levels were less than the comparable U.S. values and were no lower than those of elderly subjects and patients without cancer. In two cancer patients blood selenium levels were decreasing, and the lowest values were obtained for five cancer patients, and two noncancer patients after a long period of inanition. These low values were similar to values for patients on parenteral nutrition with negligible intakes. Robinson *et al.* (1979) have suggested that the low selenium status of cancer patients was more likely a consequence of their illness than the cause of cancer. The lower blood selenium levels were associated with lower serum albumin and glutathione peroxidase activities.

8.1.9 Selenium as a Carcinogen

The carcinogenic potential has been claimed by a few researchers, but challenged by others. In general, flaws in the experimental design or the nonphysiological nature of feeding markedly toxic amounts for a long period of time have clouded the results. In an experiment to determine the lowest level of selenium to cause chronic toxic effects Nelson *et al.*

(1943) fed rats 5–10 ppm of selenium as seleniferous wheat or corn, or 10 ppm of ammonium potassium selenide. Seventy-five per cent of the animals that survived the first three months on a seleniferous diet developed liver cirrhosis. In addition, 15 out of 53 rats that survived 18–24 months on the diet developed adenomas or adenomatoid hyperplasia. In control rats, the incidence of spontaneous hepatic tumors was less than 1%. Nelson *et al.* (1943) experienced difficulty in distinguishing between hyperplasia and true carcinomas, because it was difficult to distinguish just when the borderline between nonmalignant and malignant tumor has been passed, and also just when hyperplasia has passed into tumor. Shapiro (1972) has postulated that these changes may have represented one phase of hepatic regeneration rather than neoplasia. The inability of these "tumors" to metastasize, implying a lack of malignancy, was also noted.

A group of Soviet researchers have also claimed that selenium induced hepatic tumors (Volgarev and Tscherkes, 1967). In their first experiment they noted that 10 out of 23 male rats fed 4.3 ppm selenium developed tumors, while 5 out of 19 rats fed 8.6 ppm showed cancerous growths. Because the researchers did not mention any controls, no reliable conclusions can be made. In a second experiment they were unable to support their prior claim. In one group 5 out of 60 rats given selenium as selenate developed tumors and zero out of 100 rats on the same selenium treatment developed tumors in the second group. Volgarev and Tscherkes (1967) noted that the normal spontaneous cancer rate for the strain of rats used was 0.5% and this rate was used as a basis for comparison with the selenate-treated rats. Because of this lack of proper controls, no conclusions can be drawn from this report.

Schroeder and Mitchener (1971) have also reported on an experiment designed to show the carcinogenicity of selenium. In this experiment 418 Long–Evans rats were divided into four groups, and then further subdivided by sex. The experiment developed two problems. First of all, the groups received 2 ppm selenite, selenate, or tellurite. The initial high mortality rate among the selenite-treated males (50% mortality at 58 days) led the investigators to substitute selenate for selenite. In addition, after 12 months, the investigators raised the selenate (males and females) and selenite (females) treatments from 2 to 3 ppm for the experiment. A second experimental difficulty was a virulent pneumonia that struck the colony of rats when they were 21 months old. Penicillin brought it under control after three weeks, but losses were immense: 36% male controls, 49% selenate males, 37% female controls, and 15% selenate females. The investigators used the surviving group in spite of possibly creating statistical bias in their experiment. The following incidence of tumors was reported: 31% controls, 62% selenate, 12% selenite (females), and 36% tellurite. From these results, Schroeder and Mitchener (1971) concluded that selenium was carcinogenic and that selenium predominently induced mammary and subcutaneous fibrous cancers and not hepatic carcinomas.

In addition to the two experimental problems the investigators could have drawn a different conclusion about the carcinogenicity of selenium. First of all the 31% incidence in the controls and 12% in the selenite is statistically significant, indicating that selenite significantly reduced cancer incidence in the rats. The 62% incidence and the 31% incidence are not statistically significant. Secondly, Schroeder and Mitchener (1971) failed to take into account the importance of age as a contributing factor in the development of cancer. In the selenate-fed male rats 50% mortality was observed at 962 days, and at 1014 days in the selenate-fed female rats. However, 50% mortality in the controls occurred much earlier at only 853 days in males and 872 days in females. Older rats have a greater cancer incidence and it appears that this may have been a factor in their experiment. No attempt to age adjust was made.

Schroeder and Mitchener (1972) have also evaluated the effects of 3 ppm of selenate and selenite in the drinking water on 427 Swiss mice. They reported that both selenate and selenite were ineffective tumor-inducing agents. Harr *et al.* (1967) have studied extensively the effect of chronic toxicity in rats. They reported no excess of neoplasms in rats fed selenite and selenate at levels up to 16 ppm, even though hepatic toxicity was observed.

Seifter *et al.* (1946) have observed multiple thyroid adenomas and "adenomatous hyperplasia of liver" in rats fed bis 4-(acetoaminophenyl) selenium dihydroxide for 105 days. This compound also has inherent goitrogenic activity.

The National Cancer Institute Carcinogenesis Test Program (1980) has bioassayed SeS for possible carcinogenicity. SeS was administered by gavage to rats and mice of each sex for 105 weeks. The animals were administered SeS suspended in 0.5% aqueous carboxymethylcellulose seven days/week. The same vehicle without SeS was administered to the control animals. The animals who were administered SeS had decreased body weight, and increased tumor formation was observed in female mice and in rats of either sex. Dosed rats and female mice had an increased incidence of hepatocellular carcinomas and adenomas. Female mice who were dosed with SeS also had an increased incidence of alveolar/bronchiolar carcinomas and adenomas. Because large toxic amounts were administered by gavage, it is unlikely this experiment is of physiological value.

8.2 Selenium and Mutagenesis

Selenium appears to have both antimutagenic and mutagenic activities. Ordinarily selenium is present in blood at concentrations of about $1.25-2.50 \times 10^{-6} M$. Hopefully experiments will be designed using this physiological concentration as a starting point. Otherwise toxic or non-physiological levels might confound the interpretation of the results.

8.2.1 Antimutagenicity

Sodium selenite and seleno-amino acids interfere with the crossing over in barley (Walker and Ting, 1967). The authors provide cytological evidence for the deformation of the chromatin content of the meiocyte as a result of selenium treatment. Their observations of normal pairing discount the possibility of an effect on the ability of homologs to synapse. The latter two observations lead to the suggestion that selenium may reduce crossingover by a relaxation of the chromosomal protein which may lead to a relaxation stress in the chromosomal fibril. The selenohydryl group is less reactive than the sulfhydryl group (Nickerson *et al.* 1956); hence this and associated differences in bond strengths and distance might introduce alterations in the physicochemical properties of selenosubstituted proteins.

Similarily, selenocystine and selnomethionine also interfere with the crossover distribution along the X chromosome of *Drosophilia melanogaster* (Ahmed and Walker, 1975). These results were attributed in proteins by the error incorporation of selenocystine residues which supported a mechanism of preexchange DNA breakage induced by stress in an associated protein.

Sodium selenite also acts as an antimutagen if applied together with some known mutagens in *Salmonella* (Jacobs *et al.* 1977b; Shamberger *et al.,* 1979; Adams *et al.,* 1980; Martin *et al.,* 1981). Using the *Salmonella typhimurium* TA 1538 bacterial tester system graded decreases in mutagenicity with increasing selenium concentrations were observed for each of the three mutagens, 2-acetylaminofluorene (AAF), *N*-hydroxy-2-acetylaminofluorene (*N*-OH-AAF), and *N*-hydroxyaminofluorene (*N*-OH-AF). (Jacobs *et al.,* 1977b) Selenium decreased the mutagenicity of AAF, *N*-OH-AAF, and *N*-OH-AF to 65%, 68%, and 61% of their respective controls with mutagen alone. Shamberger *et al.* (1979a) have tested the direct alkylating agents, malonaldehyde and β-propiolactone and the antioxidants vitamin C, vitamin E, selenium and butylated hydroxytoluene (BHT) on seven mutants of *Salmonella typhimurium*, five of which mutated by a frameshift mechanism and two of which mutated through base-pair substitution. The antioxidants vitamin C, vitamin E, selenium, and BHT at three logarithmic concentrations markedly reduced mutagenesis in those strains which mutated by a frameshift mechansim. Using test strain TA 100 Adams *et al.* (1980) have tested the mutagenicity of 7,12-dimethylbenzanthracene (DMBA) and sodium nitrite alone and with selenium. Selenium addition significantly reduced the mutagencity of both the DMBA and the sodium nitrite. Martin *et al.* (1981) have tested the antimutagenic effects of selenium as sodium selenite on the mutagenicity of acridine orange and DMBA. Eight ppm of selenium reduced the number of histidine revertants caused by acridine orange and DMBA by 52%, and 74%, respectively. If the amounts of selenium added to the plates were increased, the mutagenicity of the test compounds was further suppressed.

Selenium has also been shown to reduce the mutagenicity of known mutagens in human lymphocyte cultures (Shamberger *et al.*, 1973b; Ray *et al.*, 1978). Shamberger *et al.* (1973b) incubated blood leukocyte cultures along with DMBA and the antioxidants, ascorbic acid, BHT, sodium selenite, and D,L-tocopherol. The reduction of human chromosomal breaks were as follows: ascorbic acid, 31.7%; BHT, 63.8%; sodium selenite, 42%; and D,L-tocopherol, 63.2%. More acrocentric-type chromosomal breaks (21.7%) were seen in the untreated controls than the DMBA-treated group (4.8%). In contrast the DMBA-treated groups had a higher percentage of meta breaks than the untreated controls. Ray *et al.* (1978) reported that selenium as sodium selenite, when tested with methyl methanesulfonate (MMS) or *N*-hydroxy-2-acetylaminofluorene (*N*-OH-AAF), reduced the sister-chromatid exchange (SCE) by 25%–30% and 11%–17%, respectively, below the SCE frequencies produced by the individual compounds. However, high sodium selenite concentrations (7.90×10^{-5} *M*) resulted in a threefold increase in the SCE frequency above the background level of 6–7 SCEs/cell.

Russell *et al.* (1980) have found that selenium compounds are able to induce DNA repair synthesis as a measure of DNA damage in both the isolated rat liver cell system and the Ames' *Salmonella* assay. In the liver cells, DNA repair measured by uptake of (^{3}H) thymidine was found to be greater with sodium selenite and selenate than with selenomethionine. Williams (1978) has previously reported (^{3}H) thymidine uptake by whole cells may be indicative of DNA repair synthesis. In the bacterial culture system, the repair-deficient variant was inhibited more by selenomethionine than by selenite and selenate.

Lawson and Birt (1981) have studied the repair of *N*-nitrosobis (2-oxopropyl) amine (BOP)-induced damage of pancreas DNA when the hamsters were pretreated with selenium. BOP is a potent carcinogen, preferentially producing pancreatic ductular adenocarcinoma in Syrian golden hamsters. Lawson and Birt (1981) examined the initiating phase of BOP carcinogenesis by measuring the production and repair of DNA damage in target (pancreas) and nontarget (liver and salivary gland) tissues of Syrian golden hamsters given a single subcutaneous dose of BOP (20 mg/kg). DNA damage was measured by the alkaline sucrose gradient–fluorometric method. DNA damage repair was slower in the pancreas than the liver and the salivary gland. Six weeks after dosing, there was still DNA in the pancreas which sedimented at a lower density than pancreas DNA. Selenium was fed at a basal level and added to the diet at levels of 0.1 and 5.0 ppm. DNA damage was measured in the hamsters that had been pretreated for four weeks prior to BOP administration with selenium. There was no apparent difference in the amounts of the DNA damage with any of the different treatments. However, in the hamsters fed 5.0 ppm of selenium, the DNA damage was repaired faster so that by 24 hr all the DNA sedimented at the same density as pancreas DNA from control hamsters. The enhancement of repair observed by Lawson and Birt (1981) seemed to be similar to that of Russell and Nader (1980). Even though the hamsters were fed a toxic 5-ppm level of selenium, the hamsters did not exhibit any overt toxic reactions and maintained a growth rate which was comparable with the control hamsters.

Norppa et al. (1980a; 1980b) have examined the potential mutagenic effect of sodium selenite in vivo. Norppa et al. (1980a) have grown lymphocyte cultures from nine neuronal ceroid lipofuscinosis (NCL) patients who received sodium selenite injections or tablets (0.004–0.05 mg Se/kg body weight daily) for 1–13.5 months and five healthy persons who were given sodium selenite tablets (0.025 mg Se/kg body weight daily) for two weeks. Whole-blood lymphocytes grown in a 72-hr culture from the nine NCL patients and five healthy persons showed no more cells with chromosomal aberrations than those from five other patients and two normal individuals.

Norppa et al. (1980b) have also studied the aberrations in mouse bone marrow and primary spermatocytes. Sodium selenite was injected intraperitoneally (0.8 mg Se/kg body weight) into 12 male NMR1 mice. After 24 hr, cytogenetic preparations of bone marrow and testis were made. The bone marrow of the sodium selenite-treated mice contained no more cells with structural chromosomal aberrations than that of the controls. There was no difference between the two groups in the numbers of metaphase-diakinesis stages of primary spermatocytes.

Kuhnlein et al. (1981) have measured fecal mutagenicity from six healthy male volunteers (24–33 years) who consumed low and then high selenium diets. A formula diet containing 19–24 mg Se/day was fed for 45 days followed by a similar diet containing 203–224 mg Se/day for 24 days. Aqueous extracts of feces from days 1–3, 43–45, and 67–69 were assayed for mutagenicity using *S. typhimurium* TA 100 and TA 98. The reduction in fecal mutagenicity for the three periods with TA 98 was significant with 20.5, 3.1, and 4.7 revertants. The mean number of revertants corrected for the blank with TA 100 was 9.4, 7.8, and 8.8. TA 98 is mainly affected by frameshift mutations. This observation is consistent with those previously made by Shamberger *et al.* (1979), who observed that selenium and other antioxidants mainly affected frameshift mutagenesis.

Rosin (1981) has found that sodium selenite (1–15 mmol/plate) completely suppressed spontaneous mutagenesis at two independent loci in both wild (YO-300-IC) and nine mutator isogenic strains of *Saccharomyces cerevisiae*. The two loci which were studied were the *his* 1-7, a missence mutation, and *lys* 1-1, a supersuppressible mutant of the amber variety. The amount of suppression of spontaneous reversion of prototrophy at these two loci depended on the concentration of sodium selenite, the yeast strain that was studied, and the loci being studied. Almost 30-fold greater amounts of sodium selenite were required to suppress the frequency of spontaneous reversion at the histidine locus compared to quantities necessary to show a similar inhibition of lysine spontaneous reversion rates. The histidine and the lysine locus also responded differently to the presence of sodium selenide or sodium selenate. Sodium selenide at 3 μmol/plate completely inhibited spontaneous mutagenesis at the lysine locus for strain YO-800-IC (MUT 1-1). Complete suppression of histidine reversion occurred at 30 μmol/plate. Sodium selenate suppressed the spontaneous mutagenesis at the lysine, but not the histidine locus. The results of Rosin (1981) demonstrate the complexity of the effects that environmentally added components can have on mutagenesis.

8.2.2 Mutagenicity

Nakamuro *et al.* (1976) have tested five selenium compounds, Na_2SeO_4, H_2SeO_4, Na_2SeO_3, H_2SeO_3, and SeO_2, for their capacity to induce chromosome aberrations in cultured human leucocytes and for their reactivity with DNA by a rec-assay system and inactivation of transforming activity in *Bacillus subtilis*. In general, the concentration of the five were at toxic levels and ranged from 1.3 to 5.3×10^{-4} *M*. The chromosome-breaking activity was significantly higher for four-valent rather than with six-valent selenium: H_2SeO_4, and Na_2SeO_4. The rec assay of Kada (1972), which uses *B. subtilis* with different DNA recombination capacities, suggested that damage to DNA was produced by selenites but not by selenates. In addition, the reactivity of selenites with DNA was also indicated by a significant loss of transformation of the tryptophan marker of *B. subtilis* DNA treated with H_2SeO_3 and SeO_2. Noda *et al.* (1979) have demonstrated weak base-pair substitution mutagenic activity for both selenate and selenite against *S. typhimurium* TA 100, as well as a slight modifying effect when using recombination repair positive and negative strains of *Bacilus subtilis*. SeO_2 at 0.01 M was more inhibitory in the recombination-repair-deficient (rec^-) than with wild bacteria (rec^+) indicating that this chemical is damaging cellular DNA (Kanematsu *et al.*, 1980).

Ray and Altenburg (1978) have observed that exposure of whole-blood cultures to high sodium selenite concentrations (7.90×10^{-6} to $1.58 \times 10^{-5} M$) resulted in a three-to four-fold increase in the observed average number of sister-chromatid exchanges (SCE). Analysis of different whole-blood components showed that the presence of red blood cells, and specifically red blood cell lysate, was a necessary component for sodium selenite SCE induction in purified lymphocyte cultures. The SCE frequencies of xeroderma pigmentosum as well as those of normal human lymphoblastoid cell lines were found to be unaffected by sodium selenite concentration that produced elevated SCE frequencies in whole-blood cultures. If the latter two cell types were incubated with sodium selenite and RBC lysate, there were increases in the SCE frequencies that were comparable to those increases observed in sodium selenite-exposed whole-blood cultures.

Norppa *et al.* (1980c) have studied the effect of sodium selenite on the aberrations and sister-chromatid exchanges in Chinese hamster bone marrow chromosomes. A clear rise in the number of cells with aberrations (13.0%–30.5%) or in the mean number of SCE's/cell (6.7–11.4) was observed when compared to the two control groups (0.9% and 1.0% of cells with aberrations and 3.4–4.4 SCE's/cell. The increase was only seen in the hamsters treated with 3, 4, or 6 mg Se/kg of body weight. The general toxicity of selenium at these high doses, which are near the LD_{50} of selenium, may contribute to the manifestation of observed chromosomal damage. Schwarz (1976) has observed that the LD_{50} of selenium for mice was about 4.1 mg/kg body weight. The lack of chromosomal changes at selenium doses below 3 mg/kg body weight suggested that a mechanism may exist *in vivo*, which prevents the harmful effects of selenium at low concentrations.

Lo *et al.* (1978) have added sodium selenite at doses varying 8×10^{-5} to $3 \times 10^{-3} M$ to cultured human fibroblasts and have induced DNA fragmentation, DNA repair synthesis, chromosome aberrations, and inhibition of mitosis. The response of DNA repair-deficient xeroderma pigmentosum (XP) fibroblasts was similar to that of control cells. The capacity of selenite to induce chromosome aberrations, DNA repair, and a lethal effect was enhanced when a mouse liver S-9 microsomal fraction was incubated with the fibroblasts. XP cells behaved as control cells when they were treated with activated selenite. If sodium selenate at doses ranging from 8×10^{-5} to 3×10^{-3} M were incubated with an S-9 preparation, no activation was observed as evidenced by no significant increase in the frequency of chromosome aberrations. One difficulty in uncovering the mutagenic action of selenium compounds is their high toxicity and the relatively narrow range of concentrations that lead to a mutagenic effect, rather than a lethal one. The use of different end points may lead to different results and interpretations. For example, sodium selenite but not sodium selenate gave a positive result when recombination-deficient $B.\ subtilis$ was used as a test organism in mutagenesis studies (Nakamuro $et\ al.$, 1976), whereas selenate but not selenite gives positive results in the Ames test which used an indicator strain for base-pair substitution (Lofroth and Ames, 1977). The results presented by Lo $et\ al.$ (1978) for fibroblasts of normal individuals and DNA repair-deficient XP patients were similar to those reported by Nakamuro $et\ al.$ (1976).

Whiting $et\ al.$ (1980) have observed that glutathione strongly enhanced the induction of unscheduled DNA synthesis (UDS) in cultured human cells by inorganic selenium compounds. In the presence of 10^{-3} M glutathione, high levels of UDS (74–114 grains per nucleus) were observed in cells treated with selenate at 10^{-5}–10^{-3} M selenite at 10^{-5}–3×10^{-4} M, and selenide at 10^{-5}–10^{-3} M. Glutathione also enhanced the clastogenic and cytotoxic effects of selenite and selenate in Chinese hamster ovary cells. Glutathione alone decreased DNA damage and toxicity in both the UDS and the chromosome aberration assays. In the absence of glutathione, the three inorganic selenium compounds induced low levels of UDS and moderate frequencies of chromosome aberrations. Whiting $et\ al.$ (1980) also examined the effect of three organic selenium compounds on the induction of UDS. No UDS was detected in cells treated with selenocystamine or selenomethionine, with or without added glutathione. However, selenocystine at 10^{-4}–10^{-3} M induced a low level of UDS which was enhanced by glutathione.

Whiting $et\ al.$ (1980) have demonstrated that reduction is involved in the conversion of selenium compounds to mutagenic forms. The active mutagens may be selenols GS–Se$^-$ from inorganic selenium and R–Se$^-$ from organic selenium compounds. The metabolism of inorganic selenium compounds in mammalian tissues has been shown to proceed primarily by interaction with glutathione and glutathione reductase (Hsieh and Ganther, 1975). Selenate, selenite and selenide might be interrelated by the following reactions:

$$HSeO_4^- + 2GSH \xrightarrow{\text{slow}} HSeO_3^- + GSSG + H_2O \qquad (8\text{-}1)$$

$$HSeO_3^- + 4GSH \longrightarrow GSSeSG + GSSG + OH^- + 2H_2O \ (8\text{-}2)$$

$$GSSeSG + GSH \rightleftharpoons GSSe^- + GSSG + H^+ \qquad (8\text{-}3)$$

$$GS\text{---}Se\text{---}SG + NADPH \xrightarrow{\text{glutathione reductase}} GSSe^- + GSH + NADP^+ \qquad (8\text{-}4)$$

$$GS\text{---}Se^- + H_2O \rightleftharpoons Se^0 + GSH + OH^- \qquad (8\text{-}5)$$

$$GS\text{---}Se^- + NADPH + H_2O \xrightarrow{\text{glutathione reductase}} HSe^- + GSH + NADP^+ + OH^- \quad (8\text{-}6)$$

$$GS\text{---}Se^- + GSH \rightleftharpoons HSe^- + GSSG \qquad (8\text{-}7)$$

$$HSe^- + (O) \longrightarrow Se^0 + OH^- \qquad (8\text{-}8)$$

$$HSe^- + (O) + 2GSH \longrightarrow HSe^- + GSSG + H_2O \qquad (8\text{-}9)$$

Selenate is slowly reduced to selenite by glutathione and other sulfhydryl compounds [equation (8-1)], while reduction of selenite [equation (8-2)] and then further reactions occur in seconds. Selenite reduction proceeds via the selenotrisulfide (GS–Se–SG), which is stable and can be isolated (Ganther, 1974), and the selenopersulfide (GSSe$^-$ at neutral pH), which is highly reactive. If excess sulfhydryl is present, [equation (8-9)] selenide is protected against oxidation [equation (8-8)].

Whiting et al. (1980) have explained the DNA damage induced by the inorganic selenium compounds on the basis of the above equations. Selenide cannot be oxidized in vivo beyond selenotrisulfide (GS–Se–SG), while selenate and selenite can be reduced as far as selenide. In addition, the reactive compounds selenide (HSe$^-$) and the selenopersulfide (GS–Se–) are common products of metabolism of the inorganic selenium compounds and may be the ultimate mutagens. These compounds can be easily oxidized and may produce reactive free radicals which could damage DNA. The organic selenium compound selenocystine is reduced to selenocystine by glutathione and this selenol may also be mutagenic.

Schut and Thorgeirsson (1979) have tested the effect of selenite on the mutagenicity of N-hydroxy-acetylaminofluorene (N-OH-AAF) on the Salmonella tester strain TA 1538. Sodium selenite at concentrations up to 0.6×10^{-3} did not affect the N-OH-AFF mutation frequency but was highly toxic to the bacteria. Concentrations of 0.8×10^{-3} to 2×10^{-3} M sodium selenite, which killed 67%–94% of the bacteria, increased N-OH-AAF mutation frequency up to fivefold.

Tkeshelashvili et al. (1980) have studied the effect of three metals, arsenic, selenium, and chromium, on the accuracy of DNA synthesis in vitro. Neither arsenic or selenium altered the accuracy of the fidelity assays under the normal conditions of magnesium activation, nor did they affect the mutagenicity of manganese. Chromium in the form of Cr(III) as well as Cr(VI) diminished the fidelity by which Escherichia coli DNA polymerase 1 copies polynucleotide templates.

8.3 Selenium and Immunity

Many nutritional factors may increase or decrease immunocompetence in experimental animals. Among the nutritional factors affecting immunity are intakes of calories, proteins, vitamins, and minerals. The essential amino acids, the essential fatty acids, minerals, and the vitamins are all required for proper humoral immunity.

8.3.1 Effect of Selenium on Humoral Immunity

The first report of an enhancement of humoral immunity by selenium was made by Berenshtein (1972). In this experiment Berenshtein (1972) administered thyroid vaccine to 56 rabbits together with sodium selenite (0.05 mg/kg sc) and vitamin E (0.03 mg/kg) before or during immunization with the thyroid vaccine. Rabbits given the selenium and vitamin E and the vaccine had higher antibody (Ab) titers than in control rabbits given vaccine alone. Vitamin E alone had no effect on antibody titers.

Spallholz *et al.* (1973a) have investigated the effects of dietary selenium upon the formation of antibody titers in mice sensitized to sheep red blood cells (SRBC) and upon the number of spleen plaque-forming B cells, four and seven days postsensitization. Selenium as sodium selenite added to commercial diets of mice increased titers of circulating antibodies to the SRBC antigens. Especially high anti-SRBC antibody titers (IgM) were observed from a group of mice receiving 2.8 ppm Se in their commercial diets. In addition, higher numbers of plaque-forming cells (B cells) were observed, especially from mice receiving about 1–3 ppm of dietary selenium.

In another experiment Spallholz *et al.* (1973b) fed mice Purina chow diets with supplemental graded levels of selenite (0–8.5 ppm Se). Anti-SRBC immunoglobulins IgM and IgG were titered 4, 7, and 14 days postimmunization. The higher IgM titers which had been observed on days 4 and 7 postsensitization had fallen by day 14 and were replaced by higher IgG titers. The antibody titers obtained following sensitization with SRBC antigens or tetanous toxoid were dependent upon the amount of Se administered by injection or supplemented to diets.

In another later study, Spallholz *et al.* (1974) used over 4000 mice and studied the effect of the route of administration (ip, sc, or intradermal) as well as the usage of Se and vitamin E (0–100 mg Se and/or 0–5 mg vitamin E) upon the primary and secondary immune responses to SRBC antigens or tetanus toxoid (TT). Synergism between Se and vitamin E in increasing antibody titers was observed. Anti-TT titers were not increased in the primary immune response, but were enhanced in both the primary and secondary immune responses. Measurement of the primary immune response to both antigens was measured over a 21-day period while following the second immunization–sensitization the secondary response was measured over a 21-day period.

Norman and Johnson (1976) have reported on field experiments where MuSe^R (Burns-Biotec Pharmaceutical Company, Omaha, Nebraska) was twice administered to calves (25 cc). Vaccination followed with a commerical *L. pomona* vaccine for protection against leptospirosis. After six weeks the calves were bled and titered for anti-*L. pomona* antibodies. The data indicated more than a doubling of the average leptospirosis titers in calves given MuSe^R.

Aleksondrowicz (1977) has studied the stimulation of antibody production in mice and rabbits in response to bacterial and mycotic antigens in mice fed 50 mg of selenite/kg body weight/day in a standard ration. If the selenite was increased up to 2 mg/kg body weight/day, no stimulation of antibody synthesis was observed.

Shakelford and Martin (1980), using retired male breeder mice (247 days old) given 1, 2, or 3 ppm Se or sodium selenite in their drinking water, found that there was a statistically significant increase of the anti-SRBC Ab titers of mice receiving 1 ppm Se in the drinking water over the control mice. Mice receiving 3 ppm Se in their drinking water had titers which were significantly decreased from the control mice.

Koller *et al.* (1979) has observed that selenium counteracts the immunosuppression that is produced by methylmercury. Mice fed 1, 5, and 10 ppm of methylmercury plus 6 ppm of selenium significantly increased antibody synthesis. Because methylmercury singly depresses antibody synthesis and the enhancement was greater than the increase by selenium alone, synergism between methylmercury and selenium occurred. These results indicate that environmental contaminants may not act in combination as anticipated.

Nutritional deficiencies in vitamin E and/or selenium caused impaired immune function, as measured by the humoral response to ovine erythrocytes (SRBC) by young chicks, but only at low antigen doses (Marsh *et al.*, 1981). In the two-week-old chick optimum immune function required both vitamin E and selenium; however, at three weeks of age, either vitamin E or selenium was sufficient for optimum immune function. At this stage of development, selenium appeared to be capable of replacing vitamin E with regard to the immune system. Synergism of vitamin E in the presence of adequate selenium was not observed. The amounts of vitamin E and selenium were within nutritional and nonpharmacological levels. In contrast, high dietary selenium produced significant immune suppression in male but not female chicks.

Sheffy and Schultz (1978) have used dogs as their experimental animals and found that the humoral response could be influenced by selenium and vitamin E. In addition, vaccination with a canine distemper infectious hepatitis virus vaccine resulted in lower antibody titers in animals deficient in selenium and vitamin E than in control animals. In the vitamin E- and selenium-deficient animals there was a delayed appearance of measurable antibody titers following immunization. In these dogs the primary immune response to SRBC was unaffected, whereas it was the secondary response, the IgG anti-SRBC antibody levels, that were reduced in the deficient animals.

8.3.2 Cell-Mediated Immunity

Measurements of cell-mediated immunity are generally made by delayed hypersensitivity and graft-vs.-host reactions *in vivo* as well as by mixed lymphocyte culture, mitogen stimulation, and assessment of helper, suppressor, and cytotoxic activity of T cells *in vitro*. Sometimes antibody-dependent cell-mediated cytotoxicity, which requires K cells (surface Fe receptors) for cytotoxic expression, is used. Cell-mediated activity is also conveyed by soluble products of T lymphocytes (lymphokines), of which there are many.

Martin and Spallholz (1977) have fed female Hartley guinea pigs a commercial diet supplemented with 0, 1, 3, 5, or 7 ppm of selenium of four weeks. At that time each animal was sensitized intradermally with a 0.1% solution of dinitrochlorobenzene (DNCB) in propylene glycol. After ten days each animal was challenged by a serial dilution of DNCB (0.1%–0.003%) in ethanol. Erythema and induration of the second challenge was semiquantitatively assessed 24 and 48 hr later. The guinea pigs fed diets with 1–3 ppm selenium were more sensitive than the controls of the DNCB. The group of guinea pigs receiving 7 ppm of selenium showed some increase of increased hypersensitivity, but this level of selenium was toxic with two of four guinea pigs dying and the remaining two showing much retarded growth.

High levels of glutathione peroxidase had been demonstrated in peritoneal exudate cells (Serfass and Ganther, 1976). Phagocytic cells of selenium-deficient rats were capable of ingestion of yeast cells *in vitro*, but were unable to kill them. Boyne and Arthur (1979) have confirmed the observations obtained from selenium-deficient rats in selenium-deficient cattle. When the neutrophils were isolated from selenium-deficient rats, they were able to phagocytize *C. albicans*, but were less able to kill the ingested cells. In the neutrophils of the selenium-deficient animals glutathione peroxidase activity was not detectable. In addition, reduced numbers of peritoneal macrophages were also noted in the exudates of rats.

The lymphocytes of vitamin E- and selenium-deficient dogs were found by Sheffy and Schultz (1978) to be completely suppressed and unresponsive to stimulation with the mitogens, concanavalin a, phytohemagglutinin, pokeweed mitogen, or streptolysin O.

The addition of 50% mg of Na_2SeO_4/kg body weight/day to the diets of mice and rabbits has been reported by Aleksondrovicz et al. (1977) to not only enhance antibody formation, but the selenium also shortened skin allograft rejection times.

8.3.3 Nonspecific Immune Effects of Selenium

There have been several reports of nonspecific immune effects of selenium. Properdin, complement, and lysozyme, which are indexes of natural immunity, have been reported by (Berenshtein, 1973) to be increased in the serum of rabbits administered 0.05 mg/selenium/kg a day over a period of 92 days. Leucocytes, phagocytic activity, and the bactericidal indexes were also increased in rabbits receiving subcutaneous selenite (Berenshtein, 1975). Berenshtein and Zdravoukhi (1976) have reported increased serum levels of properdin, lysozyme, and phagocytic activity of leucocytes following 14 daily subcutaneous injections of sodium selenite or sodium molybdate, (0.05 mg/kg) to mice.

Desowitz and Barnwell (1980) have demonstrated that selenium administered in drinking water potentiates a protective effect of a killed *Plasmodium berghei* vaccine for Swiss-Webster mice. In addition, they found that when the *P. berghei* antigen combined with the adjuvant dimethyl dioctadecyl ammonium bromide (DDAB) conferred a significantly high level of protective immunity. An additive effect was observed in that the greatest degree of protection was found in the group of mice maintained on selenium and vaccinated with antigen–DDAB. The challenging infection was survived by almost all of the animals treated in this manner.

Mulhern et al. (1981) have divided C57BL46J mice into four groups of six mice and fed them graded levels of selenium. *In vitro* spleen lymphocyte stimulation was performed on all the groups with mitogens that selectively activate B cells and specific T cell subpopulations. In addition, mixed lymphocyte reactions, cell-mediated lympholysis reactions, and direct plaque-forming cell response to sheep erythrocytes were determined on the spleen cells from all groups of mice. Significant differences were not observed in assays among any of the first-generation animals.

Offspring of the mice on the 0.0-ppm selenium diet had a reduced direct plaque-forming cell response to sheep erythrocytes when this group was compared to either the age-matched chow-fed animals and the first-generation 0.0-ppm selenium-torula-reared animals. The results of Mulhern *et al.* (1981) indicated that even though the different selenium diets do not affect the immune response of the first generation animals, they do affect their offspring.

8.4 Selenium and Dental Caries

The initial experiment on the effect of selenium on dental caries in rats when the selenium was consumed during the time of tooth formation was conducted by Buttner (1963). In rats development of the molar teeth begins in the latter part of pregnancy and the crowns are completed at the time of weaning, about 20 days after birth. Sodium selenite was added in the amounts of 2.3 and 4.6 ppm to the drinking water of pregnant rats. Their offspring received the same concentrations during lactation and until the end of the experimental period. The results indicated that the ingestion of selenium by the offspring during tooth development significantly increased the incidence of dental caries in proportion to the amount present in the water.

The amounts of selenium used in this experiment were toxic because there was a reduced number of young born. In addition, the selenium-treated rats lost considerable weight, which is a typical symptom of chronic selenium toxicity resulting from reduced appetite and food intake. In a second study, water containing 4 ppm of selenium as sodium selenite was given to rats starting with lactation, thereby omitting part of the period of tooth development (Navia *et al.* 1968). With these experimental conditions, there was a statistically significant increase in caries, but they were limited to the occlusal surface of the teeth in the selenium-treated animals. After the same amount of selenium was mixed with the diet, there was only a numerical increase of caries. Food consumption remained similar to that of the controls.

Bowen (1972) has shown a significant increase in the incidence of dental caries when selenium was added to the diet of monkeys. Increased caries were found in the molar teeth which were undergoing formation when increased levels of selenate were in the diet. Britton *et al.* (1980) have studied the effect of moderate amounts of dietary selenium on dental caries during tooth development. Ten-day pregnant rats received either distilled water or water containing 0.8 ppm or 2.4 ppm selenium, as sodium selenite or selenomethionine until the pups were weaned. Buccal caries were counted in the pups after they were fed a MIT-200 diet for seven weeks. Moderate levels (0.8 ppm) of selenium during tooth development significantly reduced caries in male rats when compared to control rats or to rats receiving high levels of selenium (2.4 ppm).

The results of three independent epidemiological studies among children in certain Western states indicate that the consumption of increased dietary levels of selenium during the period of tooth development may be related to the incidence of dental caries (Hadjimarkos, 1968).

Dental examinations given to about 2000 Oregon children 14–16 years of age revealed that those born west of the Cascade Mountains had a significantly higher prevalence of caries compared to children living east of the mountain range (Hadjimarkos, 1956). The hardness of the water supplies, the fluoride content of water supplies, or the dietary habits of the children were not factors responsible for this difference in caries susceptibility.

Because urinary selenium reflects dietary intake, 24-hr urine specimens were collected from 108 male high school children who were lifelong residents in counties east (Klamath) with low caries rates and the high caries rates of Jackson and Josephine Counties west of the Cascade Mountains (Hadjimarkos and Bonhorst, 1958). Further, samples of eggs, milk, and water produced and consumed locally were collected from 74 farms in the same counties (Hadjimarkos and Bonhorst, 1961). The results indicated that there is a direct relationship between the prevalence of caries and concentration of selenium in urine, eggs, and milk. The amount of selenium in the urine of children with high caries rates (Jackson and Josephine Counties) was twice as much as the children with low caries rates (Klamath County). There was also a tenfold difference in the selenium content of milk and eggs between the counties where high and low caries rates were found among the children.

The findings in Oregon on the direct association between levels of dietary selenium and the prevalence of dental caries were corroborated later by two other epidemiological studies. Children living in the high selenium states of Wyoming, Montana, Oregon, and South Dakota were shown to be more susceptible to dental caries than those in low-selenium areas (Ludwig and Bibby, 1969).

Selenium has been shown to be a normal constituent of human teeth (Nixon and Myers, 1970). Curzon and Crocker (1978) have studied whole enamel samples taken from 451 teeth. They chemically analyzed the teeth for 30 trace elements. The results of these determinations were related to dental caries by multiple linear regression analysis. Besides the expected age and sex effects, the eight elements which were found to have significantly large sample relationship to tooth decay were as follows:

negative: F, Al, Fe, Se, and Sr; positive: Mn, Cu, and Cd. The significant negative partial relationship of selenium to caries indicates that increased concentrations of selenium in enamel are associated with low caries. In general, there were low concentrations of selenium in the enamel. Hadjimarkos and Bonhorst (1959) have also observed relatively low selenium levels in enamel. Navia (1970), using ^{75}Se in rat diets, found only low concentrations of selenium in the mineralized portions of the teeth. Bowen (1972) has found significantly higher concentrations of selenium in plaque in a high-caries group of monkeys. Selenium may, therefore, have different effects depending upon whether high concentrations of the element occur in enamel or in the oral environment, such as the plaque.

Shearer (1975) has studied the mechanism for the uptake of dietary inorganic and organic selenium by the fully developed, mature molar teeth and the developing molar teeth of their pups. Pregnant rats received drinking solutions containing either 0.02 ppm selenomethionine plus 0.045 µCi of ^{75}Se-selenomethionine per ml or 0.2 ppm Na_2SeO_3 plus 0.043 µCi per ml from day 10 of pregnancy until parturition. After 13 days postpartum, the uptake of dietary ^{75}Se into developing molar teeth was much more extensive than the postdevelopmental uptake onto mature teeth. Molar teeth that were developing incorporated more ^{75}Se from dietary selenomethionine than from selenite, as did many of the other hard and soft tissues studied. The protein fraction of the enamel and dentine contained the major portion of ^{75}Se. When the selenium in the enamel and dentine was dialyzed it was observed that the selenium existed in at least three forms: loosely bound selenium, proteinaceous selenotrisulfides, and proteinaceous stable selenium. These experiments as well as the results of a previous study (Shearer, 1973) indicate that selenium is incorporated into developing proteins as selenotrisulfides or seleno-amino acids. The presence of high levels of selenotrisulfides or seleno-amino acids in the protein matrix of developing teeth may change the integrity of the protein matrix around which mineralization occurs. Improper enamel may be formed by selenium's interference with crystal nucleation, growth, or structure and this enamel could be more susceptible to dental caries.

Shearer and Ridlington (1976) have studied the interaction of dietary fluoride and selenium on the fluoride uptake into calcified molar enamel. Rats were provided drinking water containing 50 ppm F, as NaF, alone or plus 1 or 3 ppm as one of four selenium compounds: $NaSeO_3$, Na_2SeO_3, D,L-selenomethionine, or D,L-selenocystine. No evidence was found that fluoride interacted with any of the four selenium compounds.

Bawden and Hammarstrom (1977) have used whole body autoradiographic methods to study the uptake pattern of ^{75}Se into the developing enamel of rat molars. Four hours after injection, low concentrations of tracer were found in the secretory ameloblasts and higher concentrations were present in newly deposited enamel matrix. In the postsecretory ameloblasts or the maturing enamel no uptake was observed. However, 24 hr after injection, most of the ^{75}Se had left the enamel matrix and low concentrations were found in the enamel organ. This uptake pattern suggested that the ^{75}Se was secreted in the matrix by the ameloblasts, or that ^{75}Se diffused between the cells into the newly deposited matrix.

Crisp *et al.* (1979) have studied explants of developing molar teeth from eight-day-old rats. The explants were cultured either in a medium containing ^{75}Se under control conditions or following chemical inhibition by 2,4-dinitrophenol, heat inactivation, or removal of the enamel organ. Comparisons were made between the autoradiographic patterns of ^{75}Se uptake in the enamel organ and the enamel. The results indicate that the selenium is secreted into the matrix rather than diffusing into the matrix following matrix deposition. Johnson and Shearer (1979) have measured selenium concentrations in successive layers of bovine incisor enamel. Selenium concentrations showed no concentration gradient and no increase with age. In addition, selenium was taken up into the continuously growing rat incisor in proportion to the amount of selenium in the diet. This observation supports evidence that selenium is caries promoting only if ingested during tooth development.

8.5 The Anti-Inflammatory Properties of Selenium

Roberts (1963a), while screening potential antitumor compounds in rats by the Selye granuloma pouch assay (GPA), found that after ashing some inorganic substances from specific liver fractions had anti-inflammatory properties. Following further tests, the major elements of the ash were found to be biologically inactive. The minor components of the ash were then examined and selenium was identified as a minor constituent of both the ash and liver fractions. Selenium was found to be an active component of the ash when assayed as sodium selenite in the GPA (Roberts, 1963b). A complete titration of sodium selenite activity in the GPA beginning at 2 mg Se/kg body wieght to 0.6 mg Se/kg body weight showed that reduction in the exudate volume was closely related to the amount of selenium administered to rats. Similar experiments using several organoselenium compounds showed them to be either not biologically inactive in the GPA or much less effective as an equivalent amount of selenium as sodium selenite. All of the tests were performed by intraperitoneal injections of the selenium compounds or ash fractions.

Spallholz (1981) has cited some unpublished work of Roberts and

Schwarz in which they have observed a potentiating effect of vitamin E with selenium in the GPA. When vitamin E was administered as D,L-α-tocopherol (12.5 mg) with selenite, there was about a fourfold potentiation of the effect of selenite. Vitamin E alone at concentrations of 0.8 to 12.5 mg per rat was without measurable effect in reducing the exudate.

The anti-inflammatory characteristics of selenium have recently been patented for human use (Levitt Research Laboratories Inc., 1979). Administration of 0.05–1.0 mg Se was claimed to decrease inflammation of body tissues. In addition, administration of 6.5 mg of selenite for seven days diminished subcutaneous bleeding and swelling of an auto accident victim.

8.6 Selenium and Heart Disease

8.6.1 Animals

Electrocardiograms of rats maintained on selenium-deficient diets show the early development of a characteristic abnormality (Godwin, 1965). The amplitude of the T wave increases until it reaches very high proportions. The initial rise is detectable within a week of when the animals were weaned. The heart rate does not at first slow down, but when the animal becomes moribund there is a marked bradycardia. At this time there is an elevated S–T segment and the P wave becomes superimposed on the descending T wave.

In lambs, progressive development of a characteristic electrocardiogram abnormality was seen, after the animals had been on the deficient diet for a few weeks (Godwin and Fraser, 1976). Shortly before death the electrocardiographic pattern became grossly abnormal in some cases, a rise in the T wave giving way to an elevated S–T segment which is similar to that seen frequently in myocardial infarction in humans. Paralleling these changes, at least in time, there was a marked fall in blood pressure in the limbs, but present to a lesser degree in the rest of the systemic circulation. These observations taken together may indicate that the fundamental change occurring in selenium deficiency may be circulatory failure. In addition to the heart changes seen in rats and sheep severe myopathy of the heart can occur in many other animal species such as pigs (mulberry heart disease), calves, chickens, and ducks.

Cadmium at high levels has been observed to cause hypertension in rats (Perry and Erlanger, 1977). The cadmium pressor effect caused by 2.5 and 10 ppm of cadmium in the drinking water can be reversed by 3 ppm of selenium in the drinking water. The effects of four metals, lead,

copper, selenium, and zinc, on cadmium-induced hypertension was studied in groups of rats that received (a) neither metal, (b) cadmium alone,
(c) the second metal alone, or (d) both metals. All four second metals
showed some tendency to raise systolic pressure when they were used
alone, but this effect was marked for lead and zinc. Lead was unusual
in that its pressor effect was additive to that of cadmium when both metals
were given at the same time. In contrast, copper and selenium completely
inhibited the pressor effect of cadmium, and zinc decreased the pressor
effect. Hilse *et al.* (1979) have studied the effect of sodium selenite on
two types of hypertension in rats which had not been caused by cadmium.
In one type of hypertension model the right kidney of three-month-old
male Wistar rats was loosely wrapped in a cellulose acetate capsule and
the contralateral kidney was removed two weeks later. The other type
of hypertension was induced by daily six-hr-lasing infusions of angiotensin
11 into the caudal vein of three-month-old rats for 14 days. Sodium selenite
significantly decreased the systolic and diastolic blood pressure in rats
with experimentally induced renal hypertension. The rise of blood pressure after angiotensin 11 infusion was also suppressed by sodium selenite.
Guo *et al.* (1981) has reported that sodium selenite at 0.5 mg/kg (iv)
produced hypotension in dogs. Sodium selenite also inhibited the blood
pressure increase induced by hypoxia or pituitrin.

Selenium has also been implicated as a potent antioxidant which
might retard vessel wall peroxidation which has been postulated to start
the atherosclerotic process. Glavind *et al.* (1952) have suggested that oxidation products of lipids might play an active part in the genesis of
atherosclerosis. They observed a positive correlation between the degree
of atheroma in human aortas and the extent of peroxidation of the lipids
in the plaques as measured by the thiobarbituric acid test. Harland *et al.*
(1973) have found human atheromatous plaques to contain a mixture of
racemic 9 and 13-hydroperoxy-hydroxy- and keto-acids derived from cholesteryl linoleate. These compounds resemble the products which occur
after auto-oxidation of cholesteryl linoleate. When linoleate hydroperoxide was administered to rats in a total use of 353 mg by repeated subcutaneous injection, myocardial fibrosis increased the incidence of myocardial fibrosis. If the linoleate hydroperoxide was administered
subcutaneously to rabbits in doses of 700–1000 mg, both myocardial fibrosis and aortic lesions were produced. Neither linoleate nor the degradation products of linoleate hydroperoxide produced either of these
lesions in rabbits when given by repeated subcutaneous injections in a
total dose of 750 mg.

Bieri *et al.* (1961) have found that ethoxyquin, D,L-tocopheryl acetate
and selenium dioxide significantly inhibited air autooxidation in the liver,
kidney and the heart; the effect varied with different vitamin E-free diets.

Litvitskii *et al.* (1981) have observed that treatment with sodium selenite decreased the activation of myocardial free radical lipid oxidation and inhibited the depression of cardiac contractile function observed in rats during myocardial ischemia and subsequent reperfusion. These results may be of importance for clinical practice since transitory myocardial ischemia in animals represents an experimental model of the forms of human coronary heart disease which are accompanied by transient reduction of the coronary blood flow.

Koobs *et al.* (1978) have examined human myocardium from necropsies and biopsy specimens and found evidence that mitochondria can be transformed into granules of lipofuscin. Lipofuscin has been shown to be a yellowish-brown pigment that accumulates linearly with age around the nuclei of myocardial fibers (Strehler *et al.*, 1959). The pigment is generally associated with the mitochondria, and electron micrographs or dorsal root ganglia from aged mice suggest that mitochondria give rise to the pigment (Duncan *et al.*, 1960). Lipofuscin formation from isolated mitochondria as a function of lipid peroxidation could be prevented by adding an antioxidant to the incubation medium (Chio *et al.*, 1969).

Lipofuscin has been demonstrated to be a mixture of polyunsaturated lipids and protein in a random manner that involves Schiff-base formation (Hendley, 1963). Lipofuscin is thought to be formed by the reaction of malonaldehyde, a product of lipid peroxidation, with primary amino groups of proteins, phospholipids, or nucleic acids. Lipofuscin appears to be chemically similar and has the same fluorescence whether the pigment comes from the heart or other tissues. The vitamin E content of the diet has been inversely correlated with the formation of lipofuscin in experimental animals (Tappel, 1972). Koobs *et al.* (1978) have postulated that malonaldehyde which can react with DNA (Reiss *et al.*, 1972) may block template activity. This type of nuclear damage might reduce the capacity for protein synthesis and limit mitochondrial and contractile protein replacement. This type of damage might contribute to heart failure during stress. Peroxidative damage to the myocadium appears to be cumulative and irreversible. Anderson (1973) has postulated that coronary risk factors are related to developments in food processing that have disturbed the normal ratio between unsaturated fatty acids and biologically effective antioxidants in the human diet.

Revis and Armstead (1979) have studied the effect of selenium and vitamin E on the lipid peroxidation and myocardial necrosis induced by isoprenaline treatment or coronary artery ligation in rats. The degree of myocardial damage or necrosis as measured by creatine phosphokinase or lactic acid dehydrogenase activity was reduced by selenium and vitamin E, indicating that lipid peroxidation of myocardial membranes plays an important role in myocardial necrosis. Doxorubicin (adriamycin) is an antibiotic used in the treatment of several types of cancers, but its use has been limited owing to cardiotoxic effects, which include myocardial necrosis and fibrosis (Young, 1975). The mechanism responsible for these effects are not known. Meyers *et al.* (1976a) have suggested that the cardiotoxic effects of doxorubicin may be the result of membrane lipid peroxidation as it has been demonstrated that doxorubicin induces lipid peroxidation in the hearts of the mice (Meyers *et al.*, 1976). Revis and Marusic (1978) have found that both the glutathione peroxidase and the selenium concentration are significantly reduced in the hearts of doxo-rubicin rabbits.

Van Vleet *et al.* (1980; 1981) have tested adriamycin on rabbits, pigs and dogs with and without vitamin E–selenium supplementation.The re-sults were mixed with vitamin E–selenium treatments decreasing the incidence and severity of cardiomyopathy and decreasing the cumulative mortality in rabbits (1978). No effect of the vitamin E–selenium supple-ments was observed in adriamycin-treated dogs (Van Vleet, 1980). In pigs the vitamin E–selenium supplements only reduced adriamycin-induced cardiomyopathy in one of three experiments (Van Vleet, 1981). Van Vleet *et al.* (1977) have administered vitamin E–selenium to pigs before they were given isoproterenol. The pigs lived longer, but the vitamin E–selenium did not protect against increases in plasma enzymic activity or myocardial damage.

Matthes *et al.* (1980) have found that both contractility and membrane potential were suitable viability assays of cryopreservation from rat heart muscle tissue tissue in dimethyl sulfoxide. Pretreatment of donor animals with sodium selenite improves the survival of cryopreserved auricle frag-ments as measured by the contractility and membrane potential assays. Seleno-L-methionine pretreatment of male albino rats improves the cry-oprotection of the heart muscle and also reduces the amount of malon-aldehyde (Mathes *et al.*, 1981). Pories *et al.* (1979) have studied the min-eral metabolism of healing arterial walls by measuring the accumulation of isotopes of chromium, iron, manganese, selenium, strontium, or zinc at the site of vascular repair in rats. There were marked statistically significant difference in the preferential accumulation of several of the radioisotopes in healing compared with normal aorta. Zinc appeared to be the element most involved in vascular repair, followed by selenium and chromium.

8.6.2 Humans

There have been several epidemiological studies showing an inverse relationship between environmental levels of selenium and coronary heart

disease in humans. In addition, there have been reports of severe congestive heart disease in Chinese children who live in severely selenium-
deficient areas of China. This type of congestive heart disease has been
named Keshan disease.

8.6.2.1 Epidemiological Relationship

Shamberger et al. (1975) have compared the selenium content of
various forage crops in the United States to the male and female 54–64
age-specific death rates per 100,000 for cardiovascular–renal, cerebrovascular, coronary, and hypertensive heart disease and found they were
significantly lower in the high-selenium forage area than the low-selenium
forage area. States with occasionally large amounts of selenium in the
drinking water such as Colorado, North and South Dakota, and Utah had
the lowest hypertensive death rate. This observation may be due to selenium in the drinking water or could be due to other unrecognized environmental factors.

When nine types of heart mortality were compared in 17 matched
cities located in high- and low-selenium areas, there were significantly
lower death rates due to arteriosclerotic and hypertensive heart disease
for both males and females from cities located in the high-selenium areas
(Shamberger et al., 1975).

In another epidemiological study estimated intakes of selenium, cadmium, zinc, copper, chromium, arsenic, and manganese in 25 countries
were related to the 1973 age-specific mortality from hypertensive heart
disease, cerebrovascular heart disease, and diseases of the arteries and
capillaries in those 25 countries (Shamberger et al., 1978). There was a
negative significant correlation coefficient observed with selenium and
ischemic heart disease and a positive correlation was seen with cadmium
and ischemic heart disease. Both cadmium and selenium compete for the
same sulfhydryl binding site and could be interrelated. The Cd/Se ratio
was even more significant when related to ischemia heart disease than
either cadmium or selenium alone. The greater significance may indicate
some types of interaction between cadmium and selenium. Japan and
Hong Kong, which were not included in the study with the 25 countries
because of racial differences, had low Cd/Se ratios and low incidences
of ischemic heart disease. In kidney autopsy specimens, Shamberger et
al. (1977) observed no difference in the Cd/Se tissue levels from those
who died from heart disease and those who died from other causes. There
is a possibility that kidney cadmium and selenium determinations may
not reflect changes that might have occurred in the heart. Voors et al.
(1978) have done a similar study on autopsy kidneys but related heart
weight as a degree of heart congestion to the kidney quotient of Cd/Se

and Zn. A positive correlation was observed, indicating that cadmium may increase cardiac congestion and selenium and zinc may decrease cardiac congestion.

Shamberger *et al.* (1979b) observed in another epidemiological experiment that the male and female 55–64 age-specific death rates per 100,000 for cardiovascular–renal, coronary, and hypertensive heart disease had significant inverse correlations with the mean blood bank selenium concentrations in 19 states. In addition, a positive Chi-square association was observed between blood bank selenium and forage crop selenium.

Kurkela and Jaakola (1980) have measured selenium in the water, potatoes, carrots, and other crops from two regions of Finland with relatively low and high rates of heart disease mortality. In general, an inverse association was observed between environmental selenium and heart disease mortality. Jaakola *et al.* (1980) have administrated selenium as 2 mg sodium selenate, and 140 mg of vitamin E twice daily to 32 patients with cardiac pain which was resistant to conventional therapy for several years. They found a statistically significant improvement in all parameters evaluated before the beginning of the trial and at the end of the follow-up periods. These included monthly consumption of nitroglycerin, maximal walking distance without cardiac pain, and working capacity. In spite of the high levels of selenium, toxic or side effects of selenium were not observed during the three-year trial. Villalon (1974) has observed that patients taking selenium–vitamin E capsules (Tolsem) have shown reduced anginal pain in 22 out of 24 cases.

Thomson *et al.* (1978) have measured the selenium concentrations and glutathione peroxidase activities in the blood of hypertensive patients from New Zealand. There were no significant differences in the blood selenium concentrations and glutathione peroxidase activities among four groups of hypertensive patients when compared to the controls. Because both the hypertensives and the controls resided in New Zealand, all of the patients would have relatively low selenium levels which might make it difficult to observe an effect by selenium.

8.6.2.2 Keshan Disease

Keshan disease is an endemic cardiomyopathy of unknown cause in China. The main pathological feature of this cardiomyopathy, which was first observed in 1935, is multiple focal myocardial necrosis scattered throughout the heart muscle with different degrees of cell infiltration and various stages of fibrosis. There are four types depending upon the severity: acute, subacute, chronic, and latent. With the exception of the latent type, mortality is usually high from congestive heart failure. The most susceptible people are children less than 15 years of age and women of child-bearing age in the northern provinces and children of 2–7 years of age in the southern affected areas.

Encouraging results were obtained in trials using sodium selenite to prevent the disease by researchers at Sian Medical College in 1965 and the Chinese Academy of Medical Sciences in 1969. This led their researchers to conduct a four-year clinical trial conducted by the Keshan Disease Research Group of the Chinese Academy of Medical Sciences. In 1974 the preventative effect of sodium selenite was tested in 119 production teams of three people's communes and was extended in 1975 to 169 teams of four communes in Mianing county, Sichuan province. Children received sodium selenite tablets or placebo once a week, the dosage being: 1–5 years, 0.5 mg; and 6–9 years, 1.0 mg. Of the 36,603 selenium-treated children, 21 cases occurred during the four years of investigation. Of these, only three died and one became chronically ill up to the end of 1977; whereas in the control group of the 9642 children, 107 cases were identified of which 53 were fatal and six had insufficient heart function (Chen et al., 1980). The Keshan Research Group also measured the selenium content of scalp hair, blood and urine of the residents, as well as the staple cereals grown locally in both the affected and the nonaffected areas. They found that all affected areas were invariably very low in selenium. The dose relationship between selenium and the regional characteristics of Keshan disease suggests that it is probably a biogeochemical disease. These results may well establish that selenium is a necessary trace element for human nutrition.

Two cases of cardiomyopathy due to selenium deficiency have been identified in two patients. A 43-year-old man received parenteral alimentation for two years (Johnson et al., 1981). The hyperalimenation solutions were almost devoid of selenium. Erythrocyte selenium and glutatione peroxidase were only about 10% of the normal values. The patient exhibited ventricular fibrillation, frequent ventricular extrasystoles, as well as bursts of nonsustained ventricular tachycardia. A multigated cardiac blood-pool scan showed that the left ventricle was greatly dilated and the left ventricular ejection fraction was greatly depressed. Collipp and Chen (1981) have described a two-year-old Black girl who had typical features of Keshan disease. She was admitted to the hospital with dyspnea, tachycardia, cardiomegaly, and congestive heart failure. She was considered to have myocarditis. An electrocardiogram demonstrated inverted T waves in Leads 2, 3 and aVF and flat T waves over the entire precordium. An echocardiogram showed pericardial effusion and left ventricular hypertrophy. Blood determinations of selenium indicated a level of 3.5 μg/dl (range in 40 normal children was 7 to 16 μg/dl. After four weeks of selenium (sodium selenite) supplementation (2 mg per day by mouth), the serum selenium concentration rose to 15 μg/dl and the supplements were stopped. The patient improved steadily at home, although she still had cardiomegaly three months after discharge. Her diet before the onset of this illness was low in selenium. It consisted of grits, sausage, and beans for breakfast, a frankfurter and beans at lunch, and pork and beans with rice for supper. Her only beverages were water or Kool-Aid. Her estimated selenium intake was only 10 μg/day.

King *et al.* (1981) has studied whether supplements of selenium will reverse the abnormal deficiency state. Van Rij *et al.* (1979) have previously reported little change in the selenium level and glutathione peroxidase activity in erythrocytes after 24 days of intravenous supplementation in a patient with selenium deficiency. After total parenteral nutrition for 14 months a 39-year-old man had selenium levels of 0.042 ng/mg hemoglobin (normal 0.7) and erythrocyte glutathione peroxidase (EGP) activity of 2.86 IU/g of hemoglobin (normal 17.5 IU/g). Parenteral nutrition solutions have been shown to be very low in selenium (Phillips and Garnys, 1981; Van Rij *et al.*, 1974). The patient was supplemented for the next 12 days; selenium was given orally (100 mg four times daily for one week, then 25 mg four times daily). The EGP activity increased to 9.5 IU/g hemoglobin. After oral supplementation for five months, both the erythrocyte selenium concentration (0.6 ng/mg) and the EGP activity (20.4 IU/g hemoglobin) had returned to normal.

Cross *et al.* (1981) have measured the selenium concentration in healthy and atheromatous human aorta and found no significant differences. There was also no relation between age and concentration in the aorta. There were also no significant differences in the levels of selenium in the intima, media, and the advenita layers of the aorta.

8.7 Selenium and Aging

Lipofuscin pigments (LP) have been commonly referred to as age pigments, since that pigment concentration increases in cells with age. These pigments are believed to be Schiff-base condensation products of malonaldehyde, derivated from peroxidized unsaturated fat, with the proteins and nucleic acids. Several investigators have reported that high levels of LP accumulate in vitamin E deficiency in certain tissues and lower levels of LP result when dietary levels of vitamin E are increased. Because of the known interactions between vitamin E and selenium, it is possible that selenium has a similar effect.

In addition, there is a decline in both the humoral and cell-mediated immune responses with age (Kay, 1976). Harmen *et al.* (1977) have observed that vitamin E has enhanced humoral immune responses in

C3HeB/FeJ female mice up to 88 weeks. Selenium has also previously been shown to enhance humoral response (see Section 8.3.1). If the immune response has an effect on aging, then selenium as well as vitamin E may reduce free radical reactions, which could contribute to the decline of the immune response with age.

Burch *et al.* (1979) have examined the selenium content of various rat organs during the first 120 days of life. There was a general increase in selenium content of the kidney, liver, and the lung. No increases were observed in the brain and skin and hair. Nakaidze *et al.* (1979) have studied the effect of selenurea on the life span of *Drosophila melanogaster*. Selenourea added to the nutrient medium extended the life span by 9%.

8.8 Cystic Fibrosis

Cystic fibrosis (CF) is the most prevalent inheritable chronic disease in Caucasian children. It is present in one of 2000 births with a gene frequency of one in 20, and is inherited as an autosomal recessive disorder. The disease is characterized by elevated sodium in sweat electrolytes and the precipitation of glycoproteins present in mucous membranes. The precipitates block the tubular passages from the exocrine glands causing secondary complications such as pulmonary disease, pancreatic insufficiency, and malabsorption. About 50% of the patients do not survive beyond the age of 21 years.

Instead of a genetically inherited disease Wallach and Garmaise (1979) have suggested that CF is an acquired nutritional disease. This suggestion arose from an autopsy of a rhesus monkey which had died from unknown causes, but histologic lesions of the lungs and pancreas resembled those found in patients dying of this disease and ultrastructural lesions resembled those of selenium deficiency. Linking these two observations, he suggested that CF is primarily a consequence of a prenatal and perinatal dietary deficiency of selenium complicated by further deficiencies in zinc, copper, riboflavin, and vitamin E and aggravated by an excess of polyunsaturated fatty acids.

A useful approach to the further investigation would be an evaluation of the selenium status of CF subjects. Lloyd-Still and Ganther (1980) have collected blood samples from 20 infants and children with CF and compared the selenium levels to those found in 19 controls. Nine control female and ten males averaged 22.8 $\pm$ 0.8 and 21.8 $\pm$ 0.7 mg/dl. In contrast, the mean value for all 20 CF patients was 12.2 $\pm$ 2.5 (males 11.6 $\pm$ 2.2; females 13.2 $\pm$ 2.8). Glutathione peroxidase levels were somewhat greater in the CF patients. Castillo *et al.* (1981) have measured serum and red cell selenium as well as glutathione peroxidase activity in 32 CF patients and 21 controls. Both the red cell selenium and the glutathione peroxidase activity were similar in the CF patients and the controls. Serum selenium was about 15% lower in the CF patients and significantly lower in another group of 8 CF patients who also had a vitamin E deficiency. The decrease of selenium in the CF patients may be due to a decreased absorption of selenium. Heinrich *et al.* (1977) have observed that children with CF absorbed 60% and 71% of the ^{75}Se from an oral dose of 200 g ^{75}Se-selenomethionine labeled pork, whereas normal adults absorbed 76%–100%. Pancreatin did not improve the bioavailability of ^{75}Se-selenomethionine from labeled pork in cystic fibrosis patients.

8.9 Multiple Sclerosis

High-frequency areas are northern and central Europe into the USSR, Southern Canada, northern United States, New Zealand, and southern Australia (Kurtzke, 1980). From a large nationwide case-control series, the division between high and medium in the United States is 37° north latitude across most of the country but extending to some 39° in the east.

Wikstrom *et al.* (1976) have found whole-blood selenium content to be less in patients with multiple sclerosis (MS) than in controls in high-MS-risk areas of Finland. In the low-MS-risk areas there were no differences in blood selenium levels. Shukla *et al.* (1977) have found significant decrease in the activity of glutathione peroxidase in the erythrocytes of 24 patients with MS. There was no significant correlation between age-adjusted mortality from MS and the United States forage crop selenium, but there seemed an exciting relationship between the prevalence rates and the forage crop selenium. The highest selenium serum values have been found in Lapps from the northern part of Finland where the prevalence of MS is low (Shukla *et al.*, 1977). There is a possiblity that disturbed lipid peroxidation has a role in the pathogenesis of (MS) and that selenium and vitamin E are involved.

8.10 Cataracts

Selenium seems to have a bimodal effect on cataracts depending upon the concentration. Selenium deficiency has been shown to cause lens cataracts (Whanger and Weswig, 1975). On the other hand, Ostadalova *et al.* (1977) observed a single injection of sodium selenite on day 10 postpartum produced nuclear cataracts. White, opaque cataracts were grossly observable in the eyes of rat pups receiving sodium selenite at levels of 0.25 mg Se/kg body weight and above (Shearer *et al.*, 1980). Selenomethionine at comparable levels did not induce cataracts in the same experiment.

Bhuyan *et al.* (1981a) have observed increased lipid peroxidation (malondialdehyde) in human senile cortical and nuclear cataracts. They believe that toxic metabolites of oxygen are involved in mediating peroxidation of polyunsaturated fatty acyl residues of phospholipids in the lipid biolayers of the plasma membrane, thus causing membrane damage in cataracts. In another experiment Bhuyan *et al.* (1981b) also observed an increase of H_2O_2 in the aqueous humor of rats with selenium induced cataract. Abdullaev *et al.* (1981) have reported that the addition of 0.1% Se (as Na_2SeO_3) to a perfusion liquid caused and initial increase of a and b waves in the electroretinogram of frogs with subsequent reduction in the b wave.

8.11 Other Diseases

Shukla *et al.* (1978) have found that there is about a 38% decrease of glutathione peroxidase in patients with Batten's disease. This disease is a neurological disorder with a recessive genetic trait whose initial symptom is impairment of vision (retinitis pigmentosa) at about four years of age, followed later by seizures, progressive mental deterioration, motor regression, blindness, and cerebellar dysfunction, which leads to death between 15 to 30 years of age (Jensen *et al.*, 1977). The fatty acid patterns of serum lipids in this disease may reflect a primary or secondary deficiency in an essential fatty acid (linoleic acid) probably related to a glutathione peroxidase abnormality.

Westermarck (1977) has found low blood selenium levels in inhabitants of Finland, but even lower blood selenium values were seen in patients with acrodermatitis enteropathica, Duchenne muscular dystrophy, infantile and juvenile type of neuronal ceroid lipofuscinosis (NCL), severe mental retardation, and myocardial infarction. In addition, the vitamin E level of serum in patients with NCL as well as in subjects with severe mental retardation were low compared to normals. Sodium selenite supplementation in patients with NCL produced a transitory improvement with no toxic effects during one year of administration.

Matsumoto *et al.* (1981) have examined serum vitamin E, lipid peroxide, and glutathione peroxidase in patients with chronic pancreatitis.

Both serum vitamin E concentrations and glutathione peroxidase activities were depressed, especially in patients with chronic calcifying pancreatitis. On the other hand, serum lipid peroxide levels were elevated. The authors suggest than an elevation of the serum lipid peroxide level may be due to the lack of an antioxidative defense mechanism.

Rhead and Schneider (1976) found four times more selenium in kidney samples from children with cystinosis than do kidney samples from normals. When cultured skin fibroblasts from cystinotic patients and normal control individuals were incubated with Se-methionine, Se-cystine, Se-cystamine, Se-urea, selenite, or in the medium without added selenium, only the cystinotic fibroblasts grown in Se-urea, or selenite, contained more selenium than do the normal cells. Cystine metabolism of cystinotic cells appears to be more sensitive to certain selenium compounds than is that of normal fibroblasts. Clinically, nephropathic cystinosis is a rare metabolic disease, characterized by autosomal recessive inheritence and end-stage renal failure in the first decade of life. Perhaps selenium may prevent or reduce cystine uptake by the kidney. Rhead and Schneider (1976) also observed that ^{35}S-cystine incorporation into glutathione is inhibited by Se-cystine in both fibroblast types.

8.12 Radioselenium as a Diagnostic Agent

Blau and Bender (1962) first introduced ^{75}Se-selenomethionine into clinical practice for scintigraphic visualization of the pancreas which selectively takes up ^{75}Se-selenomethionine and incorporates the isotope into the secretory proteins. The technique was then used by numerous other clinicians. However, due to the overlap of the pancreatic scan by the liver shadow there can be false negative and false-positive scans, and the diagnostic reliability of the ^{75}Se-selenomethionine pancreatic scintigraphy is widely debated.

To improve the diagnostic effectiveness of scintigraphy various procedures have been suggested such as a high-protein breakfast associated with intravenous cholecystokinin (CCK PZ), a high-protein meal followed by oral glutamic acid, morphine, glucose, amino acid solutions, secretin, meritene, and (CCK PZ). Several other clinicians found no difference in scan quality with or without additional preparations. Neither the modification of the scanning procedures nor the application of new compounds could further improve the results of pancreatic scintigraphy and the technique is now—because of its poor accuracy—widely abandoned.

Pointer and Kletter (1980) have performed radioselenium–secretin–pancreozymin test on 22 normal subjects and in 23 patients with pancreatic exocrine insufficiency. Considerable overlap could be observed in the radioactivity from samples of duodenal juice from normals/ and patients with pancreatic insufficiency. McColl *et al.* (1979) have measured the radioactivity of parotid juice and duodenal juice following ^{75}Se-selenomethionine administration and a Lundh test meal. Patients with exocrine pancreatic insufficiency, patients with pancreatitis, and normals were tested. In chronic pancreatitis both the parotid and the pancreatic secretion were impaired, but in patients with pancreatic carcinoma only the pancreatic excretion was impaired. The authors suggest that the combined pancreatic/parotid radioselenium test may be useful in differentiating between chronic pancreatitis and pancreatic carcinoma. DiGiulio and Morales (1969) have studied the value of the selenomethionine ^{75}Se scan in localizing parathyroid adenomas. In both studies, only about 60% of the abnormal parathyroid glands were diagnosed by this procedure. Size seemed to be the most important factor contributing to visualization of the parathyroid adenomas. Adenomas larger than 2 g were consistently visualized, while those weighing less than 1 g were commonly not detected by the scan.

^{75}Se-methionine has been used as a label of the platelet population, in order to quantitate the platelet production (Evatt *et al.*, 1976). Thrombocytopoiesis in humans and animals can be quantified by measuring the rate of incorporation of this tracer in the circulating platelets, either when radioactivity is at a maximum or at the 24th hour after its infusion.

^{75}Se-methionine has been utilized as a biosynthetic label of lymphocyte surface antigens which can be precipitated with monoclonal antibodies (Dosseto *et al.*, 1981). The same method can be used to label immunoglobulins produced by hybridomas and to determine the nature of the secreted light chains. Crowley *et al.* (1975) have used ^{75}Se cortisol in a competitive binding assay for cortisol. In addition ^{75}Se-labeled methotrexate has been used as a radiolabel in a radioimmunoassay for methotrexate (Aherne *et al.*, 1978).

References

Abdullaev, G. B., Gasanov, G. G., Ragimov, R. N., Teplyakova, G. V., Mekhtiev, M. A., and Dzhafarov, A. I., 1973. Selenium and tumor growth under experimental conditions, *Dokl. Akad. Nauk Azerb. SSR*, 29:18–24.
Abdullaev, G. B., Gasanov, G. G., Kulieva, E. M., Dzhafarov, A. I., and Perelygin, B. B., 1981. Effect of selenium on the electroretinogram of isolated retina of cold-blooded animals, *Dokl. Akad. Nauk Az.* SSR 37:29–34.
Adams, G., Martin, S., and Milner, J., 1980. Effects of selenium on the Salmonella/microsome mutagen test system, *Fed. Proc. Fed. Am. Soc. Exp. Biol.* 39:790.

Aherne, W., Piall, E., and Marks, V., 1978. Radioimmunoassay of methotrexate: Use of ^{75}Se-labelled methotrexate, *Ann. Clin. Biochem.* 15:331–334.

Ahmed, Z. U., and Walker, G. W. R., 1975. The effects of urethane, sodium monohydrogen arsenate and selenocystine on crossing-over in *Drosophilia melanogaster, Can. J. Genet. Cytol.* 17:55–66.

Aleksondrovicz, J., 1977. Effects of food enrichment with various doses of sodium selenite on some immune response in laboratory animals, *Rocz. Nauk. Zootech.* 4:113–126.

Anderson, T. W., 1973. Nutritional muscular dystrophy and human myocardial infarction, *Lancet* 2:298–302.

Balanski, R. M. and Hadsiolov, D. H., 1979. Influence of sodium selenite on the hepato-carcinogenic action of diethylnitrosamine in rats, *Comptes Rendus de l'Academie Bulgare Des Sciences*, 32:697–698.

Banner, W. P., Tan, Q. H., and Zadeck, S., 1981. Studies of the mechanism by which selenium inhibits tumor formation, *Proc. Am. Assoc. Cancer Res.* 22:115.

Bawden, J. W., and Hammarstrom, L. E., 1977. Autoradiography of selenium-75 in developing rat teeth and bone, *Caries Res.* 11:195–203.

Berenshtein, T. F., 1972. Effect of selenium and vitamin E on antibody formation in rabbits, *Zdrawookhr Boloruss* 18:34–41.

Berenshtein T. F., 1973. Stimulation of a nonspecific immunity in immunized rabbits by sodium selenite, *Ser. Biyal Navek* 1:87–89.

Behrenshtein, T. F., 1975. Change in the immunological response of vaccinated rabbits following the administration of selenium and vitamin A, *Selen. Biol. Mater. Nauchen. Konf.*, 2nd, 1:94–96.

Berenshtein, T. F., and Zdravoukhi, 1976. Stimulation of the activity of nonspecific immunity factors by biological bases of molybdenum and selenium, *Belowes* 3:67–68.

Bhuyan, K. C., Bhuyan, D. K., and Podos, S. M., 1981a. Evidence of increased lipid peroxidation in cataracts, IRCS Medical Science, *Pathology* 9:126–127.

Bhuyan, K. C., Bhuyan, D. K., and Podos, S. M., 1981b. Cataract induced by selenium in rat: II. Increased lipid peroxidation and impairment of enzymatic defense against oxidative damage, IRCS Medical Science, *Pharmacology* 9:195–196.

Bieri, J. G., Dam, H., Prange, I., and Sondergaard, E., 1961. Effect of dietary selenium dioxide, cystine, ethoxyquin and vitamin E on lipid autoxidation in chick tissues, *Acta Physiol. Scand.* 52:36–43.

Blau, M., and Bender, M. A., 1962. ^{75}Se-selenomethionine for visualization of the pancreas by isotope scanning, *Radiology* 78:974–979.

Bogden, J. B., Kemp, F. W., Buse, M., Thind, I. S., Louria, D. B., Forgacs, J., Llanos, G., and Terrones, I. M., 1981. Composition of tobaccos from countries with high and low incidences of lung cancer. I. Selenium, Polonium-210, *Alternaria*, tar and nicotine, *J. Nat. Cancer Inst.* 66:27–31.

Bowen, W. H., 1972. The effect of selenium and vanadium on caries activity in monkeys (*Macaca irus*), *J. Irish Dent. Assoc.* 18:83–89.

Boyne, R., and Arthur, J. R., 1979. Alterations of neutrophil functions in selenium deficient cattle, *J. Comp. Path.* 89:151–158.

Bradley, C. J., Kledzik, G. S., and Meites, J., 1976. Prolactin and estrogen dependency of rat mammary cancers at early and late stages of development, *Cancer Res.* 36:319–324.

Britton, J. L., Shearer, T. R., and DeSart, D. J., 1980. Cariostasis by moderate doses of selenium in the rat model, *Arch. Envir. Health*, 35:74–76.

Broghamer, W. L., McConnell, K. P., and Blotcky, A. J., 1976. Relationship between serum selenium levels and patients with carcinoma, *Cancer* 37:1384–1388.

Broghamer, W. L., McConnell, K. P., Grimaldi, M., and Blotcky, A. J., 1978. Serum selenium and reticuloendothelial tumors, *Cancer* 41:1462–1466.

Burch, R. E., Sullivan, J. F., Jetton, M. M., and Hahn, H. K. J., 1979. The effect of aging on trace element content of various rat tissues: 1. Early stages of aging, *Age* 2:103–107.

Buttner, W., 1963. Action of trace elements on the metabolism of fluoride, *J. Dent. Res.* 42:453–460.

Calautti, P., Moschini, G., Stievano, B. M., Tomio, L., Calzavara, F., and Perona, G., 1980. Serum selenium levels in malignant lymphoproliferative diseases, *Scand. J. Haematol.* 24:63–66.

Capel, I. D., and Williams, D. C., 1979. Selenium and glutathione peroxidase in breast cancer, *Obstet. Gynecol.* 7:425.

Castillo, R., Landon, C., Eckhardt, K., Morris, V., Levander, O., and Lewiston, N., 1981. Selenium and vitamin E status in cystic fibrosis, *J. Pediat.* 99:583–585.

Chen, X., Yang, G., Chen, J., Chen, X., Wen, Z., and Ge, K., 1980. Studies on the relations of selenium and Keshan disease, *Biol. Trace Elem. Res.* 2:91–107.

Chio, K. S., Reiss, U., Fletcher, B., and Tappel, A. L., 1969. Peroxidation of subcellular organelles: Formation of lipofuscin like fluorescent pigments, *Science* 166:1535–1536.

Clayton, C. C., and Baumann, C. A., 1949. Diet and azo dye tumors: Effect of diet during a period when the dye is not fed, *Cancer Res.* 9:575–582.

Collipp, P. J., and Chen, S. Y., 1981. Cardiomyopathy and selenium deficiency in a two-year-old girl, *New Eng. J. Med.* 304:1304–1305.

Crisp, F. D., Deaton, T. G., and Bawden, J. W., 1979. *In vitro* study of selenium-75 distribution in developing rat molar enamel, *Caries Res.* 13:313–318.

Cross, J. D., Rale, R. M., and Smith, H., 1981. Bromine and selenium in human aorta, *J. Clin. Path.* 34:393–395.

Crowley, M. F., Garbien, K. J. T., and Tuttlebee, J. W., 1975. Assessment of a simple competitive protein-binding technique for plasma cortisol assay, *Ann. Clin. Biochem.* 12:66–69.

Curzon, M. E. J., and Crocker, D. C., 1978. Relationship of trace elements in human tooth enamel to dental caries, *Arch. Oral Biol.* 23:647–653.

Cutler, M. G., and Schneider, R., 1974. Linoleate oxidation products and cardiovascular lesions, *Atherosclerosis* 20:383–394.

Daoud, A. H., and Griffin, A. C., 1980. Effect of retinoic acid, butylated hydroxytoluene, selenium and sorbic acid on azo-dye hepatocarcinogenesis, *Cancer Lett.* 9:299–304.

Desowitz, R. S., and Barnwell, J. W., 1980. Effect of dimethyl dioctadecyl ammonium bromide on the vaccine-induced immunity of Swiss-Webster mice against malaria (Plasmodium berghei), *Infect. Immun.* 27:87–89.

Dickson, R. C., and Tomlinson, R. H., 1967. Selenium in blood and human tissues, *Clin. Chem. Acta* 16:311–321.

DiGiulio, W., and Morales, J. O., 1969. The value of the selenomethionine Se[75] scan in preoperative localization of parathyroid adenomas, *JAMA* 209:1873–1880.

Dosseto, M., Rohner, C., Pierres, M., and Gordis, C., 1981. Biosynthetic incorporation of {[75]Se} selenomethionine: A new method for labelling, lymphocyte membrane antigens, *J. Immunol. Meth.* 41:145–153.

Duncan, D., Nall, D., and Morales, R., 1960. Observations on the fine structure of old age pigment, *J. Gerontol.* 15:366–372.

Dzhioev, F. D., 1978. Effect of selenium, phenobarbital, teturam, and carbon tetrachloride on the carcinogenic effect of diethylnitrosamine and 1,2-dimethylhydrazine, in *Kantserog N-Nitrozonsoedin: Deistvie, Obraz., Opred., Mater Simp.*, 3rd ed., G. O. Loogna (ed.), Tallinn, USSR, pp. 51–53.

Evatt, B. L., Spivak, J. L., and Levin, J., 1976. Relationships between thrombopoiesis and erythropoiesis: With studies of the effects of preparations of thrombopoietin and erythropoietin, *Blood* 48:547–558.

Exon, J. H., Koller, L. D., and Elliott, S. C., 1976. Effect of dietary selenium on tumor induction by an oncogenic virus, *Clin. Toxicol.* 9:273–279.

Gillett, A. S., and Wakeley, C. P. G., 1922. Selenium in the treatment of malignant disease, *Brit. J. Surg.* 9:532–539.

Glavind, J., Hartmann, S., Clemmesen, J., Jessen, K. E., and Dam, H., 1952. Studies on the role of lipoperoxides in human pathology, Part 2. The presence of peroxidized lipids in the atherosclerotic aorta, *Acta Pathol. Microbiol. Scand.* 30:1–6.

Godwin, K. O., 1965. Abnormal electrocardiograms in rats fed a low selenium diet *Q. J. Exp. Physiol.* 50:282–288.

Godwin, K. O., and Fraser, F. J., 1966. Abnormal electrocardiograms, blood pressure changes and some aspects of the histopathology of selenium deficiency in lambs *Q. J. Exp. Physiol.* 51:94–102.

Greeder, G. A., and Milner, J. A., 1980. Factors influencing the inhibitory effect of selenium on mice inoculated with Ehrlich ascites tumor cells, *Science* 209:825–827.

Griffin, A. C., and Jacobs, M. M., 1977. Effects of selenium on azo dye hepatocarcinogenesis, *Cancer Lett.* 3:177–181.

Guo, J., Li, G., Zhao, W., Tang, F., Zhang, X., and Zhang, C., 1981. Cardiovascular effects of sodium selenite, *Chung kuo Yao Li Hsueh Pao* 2:93–97.

Hadjimarkos, D. M., 1956. Geographic variations of dental caries in Oregon, *J. Pediatr.* 48:195–201.

Hadjimarkos, D. M., 1968. Effect of trace elements on dental caries, in *Advances in Oral Biology*, Vol. 3, P. H. Staple (ed.), Academic Press, New York, pp. 253–292.

Hadjimarkos, D. M., and Bonhorst, C. W., 1958. The trace element selenium and its influence on dental caries susceptibility, *J. Pediatr.* 52:274–278.

Hadjimarkos, D. M., and Bonhurst, C. W., 1959. Selenium content of human teeth, *Oral Surg.* 12:112–116.

Hadjimarkos, D. M., and Bonhorst, C. W., 1961. The selenium content of eggs, milk and water in relation to dental caries in children, *J. Pediatr.* 59:265–259.

Harbach, P. R., and Swenberg, J. A., 1981. Effects of selenium on 1,2-dimethylhydrazine metabolism and DNA alkylation, *Carcinogenesis* 2:575–580.

Harland, W. A., Gilbert, J. D., and Brooks, C. J. W., 1973. Lipids of human atheroma, Part 8. Oxidized derivatives of cholesteryl lineoleate, *Biochem. Biophys. Acta* 316:378–385.

Harman, D., Heidrick, M. L., and Eddy, D. E., 1977. Free radical theory of aging: Effect of free-radical inhibitors on the immune response, *J. Am. Ger. Soc.* 25:400–407.

Harr, J. R., Bone, J. F., Tinsley, I. J., Weswig, P. H., and Yamamoto, R. S., 1967. Selenium toxicity in rats. II. Histopathology, in Symposium: *Selenium in Biomedicine*, O. H. Muth (ed.), Avi Publishing Company, Westport, Connecticut, pp. 153–178.

Harr, J. R., Exon, J. H., Whanger, P. D., and Weswig, P. H., 1972. Effect of dietary selenium on N-2 fluorenyl-acetamide (FAA)-induced cancer in vitamin E supplemented, selenium depleted rats, *Clin. Toxicol.* 5:187–194.

Harr, J. R., Exon, J. H., Weswig, P. H., and Whanger, P. D., 1973. Relationship of dietary selenium concentration; chemical cancer induction; and tissue concentration of selenium in rats, *Clin. Toxicol.* 8:487–495.

Hendley, D. D., Mildvan, A. S., Reporter, M. C., and Strehler, B. L., 1963. The properties of human cardiac age pigment. II. Chemical and enzymatic properties, *J. Gerontol.* 18:250–259.

Heinrich, H. C., Gabbe, E. E., Bartels, H., Oppitz, K. H., Bender-Gotze, C., and Pfau, A. A., 1977. Bioavailability of food iron-(^{59}Fe), vitamin B_{12}-(^{60}Co) and protein bound selenomethionine-(^{75}Se) in pancreatic exocrine insufficiency due to cystic fibrosis, *Klin. Wschr.* 55:595–601.

Hilliyard, L. A., and Abraham, S., 1979. Effect of dietary polyunsaturated fatty acids on growth of mammary adenocarcinomas in mice and rats, *Cancer Res.* 39:4430–4437.

Hilse, H., Oehme, P., Krause, W., and Hecht, K., 1979. Effect of sodium selenite on experimental hypertension in rat, *Bulg. Acad. Sci.* 5:47–50.

Hsieh, H. S., and Ganther, H. E., 1975. Acid-volatile selenium formation catalyzed by glutathione reductase, *Biochemistry* 14:1632–1636.

Ip, C., 1981a. Factors influencing the anticarcinogenic efficacy of selenium in dimethylbenzanthracene-induced mammary tumorigenesis in rats, *Cancer Res.* 41:2683–2686.

Ip, C., 1981b. Prophylaxis of mammary neoplasia by selenium supplementation in the initiation and promotion phases of chemical carcinogenesis, *Cancer Res.* 41:4386–4390.

Ip, C., 1981c. Modification of mammary carcinogenesis and tissue peroxidation by selenium deficiency and dietary fat, *Nutr. Cancer*, 2:136–142.

Ip, C., and Ip, M. M., 1981. Chemoprevention of mammary tumorigenesis by a combine regimen of selenium and vitamin A, *Carcinogenesis* 2:915–918.

Ip, C., and Sinha, D. K., 1981. Enhancement of mammary tumorigenesis by dietary selenium deficiency in rats with a high polyunsaturated diet, *Cancer Res.* 41:31–34.

Ip, C., Yip, P., and Bernardis, L. L., 1980. Role of prolactin in the promotion of dimethylbenzanthracene-induced mammary tumors by dietary fat, *Cancer Res.* 40:374–378.

Ip, C., Ip, M. M., and Kim, U., 1981. Dietary selenium intake and growth of the MT-W9B transplantable rat mammary tumor, *Cancer Lett.* 14:101–107.

Jaakkola, K., Kurkela P., Arstila, A. U., and Larni, H. M., 1981. Selenium and degenerative heart disease. Part II. Clinical studies. Selenium and vitamin E in the treatment of cardiac pain and diseases, in Mineral Elements 80. A Nordic Symposium on Soil–Plant–Animal–Man Interrelationships and Implications to Human Health, Helsinki, Finland, December 9–11, 1980, pp. 225–233.

Jacobs, M. M., Matney, T. S., and Griffin, A. C., 1977. Inhibitory effects of selenium on the mutagenicity of 2-acetylaminofluorene (AAF) and AAF metabolites *Cancer Lett.* 2:319–322.

Jacobs, M. M., Jansson, B., and Griffin, A. C., 1977. Inhibitory effects of selenium on 1,2-dimethylhydrazine and methylazoxymethanol acetate induction of colon tumors, *Cancer Lett.* 2:133–138.

Jacobs, M. M., Shubik, P., and Feldman, R., 1980. Influence of selenium on vascularization in the hamster cheek pouch, *Cancer Lett.* 9:353–357.

Jacobs, M. M., Forst, C. F., and Beams, F. A., 1981. Biochemical and clinical effects of selenium on dimethylhydrazine-induced colon cancer in rats, *Cancer Res.* 41:4458–4465.

Jansson, B., and Jacobs, M. M., 1976. Selenium—a possible inhibitor of colon and rectum cancer, in *Proceedings of the Symposium on Selenium–Tellurium in the Environment*, Industrial Health Foundation, Pittsburgh, pp. 326–340.

Jensen, E. G., Clausen, J., Melchior, J. C., and Konat, G., 1977. Clinical, social and biochemical studies on Batten's syndrome, *Eur. Neurol.* 15:203–211.

Johnson, J. R., and Shearer, T. R., 1979. Selenium uptake into teeth determined by fluorimetry, *J. Dent. Res.* 58:1836–1839.

Johnson, R. A., Baker, S. S., Fallon, J. T., Maynard, E. P., Ruskin, J. N., Wen, Z., Ge, K., and Cohen, H. J., 1981. An occidental case of cardiomyopathy and selenium deficiency, *New Eng. J. Med.* 304:1210–1212.

Kada, T., Tutikawa, K., and Sadaie, Y., 1972. In vitro and host mediated "rec assay" procedures for screening chemical mutagens; and phloxine, a mutagenic red dye detected, *Mutation Res.* 16:165–174.

Kanematsu, N., Hara, M., and Kada, T., 1980. Rec assay and mutagenicity studies on metal compounds, *Mutation Res.* 77:109–116.

Kay, M. M. B., 1976. Aging and the decline of immune responsiveness, in *Basic and Clinical Immunology*, H. H. Fudenberg (ed.), Lange Medical Publications, Los Altos, California, pp. 267–278.

King, W. W., Michel, L. Wood, W. C., Malt, R. A., Baker, S. S., and Cohen, H. J., 1981. Reversal of selenium deficiency with oral selenium, *New Eng. J. Med.* 304:1304–1305.

Koller, L. D., Issacson-Kerkvliet, N., Exon, J. H., Brauner, J. A., and Patton, N. M., 1979. Synergism of methylmercury and selenium producing enhanced antibody formation in mice, *Arch. Environ. Hlth.* 34:248–251.

Kollmorgan, R. E., Alexander S. S., and King, M., 1979. Inhibition of lymphocyte function in rats fed high-fat diets, *Cancer Res.* 39:3458–3462.

Koobs, D. H., Schultz, R. L., and Jutzy, R. V., 1978. The origin of lipofuscin and possible consequences to the myocardium, *Arch. Pathol. Lab. Med.* 102:66–68.

Kuhnlein, H. V., Levander, O. A., King, J. C., Sutherland, B., and Riskie, L., 1981. Dietary selenium and fecal mutagenicity in young men, *Fed. Proc. Fed. Am. Soc. Exp. Biol.* 40:903.

Kurkela, P., and Jaakkola, K., 1980. Selenium and degenerative heart disease. Part I. Endemic studies. Endemic studies of the etiologic relationship between selenium and myocardial infarction, in Mineral Elements 80, A Nordic Symposium on Soil–Plant–Animal–Man Interrelationships and Implications to Human Health, Helsinki, Finland, December 9–11, 1980, pp. 279–292.

Kurtzke, J. F., 1980. Geographic distribution of multiple sclerosis: An update with special reference to Europe and the Mediterranean region, *Acta Neurol. Scand.* 62:65–80.

Lawson, T., and Birt, D., 1981. BOP induced damage of pancreas DNA and its repair in hamsters pretreated with selenium, *Proc. Am. Assoc. Cancer Res.* 22:93.

Levitt Research Laboratories, 1979. Antiinflammatory compositions containing selenium compounds, U.S. Appl. No. 821, 156, *Chem. Abst.* 91:96638s.

Litvitskii, P. F., Kogan, A. K., Kudrin, A. N., and Lukyanova, L. O., 1981. Pathogenic role of lipid peroxidation activation and the protective effect of sodium selenite during ischemia and myocardial reperfusion, *Byull. Eksp. Biol. Med.* 91:271–274.

Lloyd-Still, J. D., and Ganther, H. E., 1980. Selenium and glutathione peroxidase levels in cystic fibrosis, *Pediatrics* 65:1010–1012.

Lofroth, G. and Ames, B. N., 1978. Mutagenicity of inorganic compounds in *Salmonella typhimurium*: Arsenic, chromium, and selenium, *Mutation Res.* 53:65–66.

Lo, L. W., Koropatnick, J., and Stich, H. F., 1978. The mutagenicity and cytotoxicity of selenite, "activated" selenite and selenate for normal and DNA repair-deficient human fibroblasts, *Mutation Res.* 49:305–312.

Lotliker, P. D., Enomoto, M., Miller, J. A., and Miller, E. C., 1967. Species variations in the *N*- and ring-hydroxylation of 2-acetylaminofluorene and effects of 3-methylcholanthrene pretreatment, *Proc. Soc. Exp. Biol. Med.* 125:341–346.

Lotliker, P. D., Hong, Y. S., and Baldy, W. L., 1978. Effects of 3-methylcholanthrene pretreatment on 2-acetylaminofluorene *N*- and ring-hydroxylation by rat and hamster liver microsomes, *Toxicol. Lett.* 2:135–139.

Ludwig, T. G., and Bibby, B. G., 1969. Geographic variations in the prevalence of dental caries in the United States of America, *Caries Res.* 3:32–43.

Marsh, J. A., Dietert, R. R., and Combs, G. F., 1981. Influence of dietary selenium and vitamin E on the humoral immune response of the chick, *Proc. Soc. Exp. Biol. Med.* 166:228–236.

Marshall, M. V., Arnott, M. S., Jacobs, M. M., and Griffin, A. C., 1979. Selenium effects on the carcinogenicity and metabolism of 2-acetylaminofluorene, *Cancer Lett.* 7:331–338.

Martin, J. L., and Spallholz, J. E., 1977. Selenium in the immune response, in *Proceedings of the Symposium on Selenium–Tellurium in the Environment*, Industrial Health Foundation, Pittsburgh, pp. 204–225.

Martin, S. E., Adams, G. H., Schillaci, M., and Milner, J. A., 1981. Antimutagenic effects of selenium on acridine orange and 7,12-dimethylbenzanthracene in the Ames Salmonella/microsomal system, *Mutat. Res.* 82:41–46.

Matsumoto, M., Wakasugi, H., and Ibayashi, H., 1981. Serum vitamin E, lipid peroxide and glutathione peroxidase in patients with chronic pancreatitis, *Clin. Chim. Acta* 110:121–125.

Matthes, G., Hackensellner, H. A., Jentzsch, K. D., and Oehme, P., 1981. Further studies on the cryoprotection supporting efficiency of selenium compounds, *Cryo-Lett.* 2:241–245.

Matthes, G., Hackensellner, H. A., Richter, R., Oehme, P., and Jentzsch, K. D., 1980. Contractility and membrane of cryopreserved auricle fragments from rat hearts after pretreatment with selenium, *Acta Physiol. Pharmacol. Bulgarica* 6:60–64.

Mautner, H. G., and Jaffe, J. J., 1958. The activity of 6-selenopurine and related compounds against some experimental mouse tumors, *Cancer Res.* 18:294–298.

McColl, K. E. L., Brodie, M. J., Whitesmith, R., Gray, K. F., and Thomson, T. J., 1979. Parotid salivary gland function in patients with exocrine pancreatic insufficiency, *Acta Hepato-Gastroenterol.* 26:407–412.

McConnell, K. P., Broghamer, W. L., Blotcky, A. J., and Hurt, O. J., 1975. Selenium levels in human blood and tissues in health and in disease, *J. Nutr.* 105:1026–1031.

McConnell, K. P., Jagar, R. M., Higgins, P. J., and Blotcky, A. J., 1977. Serum selenium levels in patients with and without breast cancer, *Proceedings of the 18th Meeting of the American College of Nutrition*, Houston, Texas, J. Van Eys, M. S. Seelig, and B. L. Nichols, (eds), S. P. Medical and Scientific Books, New York, pp. 195–198.

McConnell, K. P., Jager, R. M., Bland, K. I., and Blotcky, A. J., 1980. The relationship of idetary selenium and breast cancer, *J. Surg. Oncol.* 15:67–70.

Medina, D., and Oborn, C. J., 1981. Differential effects of selenium on the growth of mouse mammary cells in vitro, *Cancer Lett.* 13:333–344.

Medina, D., and Shepherd, F., 1981. Selenium-mediated inhibition of 7,12-dimethylbenzanthracene-induced mouse mammary tumorigenesis, *Carcinogenesis* 2:451–455.

Meyers, C. E., McGuire, W. P., Liss, R., Ifrim, I., Frotzinger, K., and Young, R. C., 1976a. Adriamycin: the role of lipid peroxidation in cardiac toxicity and tumor response, *Science* 197:165–167.

Milner, J. A., and Hsu, C. Y., 1981. Inhibitory effects of selenium on the growth of L1210 leukemic cells, *Cancer Res.* 41:1652–1656.

Mulhern, S. A., Morris, V. C., Vessey, A. R., and Levander, O. A., 1981. Influence of selenium and chow diets on immune function in first and second generation mice, *Fed. Proc. Fed. Am. Soc. Exp. Biol.* 40:935.

Nakaidze, N. S., Kuzmin, V. I., Rozantsev, E. G., and Obukhova, L. G., 1979. Effect of some antioxidants on the life span of *Drosophilia melanogaster*, *Izv. Akad. SSSR Ser. Biol.* 6:926–929.

Nakamuro, K., Yoshikawa, K., Sayato, Y., Kurata, H., Tonomura, M., and Tonomura, A., 1976. Studies on selenium-related compounds, V. Cytogenetic effect and reactivity with DNA, *Mutation Res.* 40:177–184.

National Cancer Institute Carcinogenesis Test Program, 1980. Bioassay of selenium sulfide (gavage) for possible carcinogenicity, NIH Publ., NIH-80-1750, 130 pp.

Navia, J. M., 1970. Effect of minerals on the dental caries. Dietary chemicals vs. dental caries, in *Advances in Chemistry*, R. S. Harris (ed.), American Chemical Society, Washington, D. C., pp. 123–160.

Navia, J. M., Menaker, L., Seltzer, J., and Harris, R. S., 1968. Effect of Na_2SeO_3 supplemented in the diet or the water on dental caries of rats, *Fed. Proc. Fed. Am. Soc. Exp. Biol.* 27:676.

Nelson, A. A., Fitzhugh, O. G., and Calvery, H. O., 1943. Liver tumors following cirrhosis caused by selenium in rats, *Cancer Res.* 3:230–236.

Nickerson, W. J., Taber, W. A., and Falcone, G., 1956. Physiological bases of morphogenesis in fungi. 5. Effect of selenite and tellurite on cellular division of yeastlike fungi, *Can. J. Microbiol.* 2:575–584.

Nixon, G. S., and Myers, V. B., 1970. Estimation of selenium in human dental enamel by activation analysis, *Caries Res.* 4:179–187.

Noda, M., Takano, T., and Sakurai, H., 1979. Mutagenic activity of selenium compounds, *Mutat. Res.* 66:175–179.

Norman, B. B., and Johnson, W., 1976. Selenium responsive disease, *Anim. Nutr. Hlth.* 31:6.

Norppa, H., Westermarck, T., Laasonen, M., Knuutila, L., and Knuutila, S., 1980a. Chromosomal effects of sodium selenite *in vivo*. I. Aberrations and sister chromatid exchanges in human lymphocytes, *Hereditas* 93:93–96.

Norppa, H., Westermarck, T., Oksanen, A., Rimaila-Parnanen, E., and Knuutila, S., 1980b. Chromosomal effects of sodium selenite *in vivo*. II. Aberrations in mouse bone marrow and primary spermatocytes, *Hereditas* 93:97–99.

Norppa, H., Westermarck, T., and Knuutila, S., 1980c. Chromosomal effects of sodium selenite *in vivo*. III Aberrations and sister chromatid exchanges in Chinese hamster bone marrow, *Hereditas* 93:101–105.

Ostadalova, I., Babicky, A., and Obenberger, J., 1977. Cataract induced by administration of a single dose of sodium selenite to suckling rats, *Experientia* 34:222–223.

Perry, H. M., and Erlanger, M. W., 1977. Effect of a second metal on cadmium-induced hypertension, in *Trace Substances in Environmental Health*, Vol. 11, D. D. Hemphill (ed.), University of Missouri Press, Columbia, Missouri, pp. 280–288.

Phillips, G. D., and Garnys, V. P., 1981. Trace element balance in adults receiving parenteral nutrition: Preliminary data, JPEN 5:11–14.

Pointner, H., and Kletter, K., 1980 Evaluation of the [75]Se-*l*-selenomethionine test for pancreatic disease, *Digestion* 20:225–233.

Poirier, K. A., and Milner, J. A., 1979. The effect of various seleno-compounds on Ehrlich ascites tumor cells, *Biol. Tr. Elem. Res.* 1:25–34.

Pories, W. J., Atawneh, A., Peer, R. M., Childers, R. C., Worland, R. L., Zaresky, S. A., and Strain, W. H., 1979. Mineral metabolism of the healing arterial wall, *Arch. Surg.* 114:254–257.

Prowse, W. B., 1937. Carcinoma of breast with wide-spread metastases. Two cases of recovery, *Brit. Med. J.* 1:1021–1023.

Randleman, C. D., 1980. Inhibitory effects of selenium on induced rat ovarian tumors, *BIOS* 51:86–89.

Rasco, M. A., Jacobs, M. M., and Griffin, A. C., 1977. Effects of selenium on aryl hydrocarbon hydroxylase activity in cultured human lymphocytes, *Cancer Lett.* 3:295–301.

Ray, J. H., and Altenburg, L. C., 1978. Sister-chromatid exchange induction by sodium selenite: Dependence on the presence of red blood cells or red blood cell lysate, *Mutation Res.* 54:343–354.

Ray, J. H., Altenburg, L. C., and Jacobs, M. M., 1978. Effects of sodium selenite and methyl methanesulphonate or N-hydroxy-2-acetylaminofluorene co-exposure on sister chromatid exchange production in human whole blood cultures, *Mutat. Res.* 57:359–368.

Reiss, U., Tappel, A. L., and Chio, K. S., 1972. DNA-malonaldehyde reaction: Formation of fluorescent products, *Biochem. Biophys. Res. Commun.* 48:921–926.

Revis, N. W., and Armsted, B., 1979. The role of selenium and vitamin E in prevention of lipid peroxidation and myocardial necrosis induced by isoprenaline treatment of coronary artery ligation, *International Congress Series no. 491*, Florence International Meeting on Myocardial Infarction, Excerpta Medica, Amsterdam–Oxford–Princeton.

Revis, N. W., and Marusic, N., 1978. Glutathione peroxidase activity and selenium concentration in the hearts of doxorubicin-treated rabbits, *J. Molec. Cell. Card.* 10:945–951.

Rhead, W. J., and Schneider, J. A., 1976. Effect of selenium compounds on selenium content, growth and ^{35}S-cystine metabolism of skin fibroblasts from normal and cystinotic individuals, *Bioinorg. Chem.* 6:187–202.

Riley, J. F., 1968. Mast cells, cocarcinogenesis and anti-carcinogenesis in the skin of mice, *Experientia* 15:1237–1238.

Roberts, M. E., 1963a. Antiinflammation studies I. Antiinflammatory properties of liver fractions, *Toxicol. Appl. Pharm* 5:485–499.

Roberts, M. E., 1963b. Antiinflammation studies II. Antiinflammatory properties of selenium, *Toxicol. Appl. Pharm.* 5:500–506.

Robinson, M. F., Godfrey, P. J., Thomson, C. D., Rea, H. M., and van Rij, A. M., 1979. Blood selenium and glutathione peroxidase activity in normal subjects and in surgical patients with and without cancer in New Zealand, *Am. J. Clin. Nutr.* 32:1477–1485.

Rosin, M. P., 1981. Inhibition of spontaneous mutagenesis in yeast cultures by selenite, selenate and selenide, *Cancer Lett.* 13:7–14.

Russell, G. R., Nader, C. J., and Patrick, E. J., 1980. Induction of DNA repair by some selenium compounds, *Cancer Lett.* 10:75–81.

Schrauzer, G. N., Rhead, W. J., and Evans, G. A., 1973. Selenium and cancer: Chemical interpretation of a plasma "cancer test", *Bioinorg. Chem.* 2:329–340.

Schrauzer, G. N., and Ishmael, D., 1974. Effects of selenium and of arsenic on the genesis of spontaneous mammary tumors in inbred C₃H mice, *Ann. Clin. Lab. Sci.* 4:411–447.

Schrauzer, G. N., White, D. A., and Schneider, C. J., 1976. Inhibition of the genesis of spontaneous mammary tumors in C₃H mice: Effects of selenium and of selenium antagonistic elements and their possible role in human breast cancer, *Bioinorg. Chem.* 6:265–270.

Schrauzer, G. N., White, D. A., and Schneider, C. J., 1977a. Cancer mortality correlation studies-III: Statistical associations with dietary selenium intakes, *Bioinorg. Chem.* 7:23–24.

Schrauzer, G. N., White, D. A., and Schneider, C. J., 1977b. Cancer mortality correlation studies—IV: Associations with dietary intakes and blood levels of certain trace elements, notable Se-antagonists, *Bioinorganic Chem.* 7:35–56.

Schrauzer, G. N., White, D. A., and Schneider, C. J., 1978. Selenium and cancer: Effects of selenium and of the diet on the genesis of spontaneous mammary tumors in virgin inbred female C₃H/St mice, *Bioinorg. Chem.* 8:387–396.

Schrauzer, G. N., McGinness, J. E., and Kuehn, K., 1980. The effects of temporary selenium supplementation on the genesis of spontaneous mammary tumors in inbred female C₃H/St mice, *Carcinogenesis* 1:199–201.

Schrauzer, G. N., Kuehn, K., and Hamm, D., 1981. Effects of dietary selenium and lead on the genesis of spontaneous mammary tumors in mice, *Biol. Tr. Elem. Res.* 3:185–196.

Schut, H. A. J., and Thorgeirsson, S. S., 1979. Mutagenic activation of N-hydroxy-2-ace-

tylaminofluorene by developing epithelial cells of rat small intestine and effects of
antioxidants, *J. Nat. Cancer Inst.* 63:1405–1409.

Schroeder, H. A., and Mitchener, M., 1971. Selenium and tellurium in rats: Effects on
growth, survival and tumors, *J. Nutr.* 101:1531–1540.

Schroeder, H. A., and Mitchener, M., 1972. Selenium and tellurium in mice, *Arch. Environ.
Health* 24:66–71.

Schwarz, K., 1976. Essentiality and metabolic functions of selenium, *Med. Clin. N. Am.*
60:745–758.

Seifter, J., Ehrlich, W. E., Hudgma, G., and Mueller, G., 1946. Thyroid adenomas in rats
receiving selenium, *Science* 103:762.

Serfass, R. E., and Ganther, H. E., 1976. Effects of dietary selenium and tocopherol on
glutathione peroxidase and superoxide dismutase activities in rat phagocytes, *Life Sci.*
19:1139–1144.

Shakelford, J., and Martin, J., 1980. Antibody response of mature male mice after drinking
water supplemented with selenium, *Fed. Proc. Fed. Am. Soc. Exp. Biol.* 39:339.

Shamberger, R. J., 1970. Relationship of selenium to cancer. I. Inhibitory effect of selenium
on carcinogenesis, *J. Nat. Cancer Inst.* 44:931–936.

Shamberger, R. J., 1972. Increase of peroxidation in carcinogenesis, *J. Nat. Cancer Inst.*
48:1491–1497.

Shamberger, R. J., 1977. Antioxidants and Cancer. VII. Cadmium-seleniums in kidneys,
in *Proceedings of the 3rd International Symposium of Trace Element Metabolism in
Man and Animals*, M. Kirchgessner (ed.) Freising-Weihenstephen, West Germany, pp.
391–392.

Shamberger, R. J., and Bratush, C. M., 1980. Cadmium, selenium and zinc levels in kidneys,
Trace Substances in Environmental Health, Vol. 14, D. D. Hemphill (ed.), University
of Missouri Press, Columbia, Missouri, pp. 203–210.

Shamberger, R. J., and Rudolph, G., 1966. Protection against cocarcinogenesis by antioxidants, *Experientia* 22:116.

Shamberger, R. J., and Willis, C. E., 1971. Selenium distribution and human cancer mortality, *CRC Crit. Rev. Clin. Lab. Sci.* 2:211–221.

Shamberger, R. J., Baughman, F. F., Kalchert, S. L., Willis, C. E., and Hoffman, G. C.,
1973b. Carcinogen-induced chromosomal breakage decreased by antioxidants, *Proc.
Nat. Acad Sci.* 70:1461–1463.

Shamberger, R. J., Rukovena, E., Longfield, A. K., Tytko, S. A., Deodhar, S., and Willis,
C. E., 1973a, Antioxidants and cancer. I. Selenium in the blood of normals and cancer
patients, *J. Nat. Cancer Inst.* 50:863–870.

Shamberger, R. J., Andreone, T. L., and Willis, C. E., 1974. Antioxidants and cancer. IV.
Initiating activity of malonaldehyde as a carcinogen, *J. Nat. Cancer Inst.* 53:1771–1773.

Shamberger, R. J., Tytko, S. A., and Willis, C. E., 1975. Selenium and heart disease, in
Trace Substances in Environmental Health, Vol. 9, D. D. Hemphill (ed.), University
of Missouri Press, Columbia, Missouri, pp. 15–22.

Shamberger, R. J., Tytko, S. A., and Willis, C. E., 1976. Antioxidants and cancer. Part VI.
Selenium and age-adjusted human cancer mortality, *Arch. Environ. Health* 31:231–235.

Shamberger, R. J., Gunsch, M. S., Willis, C. E., and McCormack, L. J., 1978. Selenium
and heart disease. II. Selenium and other trace metal intakes and heart disease in 25
countries, in *Trace Substances in Environmental Health*, Vol. 12, D. D. Hemphill (ed.),
University of Missouri Press, Columbia, Missouri, pp. 48–52.

Shamberger, R. J., Corlett, C. L., Beaman, K. D., and Kasten, B. L., 1979a. Antioxidants
reduce the mutagenic effect of malonaldehyde and β-propiolactone, Part IX. Antioxidants and Cancer, *Mutat. Res.* 66:349–355.

Shamberger, R. J., Willis, C. E., and McCormack, L. J., 1979b. Selenium and heart disease. III. Blood selenium and heart mortality in 19 states, in *Trace Substances in Environmental Health*, Vol. 13, D. D. Hemphill (ed.), University of Missouri Press, Columbia, Missouri, pp. 59–63.

Shapiro, J. R., 1972. Selenium and carcinogenesis: A review, *Ann. N. Y. Acad. Sci.* 192:215–219.

Shearer, T. R., 1973. Comparative distribution of ^{75}Se in the hard and soft tissues of mother rats and their pups, *J. Nutr.* 103:553–559.

Shearer, T. R., 1975. Developmental and postdevelopmental uptake of dietary organic and inorganic selenium into the molar teeth of rats, *J. Nutr.* 105:338–346.

Shearer, T. R., and Ridlington, J. W., 1976. Fluoride–selenium interaction in the hard and soft tissues of the rat, *J. Nutr.* 106:451–456.

Shearer, T. R., McCormack, D. W., Desfart, D. J., Britton, J. L., and Lopez, M. T., 1980. Histological evaluation of selenium induced cataracts, *Exp. Eye Res.* 31:327–333.

Sheffy, B. E., and Schultz, R. D., 1978. Influence of vitamin E and selenium on immune response mechanisms, *Cornell Vet.* 68 Suppl. 7:89–93.

Shukla, V. K. S., Jensen, G. E., and Clausen, J., 1977. Erythrocyte glutathione deficiency in multiple sclerosis, *Acta Neurol. Scandinav.* 56:542–550.

Shukla, V. K. S., Jensen, G. E., and Clausen, J., 1978. Serum fatty acids and peroxidase abnormalities in Batten's disease, *Res. Exp. Med.* 173:27–34.

Soullier, B. K., Wilson, P. S., and Nigro, N. D., 1981. Effect of selenium on azoxymethane-induced intestinal cancer in rats fed high fat diet, *Cancer Lett.* 12:343–348.

Spallholz, J. E., 1981. Anti-inflammatory, immunologic and carcinostatic attributes of selenium in experimental animals, *Adv. Exp. Med. Biol.*, 135:43–62.

Spallholz, J. E., Martin, J. L., Gerlach, M. L., and Heinzerling, R. H., 1973a. Immunological responses of mice fed diets supplemented with selenite selenium, *Proc. Soc. Exp. Biol. Med.* 143:685–689.

Spallholz, J. E., Martin, J. L., Gerlach, M. L., and Heizerling, R. H., 1973b. Enhanced IgM and IgG titers in mice fed selenium, *Infec. Immun.* 8:841–842.

Spallholz, J. E., Heinzerling, R. H., Gerlach, M. L., and Martin, J. L., 1974. The effect of selenite, tocopherol acetate and selenite, tocopherol acetate on the primary and secondary immune responses of mice administered tetanus toxoid or sheep red blood cell antigen, *Fed. Proc. Fed. Am. Soc. Exp. Biol.* 33:694.

Spallholz, J. E., Martin, J. L., Gerlach, M. L., and Heizerling, R. H., 1975. Injectable selenium: Effect on the primary immune response of mice, *Proc. Soc. Exp. Biol. Med.* 148:37–40.

Strehler, B. L., Mark, D. D., Mildvan, A. S., and Gee, M. V., 1959. Rate and magnitude of age pigment accumulation in the human myocardium, *J. Gerontol.* 14:430–439.

Tappel, A. L., 1972. Vitamin E and free-radical peroxidation of lipids, *Ann. N. Y. Acad. Sci.* 203:12–28.

Thomson, C. D., Rea, H. M., Robinson, M. F., and Simpson, F. O., 1978. Selenium concentrations and glutathione peroxidase activities in the blood of hypertensive patients, *Proc. Univ. Otago Med. Sch.* 56:31–33.

Thompson, H. J., and Becci, P. J., 1979. Effect of graded dietary levels of selenium on tracheal carcinomas induced by 1-methyl-1-nitrosourea, *Cancer Lett.* 7:215–219.

Thompson, H. J., and Becci, P. J., 1980. Selenium inhibition of N-methyl-N-nitrosourea-induced mammary carcinogenesis in the rat, *J. Nat. Cancer Inst.* 65:1299–1301.

Thompson, H. J., and Taglaferro, A. R., 1980. Effect of selenium on 7,12-dimethylbenzanthracene-induced mammary tumorigenesis, *Fed. Proc. Fed. Am. Soc. Exp. Biol.* 39:1117.

Thompson, H. J., Soule, R. A., and Becci, P. J., 1981b. Inhibition of mammary tumorigenesis by graded levels of selenium, *Fed. Proc. Fed. Am. Soc. Exp. Biol.* 40:929.

Thompson, H. J., Meeker, L. D., and Becci, P. J., 1981a. Effect of combined selenium and retinyl acetate treatment on mammary carcinogenesis, *Cancer Res.* 41:1413–1416.

Tkeshelashvili, L. K., Shearman, C. W., Zakour, R. A., Koplitz, R. M., and Loeb, L. A., 1980. Effects of arsenic, selenium, and chromium on the fidelty of DNA, *Cancer Res.* 40:2455–2460.

van Rij, A. M., McKenzie, J. M., Robinson, M. F., and Thomson, C. U., 1974. Selenium and total parenteral nutrition, *JPEN* 3:235–239.

van Rij, A. M., Thomson, C. D., McKenzie, J. M., and Robinson, M. F., 1979. Selenium deficiency in total parenteral nutrition, *Am. J. Clin. Nutr.* 32:2076–2085.

Van Vleet, J. F., Rebar, A. H., and Ferrans, V. J., 1977. Acute cobalt and isoproterenol cardiotoxicity in swine: Protection by selenium-vitamin E supplementation and and potentiation by stress-susceptible phenotype, *Am. J. Vet. Res.* 38:991–1002.

Van Vleet, J. F., and Ferrans, V. J., 1980. Clinical observations, cutaneous lesions, and hematologic alterations in chronic adriamycin intoxication in dogs with and without vitamin E and selenium supplementation, *Am. J. Vet. Res.* 40:691–699.

Van Vleet, J. F., Greenwood, L. A., and Rebar, A. H., 1981. Effect of selenium–vitamin E on hematologic alterations of adriamycin, toxicosis in young pigs, *Am. J. Vet. Res.* 42:1153–1159.

Villalon, J. A. M., M. D. thesis, Universidad Michocana de San Nicolas de Hilago, Mexico, cited by D. V. Frost and P. M. Lish, 1975, Selenium in biology, *Ann. Rev. Pharm.* 15:259–284.

Volgarev, N. N., and Tscherkes, L. A., 1967. Further studies in tissue changes associated with sodium selenate, in Symposium: *Selenium in Biomedicine*, O. H. Muth (ed.), Avi Publishing Co., Westport, Connecticut, pp. 179–184.

Voors, A. W., Johnson, W. D., Shuman, M. S., and Blotcky, A. J., 1978. Adjusted cadmium levels in the renal cortex and heart weight at autopsy, in *Trace Substances in Environmental Health*, Vol. 12, D. D. Hemphill (ed.), University of Missouri Press, Columbia, Missouri, pp. 181–190.

Walker, G. W. R., and Ting, K. P., 1967. Effect of selenium on recombination in barley, *Can. J. Genet. Cytol.* 9:314–320.

Wallach, J. D., and Garmaise, B., 1979. Cystic fibrosis: A perinatal manifestation of selenium deficiency, in *Trace Substances in Environmental Health*, D. D. Hemphill (ed.), University of Missouri Press, Columbia, Missouri, pp. 469–476.

Watson-Williams, E., 1920. The treatment of inoperable cancer with selenium, *Brit. J. Surg.* 8:50–58.

Welsch, C. W., Goodrich-Smith, M., Brown, C. K., Greene, H. D., and Hamel, E. J., 1981, Selenium and the genesis of murine mammary tumors, *Carcinogenesis* 2:519–522.

Wester, P. O., Brune, D., and Nordberg, G., 1981. Arsenic and selenium in lung, liver and kidney tissue from dead smelter workers, *Br. J. Ind. Med.* 38:179–184.

Westermarck, T., 1977. Selenium content of tissues in Finnish infants and adults with various diseases, and studies on the effects of selenium supplementation in neuronal ceroid lipofuscinosis patients, *Acta Pharmacol. Toxicol.* 41:121–128.

Whanger, P. D., and Weswig, P. H., 1975. Effects of selenium, chromium, and antioxidants on growth, eye cataracts, plasma cholesterol, and blood glucose in selenium deficient, vitamin A supplemented rats, *Nutr. Repts. Int.* 12:345–357.

Whanger, P. D., Pederson, N. D., Hatfield, J., Weswig, P. H., 1976. Absorption of selenite and selenomethionine from ligated digestive tract segments in rats, *Proc. Soc. Exp. Biol. Med.* 153:295–297.

Whiting, R. F., Wei, L., and Stich, H. F., 1980. Unscheduled DNA synthesis and chromosome aberrations induced by inorganic and organic selenium compounds in the presence of glutathione, *Mutation Res.* 78:159–169.

Wikstom, J., Westermarck, T., and Palo, J., 1976. Selenium, vitamin E and copper in multiple sclerosis, *Acta Neurol. Scand.* 54:287–290.

Williams, G. M., 1978. Further improvements in the hepatocyte primary culture DNA repair test for carcinogens: Detection of carcinogenic biphenyl derivatives, *Cancer Lett.* 4:69–75.

Wortzman, M. S., Besbris, H. J., and Cohen, A. M., 1980. Effect of dietary selenium on the interaction between 2-acetylaminofluorene and rat liver DNA *in vivo, Cancer Res.* 40:2670–2676.

Young, D. M., 1975. Pathologic effects of adriamycin (NSC-123-127) in experimental systems, *Cancer Chemother. Rep.* 6:159–175.

Young, E. O., and Milner, J. A., 1981. Inhibition of 7,12-dimethylbenzanthracene-induced mammary tumors by selenium, *Fed. Proc. Fed. Am. Soc. Exp. Biol.* 40:949.

Synthetic Forms of Selenium and Their Chemotherapeutic Uses

9.1 Anti-infective Agents

9.1.1 Antibacterial

Because of the close resemblance of selenium to the biologically important element sulfur, a strong rationale exists for the incorporation of selenium into potential medicinal agents. In many cases selenium isoesters of sulfur-containing antibacterial agents are as effective as their analogous sulfur compounds or sometimes even more effective. Inorganic selenium compounds, when tested at the same selenium concentration as the active organic selenium compounds, have comparatively negligible antibacterial activity (Green and Bielschowsky, 1942). Many of the selenium compounds studied as potential bactericides during World War II were modeled after the sulfonamide drugs. Although no direct sulfur–selenium interchanges to give the analogous selenoamides have been accomplished, a close derivative may be found in bis(p-aminophenyl) selenide (**I**) and bis(p-aminophenyl) diselenide (**II**). (Table 9-1 lists structures **I–XV**.) The bis(p-aminophenyl) diselenide is very effective against *Brucella paramelitensis* and *Staphylococcus aureus* (Green and Bielschowsky, 1942). The diselenide is a thousandfold more effective as an antibacterial agent than bis(p-aminophenyl) disulfide. The corresponding acid is also active against *B. paramelitensis* and *S. aureus* (Green and Bielschowsky, 1942). The selenium of sulfathiazole (**III**) has shown some activity *in vitro* against *Pneumococcus* (Jensen and Schmith, 1941). The methyl alkyl selenoxide hydronitrate (**IV**) derivatives have been patented as germicides for inclusion in detergent formulations (Priestly, 1972). The monophenol phenyl selenide (**V**) derivatives also have bactericidal activity (Keimatsu, 1933). The selenium derivative of chloromycetin (**VI**) is a

Table 9-1. Antibacterial Activity of Organic Selenium Compounds

	I
	II
	III
$[CH_3Se(=O)R \cdot HNO_3]$	IV
	V
	VI
	VII
	VIII

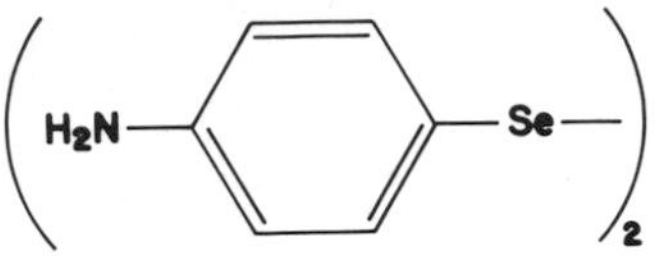

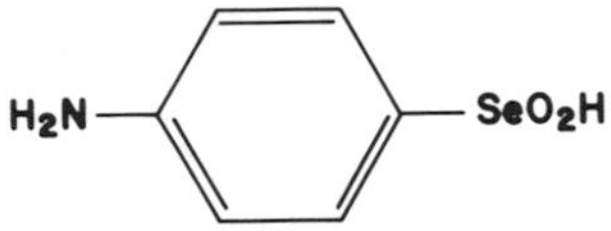

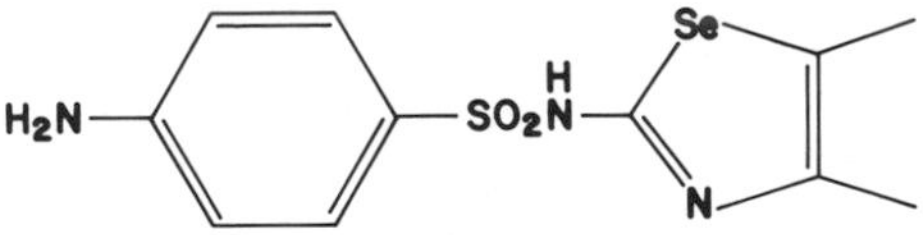

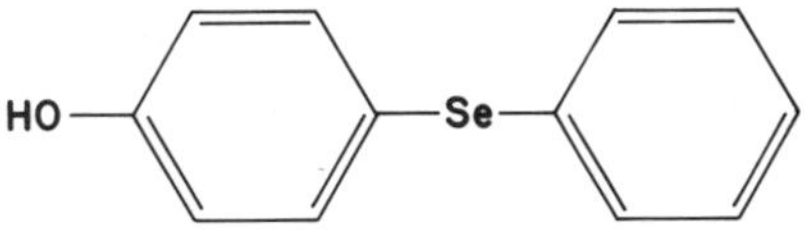

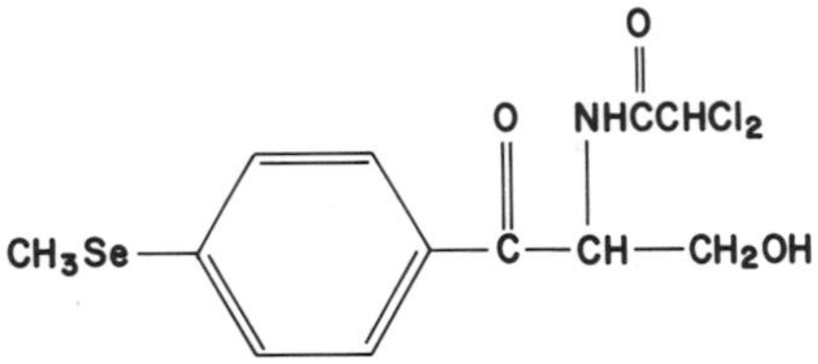

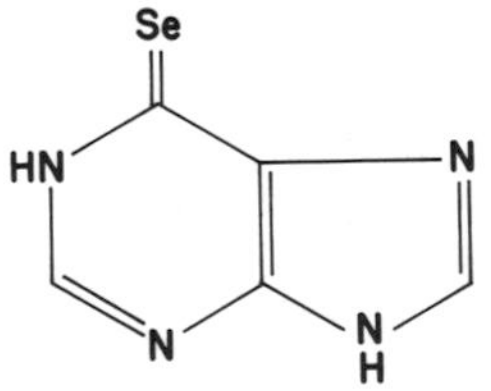

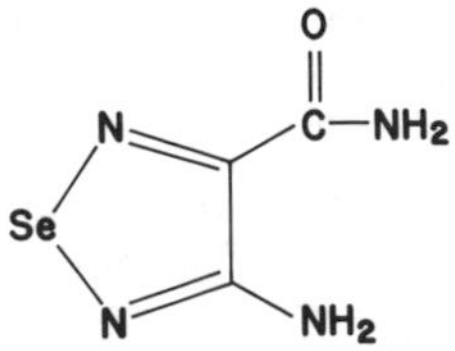

Table 9-1. (Continued)

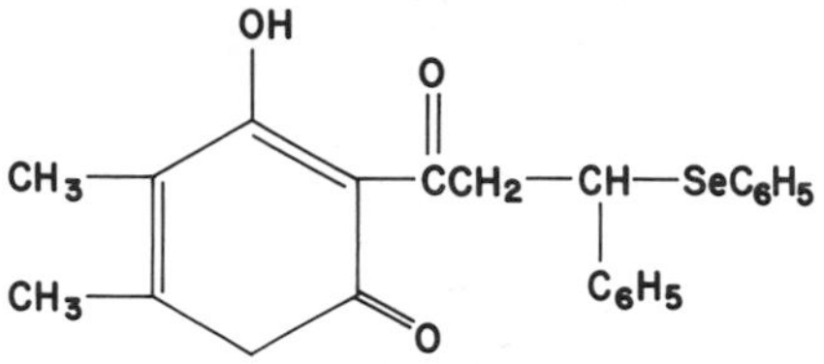

	IX
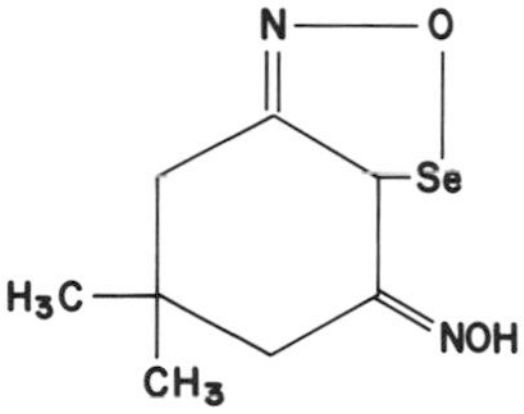	X
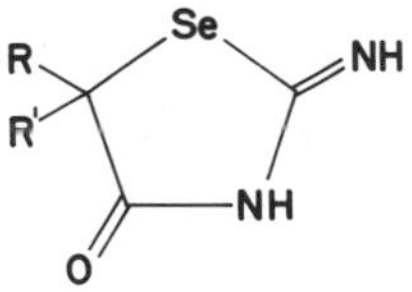	XI
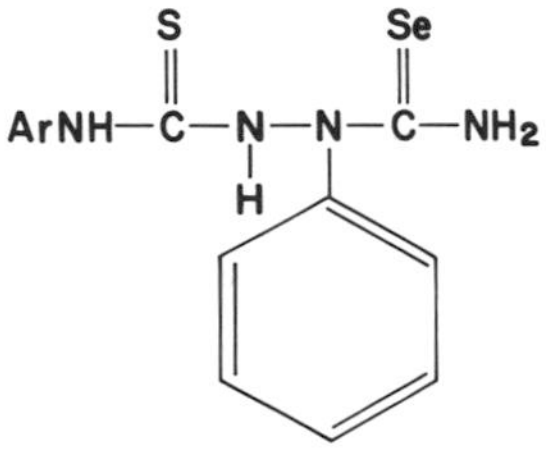	XII
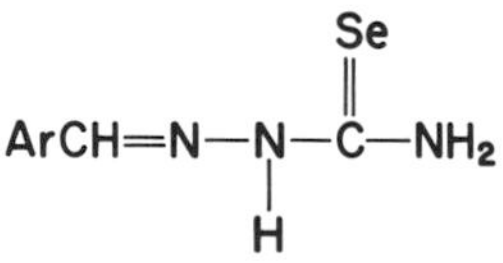	XIII
$RSeO_2H$	XIV
$Se-S$	XV

strong antibacterial agent, which is about ten times as active as the S analog (Supniewski *et al.*, 1954). Gram-positive were more sensitive than gram-negative bacteria and both the sulfur and selenium analogs were strongly active against acid-fast bacteria. The lethal doses were 50 mg/kg weight and the symptoms were identical to those of chloramphenicol. The selenium compound, but not the sulfur compound, lowered arterial pressure, decreased respiratory movement and had a diuretic effect on the cat. The selenium derivative of 6-mercaptopurine (**VII**) has been screened against ten types of bacteria and it was found to possess more antibacterial activity than 6-mercaptopurine (Mautner, 1959). One of the more promising class of seleno compounds is the 1,2,5-selenadiazole (**VIII**) family. These compounds were found to exert marked *in vitro* inhibitory activity against more than 300 species of bacteria, yeasts, and fungi (Hunt and Pittillo, 1966). The selenopyranones (**IX**) also have bactericidal activity (Kuhn and McIntyre, 1970). Selenoisonitroso-dimethyl cyclohexane (**X**) at a concentration of 1.95×10^{-5} *m* has inhibited the growth of *Mycobacterium tuberculosis* (Takeda *et al.*, 1952). The 2-iminoselenazolin-2-ones (**XI**) have shown only negligible activity against *S. aureus, Streptococcus pyogenes, E. coli*, and *Proteus vulgaris* (Comrie *et al.*, 1964). The compounds of the sulfur-selenosemicarbazide (**XII**) series have a weak effect on *S. aureus, Bacillus subtilis*, and *E. Coli*; a moderate effect on *Mycobacterium* BCG, *Mycobacterium phlei, Mycobacterium smegmatis*, and a strong effect on *M. tuberculosis* (Bednarz, 1957). The selenosemicarbazone (**XIII**) has a weak bacteriostatic effect against *S. aureus, Bacillus subtilis*, and *E. coli*; the compounds also inhibit growth in *M. phlei* and *M. smegmatis*; all compounds in this series completely inhibit the growth of *M. tuberculosis* Rv/47 (Bednarz, 1958). Seleninic acid (**XIV**) also has bactericidal activity (Aktiebolog, 1970). Selenium sulfide (**XV**) has been shown to be effective against marginal blepharitis probably through its an effect against *Staphylococcus* or other types of microbial infections (Wong *et al.*, 1956).

9.1.2 Antiviral

Selenocystamine dihydrochloride (**XVI**) (0.003 mm) inhibited particle-associated RNA-dependent RNA polymerase of several types of influenza as well as recombinant types of influenza (Oxford, 1973). The structurally related compounds L-cystine, cystamine, and *N, N'*-diacetyl selenocystamine were less active and inhibited influenza B/Lee particle-associated RNA-dependent RNA polymerase activity by 50% at concentrations of 0.3, 0.3, and 0.018 mm, respectively. Selenocystamine also inhibited the particle-associated RNA-dependent RNA polymerase activity. This inhibitory activity of 0.01 mm selelnocystamine was reversed completely by the addition of 0.06 mm β-mercaptoethanol and partially by dialysis of a virus–selenocystamine mixture. Viral infectivity wa reduced significantly after incubation with 0.03 mm selenocystamine.

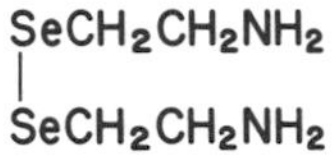

$$\begin{array}{l} SeCH_2CH_2NH_2 \\ | \\ SeCH_2CH_2NH_2 \end{array} \qquad \textbf{XVI}$$

Selcnocystine (**XVII**) inhibited influenza ribonucleic acid (RNA% polymerase activity in a cell-free assay system (Billard and Pccts, 1974). The addition of 5 mm dithiothreitol reverses the inhibitory activity. Selenocystine also significantly inhibited the deoxyribonucleic acid-directed RNA polymerases of *E. coli* and chicken embryo cells in the absence of reducing agent, but to an extent substantially less than that obtained against the viral enzyme.

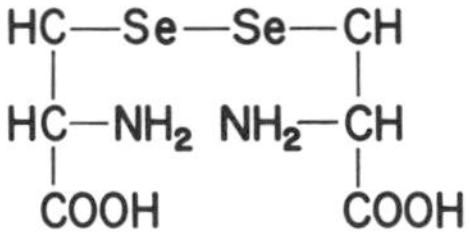

$$\begin{array}{ccc} HC\!-\!Se\!-\!Se\!-\!CH & & \\ | & | & \\ HC\!-\!NH_2 \ \ NH_2\!-\!CH & & \textbf{XVII} \\ | & | & \\ COOH & COOH & \end{array}$$

9.2 Antifungal Agents

Frequently selenium-containing compounds that have been tested for their bactericidal properties have also been screened at the same time for their potential as fungicides. Some aminoselenadiazoles have been demonstrated to possess both broad-spectrum antibacterial and antifungal activity (Shealey and Clayton, 1967) (Table 9-2, **XVIII**), this table contains structures **XVIII–XXI**. Some inorganic forms of selenium such as elemental selenium, hydrogen selenide, and selenide have been shown to be poor fungicides. The antifungal activities of phenylselenourea and 2-phenylselenosemicarbazide (**XIX**) were compared with those of their sulfur and oxygen analogs (Mautner *et al.*, 1956). It was observed that the compounds of selenium were 10–1000 times more effective on a molar basis than the sulfur compounds, whereas the oxygen analogs exhibited only negligible activity. Several of the phenylselenosemicarbazide derivatives were tested against *Trichophyton metagrophytes, Botrytis cinerea, Monilia fructagena*, and *Penicillium notatum*. Good antifungal activity was observed. Mautner *et al.* (1956) suggested that the superiority of the

Table 9-2. Antifungal Activity of Organic Selenium Compounds

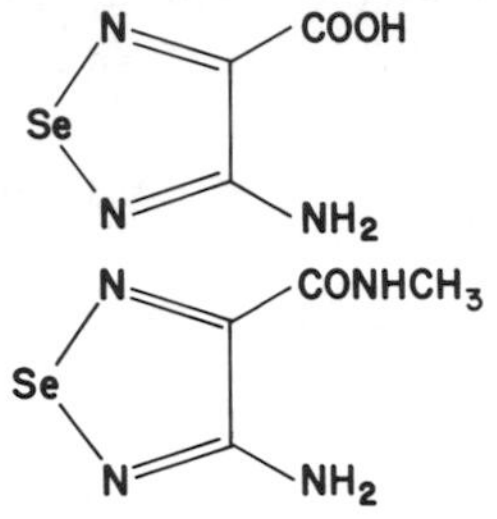

XVIII

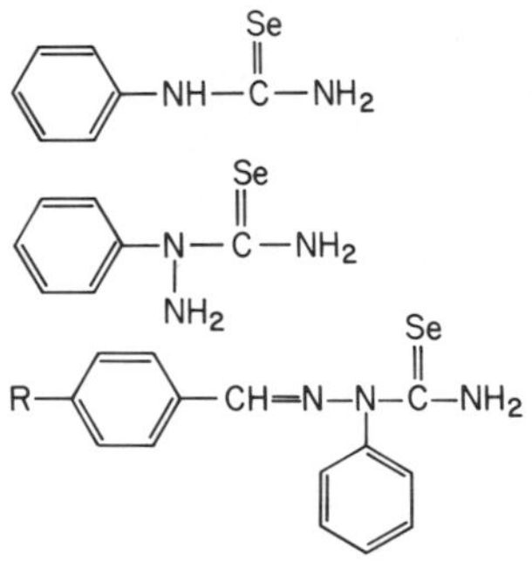

XIX

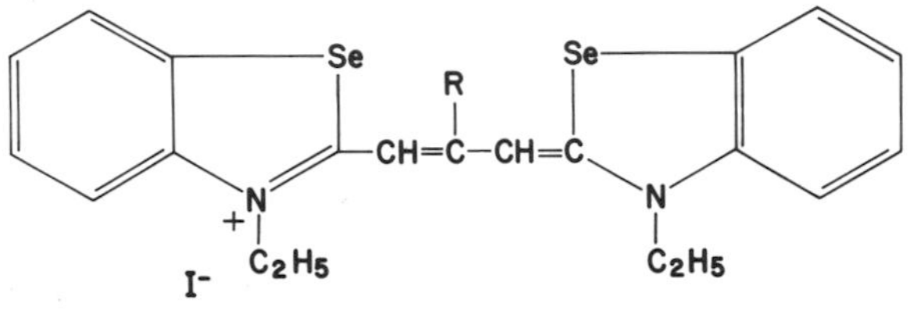

XX

Se–S   XXI

selenium compounds originated from their ability to chelate with metal ions thereby forming complexes which by themselves may be active fungicides. Thia- and selenacarbocyanine (**XX**) dyes were moderately fungicidal, but were less effective than their oxygen analogs against the *Venturia inaequalis, Botrytis cinerea*, and *Fusarium bulbigenum* test bacteria (Pianka and Hall, 1959). Neither the Se or the S analog of the keto derivative of chloromycetin had activity against *A. niger*, but the oxygen analogs were active (Supniewski *et al.*, 1954).

N-butyl-N-hexyl and N-decyl selenocyanates have been observed to possess more fungicidal activity than their isosteric thiocyanates, but their very offensive odor was mentioned as a deterrent to their use (Weaver and Whaley, 1946). A series of aryl and aralkyl selenocyanates, as well as alkyl esters of selenocyanato-acetic acid, was tested along with 450 organic sulfur compounds against ten organisms. Most of the selenium compounds possessed good fungicidal activity (Zsolnai, 1962). *Malasseia furfur* is a fungus which can cause tinea versicolor, a benign, noncontagious superficial fungal infection of the skin. Robinson and Jaffe (1956) have used a 1% selenium sulfide cream (**XXI**) on 28 patients. Treatment consisted of two daily applications to the body for two weeks and baths every three days. No recurrences were found after one year of observation. Similar results have been reported by Giordano (1963), who preferred only a single treatment.

9.3 Antiparasitic Agents

Goble *et al.* (1967) reported that selenourea (Table 9-3 **XXII**; this table contains structures **XXII–XXV**) was effective against several Lep-

Table 9-3. Antiparasitic Agents

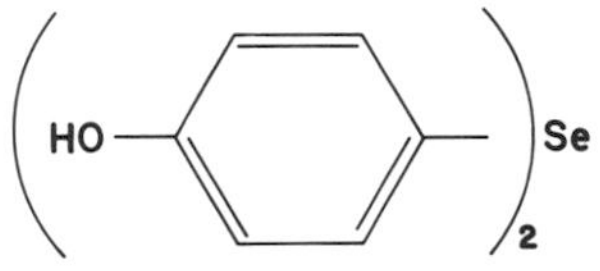

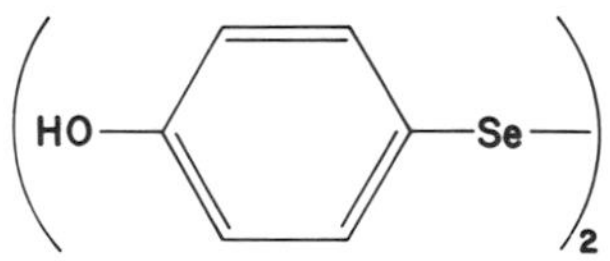

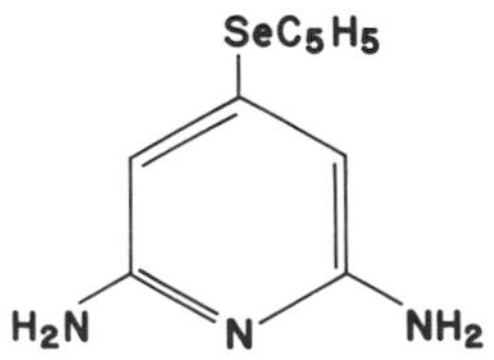

tospirosis strains including *L. australis, L. autumnalis, L. bataviae, L. canicola, L. grippotyphosa, L. icterohaemorrhagiae,* and *L. pomona.* Selenium disulfide was also almost as effective as the selenourea against the Leptospinosis species.

The compounds bis(*p*-hydroxyphenyl) selenide (**XXIII**) and bis(*p*-hydroxyphenyl) diselenide (**XXIV**) were found by Keimatsu and Yokota (1931) to possess 150 times and 100 times, respectively, greater killing potential toward *Paramecium caudatum* than phenol. In a series of 4-substituted 2,6-diaminopyridines, the 4-phenylseleno derivative (**XXV**) was among the most active against the protozoa *Tetrahymena pyriformis* and *Crithidei fasciculata* (Markees *et al.*, 1968). The dye 3,6-didamino-selenopyronine chloride has been tested as a trypanocidal drug in mice (Ehrlich and Bauer, 1915). This compound could produce only transient healing effects, which were accompanied by severe edema.

9.4 Compounds Affecting the Central Nervous System

9.4.1 Hypnotics

The major group of sulfur-containing hypnotics are the 2-thio-5,5-dialkyl barbiturates (e.g., pentothal sodium). When this type of compound is administered intravenously in small doses, it produces anesthesia of short duration (5–30 min). The synthesis of 2-selenobarbituric acid (**XXVI**) as well as of several 5-substituted and 5,5-disubstituted derivatives have been reported by Mautner and Clayton (1959) as part of a study of the relative lipid solubilities of oxygen, sulfur, and selenium compounds. 6-Selenopurine and 2-selenouracil were similarly investigated. In the case of the barbiturates tested, the sulfur and selenium analogs had similar lipid solubilities. The authors concluded that for the type of compounds tested replacement of an oxygen by a sulfur atom was effective in increasing lysis solubility with only minor changes in the steric configuration of the molecule. When the more metallic selenium replaced the sulfur, there was no increase in lipid solubility. Apparently the 2-selenobarbituates have not been tested physiologically.

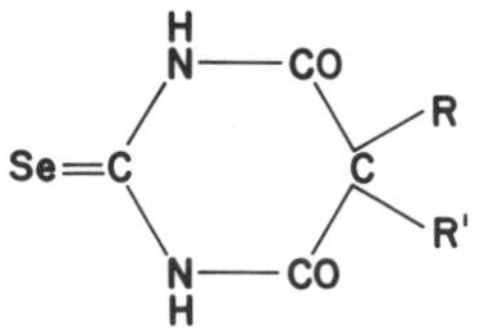

XXVI

The hypnotic activity of alphatic disulfones has been know since the 1890s. Sulfonal (**XXVII**) is one of the most active types of the ketone disulfones. However, because of the toxicity of sulfonal and its derivatives, they have been replaced by other hypnotics. In addition, attempts to synthesize the seleno compounds have not been successful.

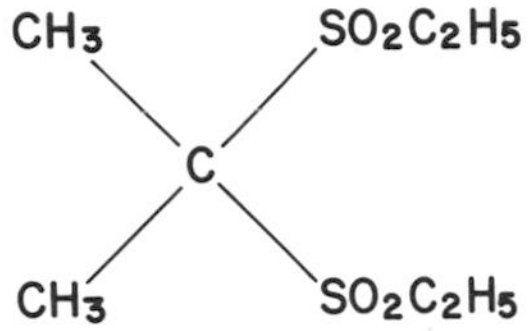

XXVII

9.4.2 Analgesics and Local Anesthetics

Hannig has been responsible for the development of numerous selenium-containing analgesics. Hannig (1963) reported the synthesis of selenium derivatives of phenylbutazone (**XXVIII**). Phenylbutazone is a potent antirheumatic, analgesic, and antipyretic which was benzoylethylated at the 4-position (**XXIX**) (Hannig, 1963). Falicaine (β-piperidineo-*p*-propoxypropiophenone) (**XXX**) was also synthesized. A series of these compounds was prepared; the propoxy group was replaced by alkylseleno (RSe) groups in which the R ranged from methyl through *n*-hexyl (**XXXI**) (Hannig, 1964). The selenium analogs were less toxic than Falicaine. Topical anesthesia experiments on the rabbit cornea showed that the activity increased as R increased in chain length. The butyl, pentyl, and hexyl homologs were more active than Falicaine, and these derivatives were also less toxic.

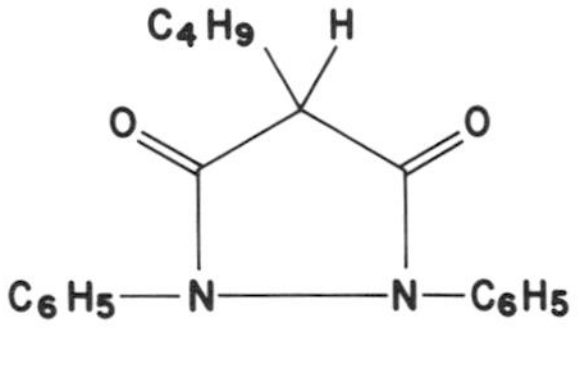

XXVIII

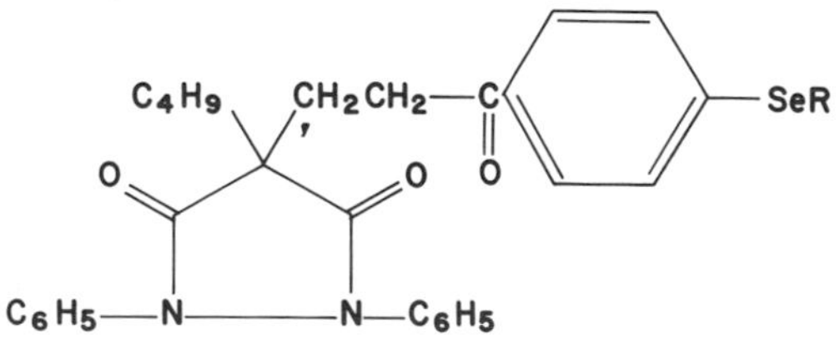

XXIX

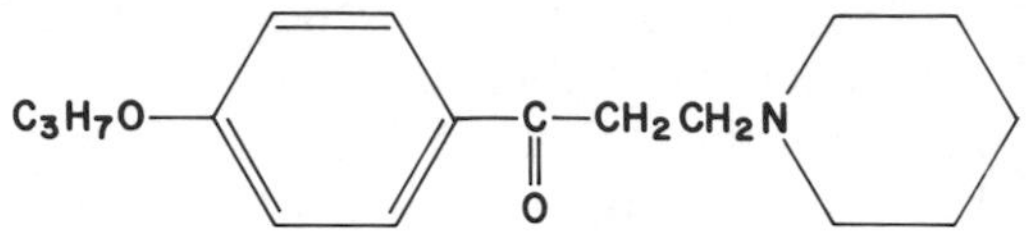

C₃H₇O— ... —C—CH₂CH₂N ...   **XXX**

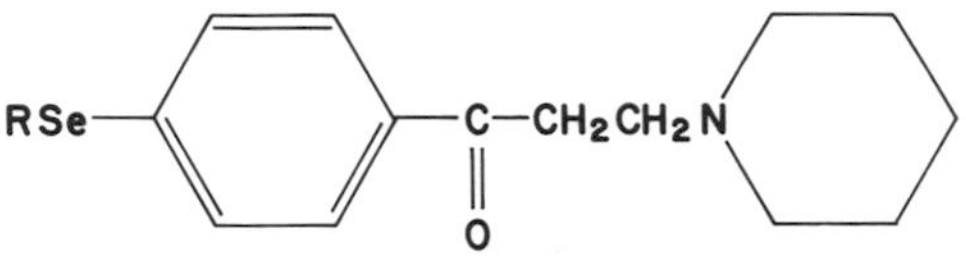

RSe— ... —C—CH₂CH₂N ...   **XXXI**

Bekemeir *et al.* (1976) have studied some *p*-alkylseleno-α,β-hexa-methyleniminopropiophenones (**XXXII**). They have noted that these compounds have slightly greater toxicity than the oxygen analog, but less toxicity than Falicaine. Derivatives of these compounds were reported to be effective against electroshock and/or pentylenetetrazole convulsions, but the derivatives have a lesser local anesthetic effect than Falicaine.

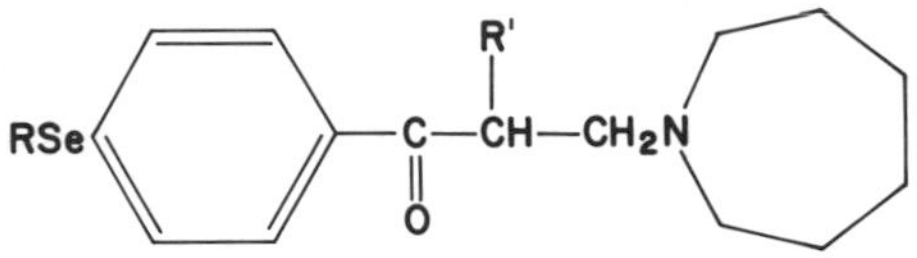

RSe ... —C—CH—CH₂N ...   **XXXII**

9.4.3 Tranquilizing Drugs

A large number of derivatives of phenothiazine derivatives such a chloropromazine (**XXXIII**) and promazine (**XXXIV**) have found use as psychopharmacologic agents as well as antiemetics, antihistamines, antihelmintics, and antimalarials.

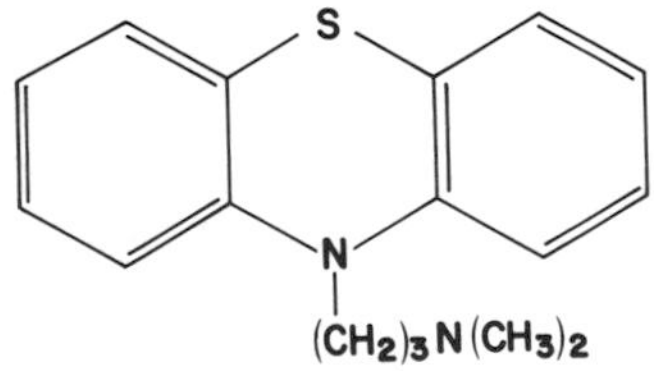

(CH₂)₃N(CH₃)₂   **XXXIII**

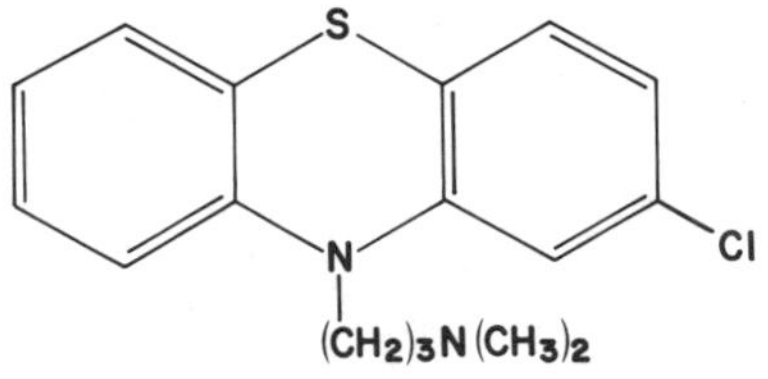

(CH₂)₃N(CH₃)₂   **XXXIV**

Muller *et al.* (1959) first prepared phenoselenazines, which have been reported by Glassman and coworkers (1960) to have less psychopharmacologic potency. The phenoselenazines nonetheless retain the characteristics of promazine and chlorpromazine in regard to their cardiovascular, adenolytic, antihistaminic, and hypoglycemic actions. The 10-substituted phenoselenazines have been reported to be as effective as central nervous system depressants (Craig, 1962).

Tricyclic psychotropic drugs that are derivatives of 6,11-dihydrodibenzo[b,e]thiepin have been studied by Protiva *et al.* (1964). These compounds contain sulfur and are effective antagonists of reserpine and also have good antidepressant activities. The seleno derivatives, the 6,11-dihydrodibenzo[b,e]selenepins (**XXXV**) (R = H or C_2H_5), and **XXXVI** were prepared by methods used for the thiepines pharmacologically (Sindelar *et al.*, 1969a). Compound **XXXVI** has particularly high antireserpine activity. Selenium analogs of substituted thioxanthenes have also been prepared by Sindelar *et al.* (1969a). This latter group of compounds has been reported to possess antiemetic properties as well as properties such as central nervous system depressant activity. (Jilek *et al.*, 1966). One of the derivatives, *cis*-2-choloro-9-(3-dimethylaminopropylidene)-selenoxanthene (**XXXVII**) has shown the greatest central nervous system depressant activity but was less effective than its sulfur analog (Sindelar *et al.*, 1969a). Compounds which were related showed antireserpine, antihistamine, and low central nervous system depressant activity.

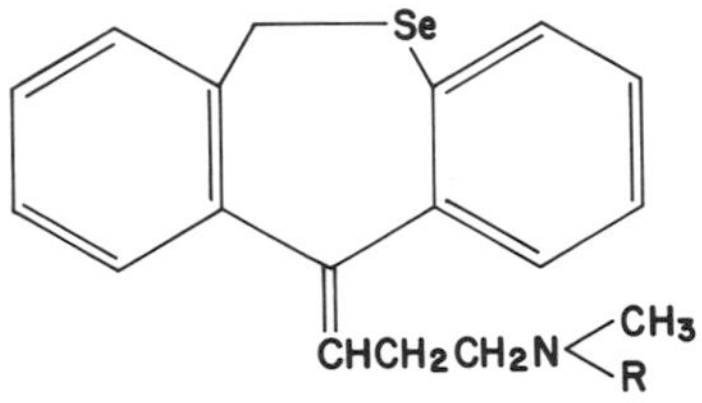

XXXV

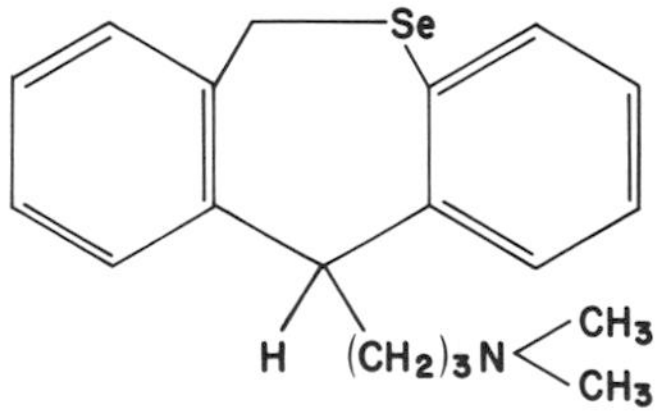

XXXVI

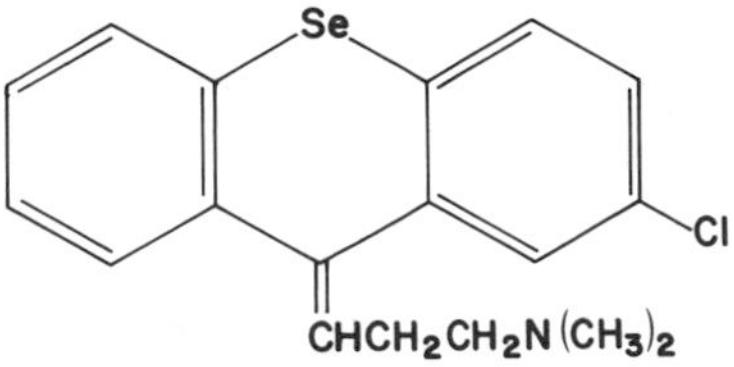

XXXVII

The 8-unsubstituted or 8-chloro-10-(4-methylpiperazino)-10,11-dih-ydrodibenzo[b,f]selenepines (**XXXVIII**) demonstrated less central nervous system depressant activity than their corresponding sulfur analogs (Sindelar *et al.*, 1969b), which were reported to be active neuroleptics (Sindelar *et al.*, 1969c). When the selenium-containing ring of **XXXVIII** is unsaturated, this compound has been found to be one of the highly active central nervous system depressants (Jilek *et al.*, 1970).

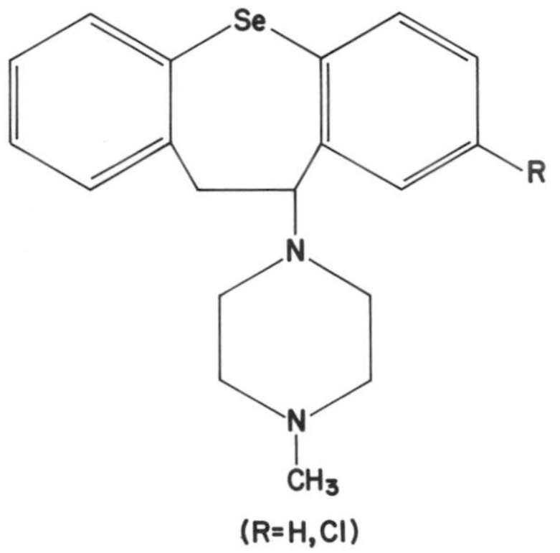

XXXVIII

(R=H,Cl)

9.5 Compounds that Affect the Autonomic Nervous System

The compound *o*-carboxybenzeneseleninic acid (**XXXIX**) has been reported to possess both adrenergic (Limongi, 1955) and cholinergic (Corbett *et al.*, 1957) blocking activity. This compound can reverse the pressor effect of epinephrine in the dog. The compound can also inhibit the action of the ephedrine on the nictating membrane of cats and can also protect mice from lethal amounts of epinephrine to 48 hr after administration.

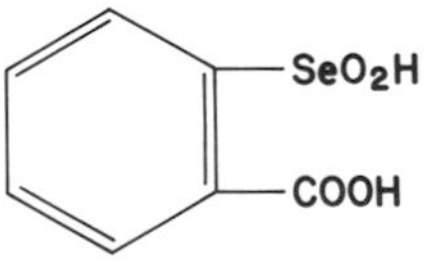

XXXIX

Transmission of nerve impulses by chemical mediation is believed to involve (1) the release by autonomic nerve endings of acetylcholine (**XL**) and its reversible combination with receptor sites; (2) acetylcholinesterase breaks down some of the acetylcholine into choline and acetic acid; (3) finally, the resynthesis of the acetycholine. Mautner and his associates have extensively studied the selenium and sulfur analogs of acetylcholine, in which the chalcogen replaces either or both of the choline oxygen atoms. The selenocholine iodide (N, R = H) and derivatives (**XLI**) (R = Ac, Bz, or PrCo) as well as selenocholine diselenide and some *N*-methyl and *N,N*-dimethyl analogs have been prepared by Gunther and Mautner (1965a). Various test systems have been used to study these compounds. These include the guinea pig ileum (Scott and Mautner, 1964), frog rectus abdominis (Scott and Mautner, 1964), phrenic nerve-stimulated rat diaphragm (Scott and Mautner, 1967), and isolated single-cell *Electrophorus electroplax* preparation (Rosenberg and Mautner, 1967).

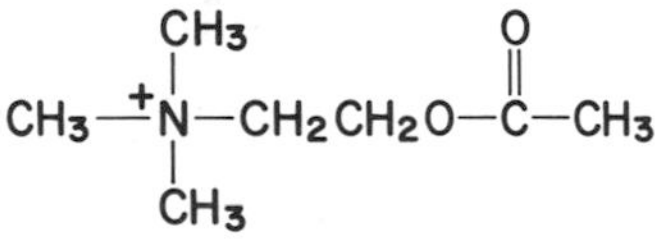

$$CH_3-\overset{\overset{\displaystyle CH_3}{|}}{\underset{\underset{\displaystyle CH_3}{|}}{N^+}}-CH_2CH_2O-\overset{\overset{\displaystyle O}{\|}}{C}-CH_3 \qquad \textbf{XL}$$

$$CH_3-\overset{\overset{\displaystyle CH_3}{|}}{\underset{\underset{\displaystyle CH_3}{|}}{N^+}}-CH_2CH_2SeR \qquad \textbf{XLI}$$

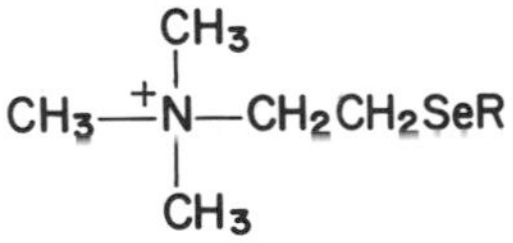

When the sulfur and selenium derivates of 2-dimethylaminobenzoate (**XLII**) (A = O, S, or Se) were compared to the oxygen derivative in regard to their ability to block squid axon electrical activity (Rosenberg and Mautner, 1967), the sulfur analog was found to have ten times and the selenium analog about 30 times the potency of the oxygen analog. Compounds had enhanced activity if the carbonyl group was replaced with a thiocarbonyl group. In addition, all compounds of the series were also tested at the synaptic junction of the electroplax. In this test system the relative activities were also in the order Se < S < O, which suggests a similar type of reactivity at the receptor sites (Mautner *et al.*, 1966).

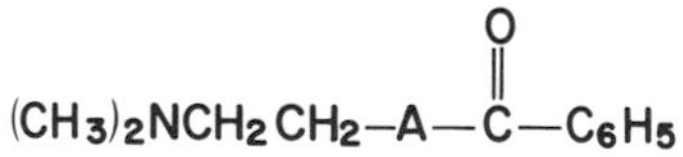

$$(CH_3)_2NCH_2CH_2-A-\overset{\overset{\displaystyle O}{\|}}{C}-C_6H_5 \qquad \textbf{XLII}$$
$$(A=O, S, Se)$$

The carbonyl oxygen atom of benzoylcholine (**XLIII**) was replaced by a thiocarbonyl functional group and the oxygen, sulfur, and selenium derivatives were tested in the single-cell electroplax preparation (Chu and Mautner, 1968) as well as in squid giant axons (Rosenberg and Mautner,

1967). The selenium analogs showed the greatest electrical inhibitory activity in both the synaptic and the axonal preparations.

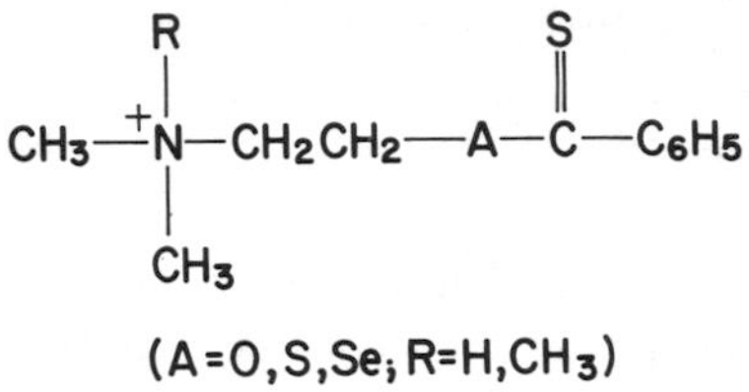

$$CH_3 - \overset{R}{\underset{CH_3}{\overset{|}{\underset{|}{N^+}}}} - CH_2CH_2 - A - \overset{S}{\overset{\|}{C}} - C_6H_5 \qquad \textbf{XLIII}$$

$$(A = O, S, Se; R = H, CH_3)$$

Chu *et al.* (1972) synthesized some derivatives of selenocarbonyl esters of acetylcholine (**XLIV**) and the selenocarboxamides (**XLV**). Several of these derivatives had considerable activity in blocking axonal conduction.

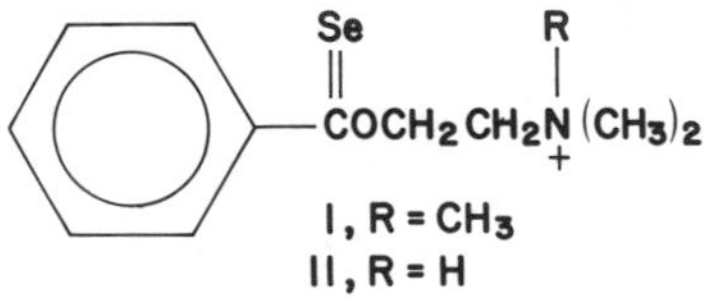

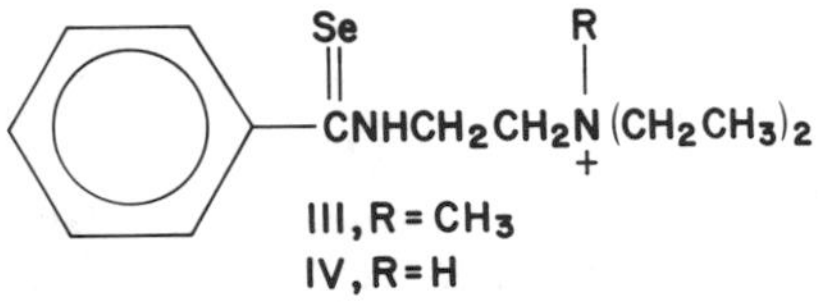

$$\overset{Se}{\overset{\|}{COCH_2}}CH_2\overset{R}{\underset{+}{\overset{|}{N}}}(CH_3)_2 \qquad \textbf{XLIV}$$

I, R = CH_3
II, R = H

$$\overset{Se}{\overset{\|}{CNHCH_2}}CH_2\overset{R}{\underset{+}{\overset{|}{N}}}(CH_2CH_3)_2 \qquad \textbf{XLV}$$

III, R = CH_3
IV, R = H

Numerous sulfur-containing derivatives of phosphoric, phosphonic, as well as pyrophosphoric acids have been found to be active inhibitors of cholinesterase (Heilbronn-Wilkstrom, 1965). Four related selenium organophosphorous compounds have been synthesized by Akerfeldt and Fagerlind (1967). These compounds exhibit potent inhibition of cholinesterase when tested in human erythrocyte cholinesterase. One of the compounds, selenophos, *o*-ethyl-Se-(2-diethylaminoethyl)ethylophosphonoselenoate (**XLVI**), was applied for 30 min in concentrations as great as $5 \times 10^{-3} M$ (Hoskin *et al.*, 1969). No blockage of the conduction was observed in the squid giant axon unless the latter was pretreated with the venom from a cotton mouth moccasin. Then selenophos caused a marked and irreversible reduction in the action potential. Selenophos has an extremely low LD_{50} of 9.21 mg/kg when administered to mice subcutaneously (Akerfeldt and Fagerlind, 1967).

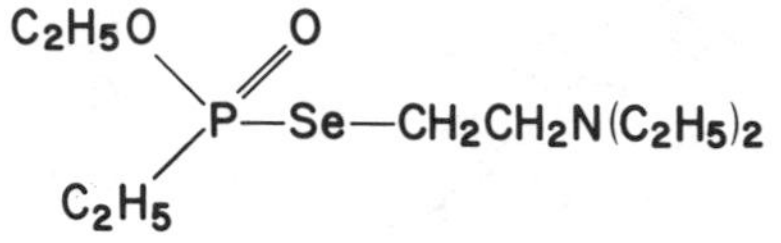

$$C_2H_5O \diagdown \underset{\underset{C_2H_5}{|}}{P} (=O) - Se - CH_2CH_2N(C_2H_5)_2 \qquad \textbf{XLVI}$$

9.6 Compounds that Affect the Circulatory System

The most potent of the four selenophene derivatives studied in a test for antiarrhethmic activity on isolated rat atria (Kudrin and Zaidler, 1968) was the compound known as Sc-5 (**XLVII**) (R = α-pyridyl, X = Cl). Se-5 was not only more effective at a lower dose than procaine amide but it also had a longer duration of action.

1-Phenyl-2-aminoethanols of the type shown in structure **XLVIII** where R^1 may be RSe or $RSeO_2$ have been shown to have peripheral vasodilating activity. Moreover these derivatives have been suggested for use in the treatment of cardiac arrhythmias (Buu-Hoi et al., 1970).

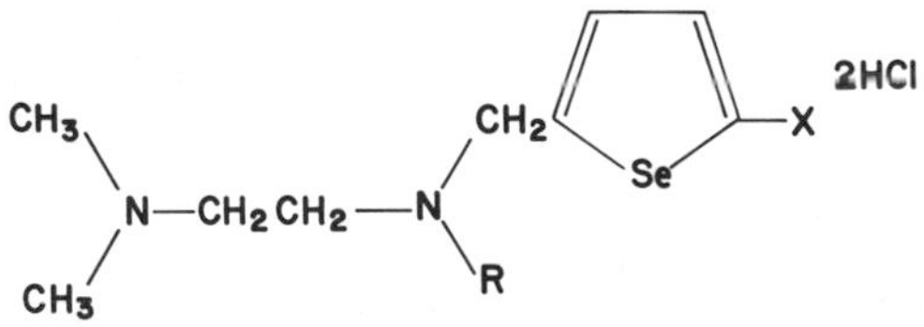

$$\underset{CH_3}{\overset{CH_3}{\diagdown}} N - CH_2CH_2 - \underset{R}{N} \diagup CH_2 \text{-(selenophene)} - X \cdot 2HCl \qquad \textbf{XLVII}$$

$$\underset{R^2}{\overset{R^1}{\diagdown}} (\text{phenyl}, R^3) - \underset{OH}{\overset{}{CH}} - \underset{R^4}{\overset{}{CH}} - N \underset{R^6}{\overset{R^5}{\diagup}} \qquad \textbf{XLVIII}$$

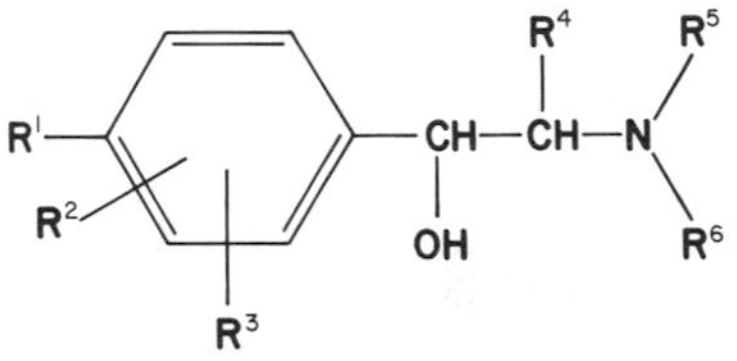

9.7 Anti-inflammatory Compounds

Roberts (1963) has reported the in vitro anti-inflammatory properties of benzylselenovaleric acid, selenocystine, and γ,γ'-diseleneovaleric acid. In this test system the compounds β,β-diselenodipropionic acid and diphenyl selenide were ineffective.

Phillips et al. (1967) found that the compound 5,5'-selenobissalicylic acid (**XLIX**) was more active than acetylsalicyclic acid. This compound was active in vitro but not in in vivo anti-inflammatory assays.

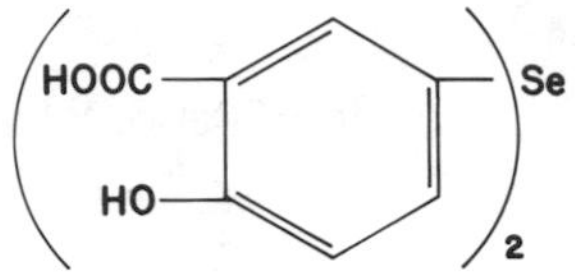

XLIX

9.8 Antihistamines

There are two important classes of sulfur-containing antihistamines. One group is the 10-substituted phenothiazines (exempliflied by the powerful drug promethazine) and the ethylenediamine derivative substituted by a 2-phenyl group. Only one selenium isomer of the pheothiazine class has been examined: 10-(3-dimethylaminopropyl) phenoselenazine (L, selenopromazine) (Rosell and Axelrod, 1963). These investigators have studied the uptake of noradrenaline^{-3}H by rat heart, and observed that L had greater antihistaminic activity than promazine. Moreover, its antiadrenalin action was about the same order of magnitude as promazine. The selenium analogs of chlorpromazine (LI) have been isolated by Miller *et al.* (1959) and were shown to have antihistaminic activity comparable to that of their sulfur analogs. Methapyrilene (LII) (thenylene) exemplified the second group of sulfur-containing antihistamines. Because of their diverse properties, selenium isomers of the methapyrilene-type compounds (XLVII) have been extensively studied by several investigators in the Soviet Union. The compounds Se-4 (XLVII) (R = *d*-pyridyl, X = H) (Chernysheva and Eliseeva, 1966), Se-5 (XLVII) (α-pyridyl, X = Cl), Se-6 (XLVII) (R = *d*-pyridyl, X = Br) (Chernysheva, 1966, and Se-1 (XLVII) (R = phenyl, X = H) were observed to be two to five times more active as antihistamines than their sulfur analogs. Moreover their length of action was 12 times longer (Chenysheva and Galbershtam, 1967). All of these compounds were found to inhibit deamination of tyramine and dopamine (Romanova *et al.*, 1968). Se-4 also demonstrated preventive action against traumatic shock at 1–5 mg/kg in dogs (Chernysheva *et al.*, 1969).

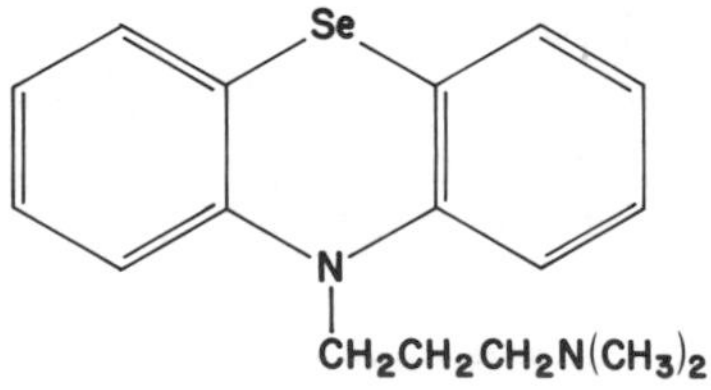

L

CH$_2$CH$_2$CH$_2$N(CH$_3$)$_2$

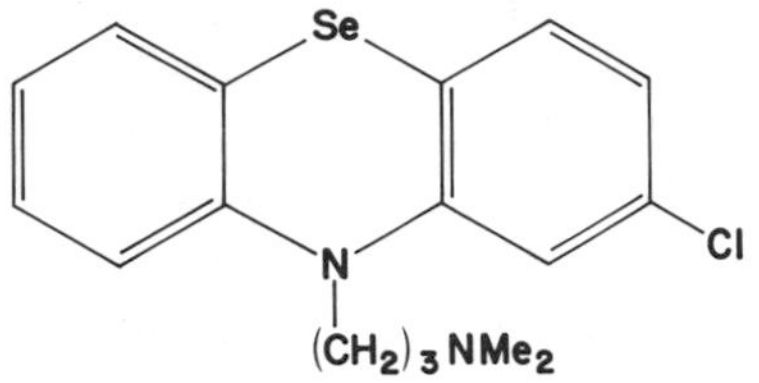

LI

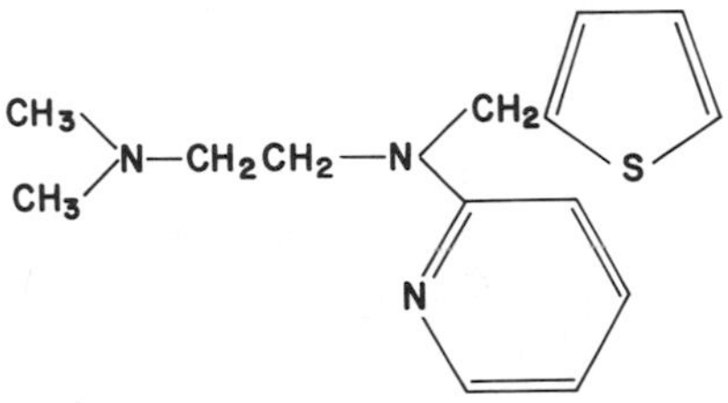

LII

Against histamine-induced cardiac impairment in the guinea pig, the compound Se-5 was found to have the greatest activity (Chernysheva and Eliseeva, 1966). Bodkov (1969) has recommended this compound for routine use in heart surgery. Se-6 was shown to be effective in the treatment of toxemia induced in pregnant rabbits by the administration of placental extracts from women with toxemia late in their pregnancy (Lotis et al., 1967). If Se-6 was administered to these rabbits, they bore fetuses with fewer abnomalities (Korzhova and Smirnova, 1969).

The selenophene derivatives have relatively low toxicity. The LD_{50} in mice, for example, for Se-1 (**XLVII**) (R = C_6H_5, X = H), equals 160 mg/kg; for Se-5 equals 340 mg/kg; and for Se-6 equals 215 mg/kg (Chernysheva and Kudrin, 1967).

9.9 Anticancer Agents

Some types of cancer have had good response to treatment by antagonists of purine and pyrimidine metabolism. One of the more important compounds thus far in the chemotherapy of certain leukemias has been the potent derivative 6-mercaptopurine (**LIII**) (6-MP) which was discovered by Elion et al. (1952). The antitumor acitivty (LePage and Howard, 1963) of 6-mercaptopurine is suggested to result from the incorporation into DNA with ultimate cell death. Mautner (1956) first prepared its selenium analog, 6-selenopurine (**LIV**) (6-SP), which was observed (Mautner and Jaffe, 1958) to inhibit mouse leukemia LIZIO as effectively as 6-MP.

The selenopurine was not as effective as 6-MP and was more toxic when administered to mice having leukemia L5178Y or sarcoma 180. Both 6-SP and 6-MP have inhibited the incorporating of formate and, to a lesser degree, adenine in the Ehrlich ascites tumor cell system (Mautner, 1958). Subsequent studies (Jaffe and Mautner, 1960) showed that 6-SP, unlike 6-MP, is unstable at body temperature, having a half-life of only 6 hr, making it a less satisfactory compound from the standpoint of stability. The antitumor activity was decreased if the Sm or Se of 6-MP or 6-SP was methylated (Mautner and Jaffe, 1958).

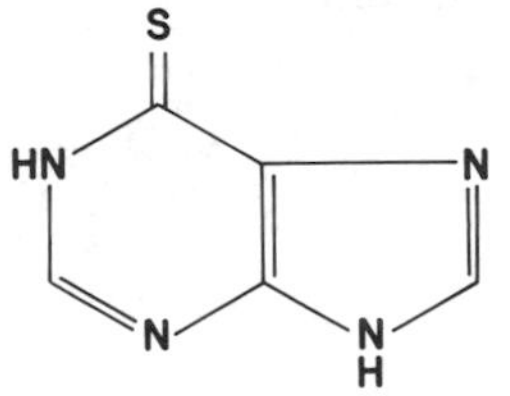

LIII

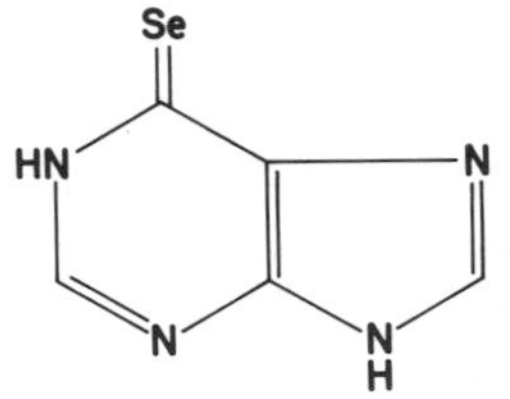

LIV

Selenoguanine (**LV**), selenocytosine (**LVI**) and its 5-methyl derivative, and diselenothymine (**LVII**) have also been synthesized by Mautner *et al.* (1963). The selenium analog of thioguanine as well as thioguanine have shown similar antitumor activities against mouse neoplasms. Townsend and Milne (1970) have synthesized 6-selenoguanosine (**LVIII**). This compound has been tested by Chu (1972), who reported that its activity was approximately equal to thioguanine in mice bearing L5178Y lymphomas. The Se-methyl derivates of selenoguanosine as well as selenoguanine were less active than thioguanine. With the Sarcoma-180 ascites tumor system in mice Ross *et al.* (1973) tested both the 6-thio and 6-seleno analogs of guanine and guanosine. The selenium-containing analogs were somewhat superior to the sulfur-containing compounds in antitumor and therapeutic index. Moreover, the toxicities of several other Se-methyl derivatives of 6-selenopurines were observed to be four to six times greater than those of the selenopurines themselves (Bergmann and Rashi, 1969). The greater toxicity may be due to the *in vivo* formation of methaneselenol. The 8-selenopurines (**LIX**) have been synthesized by

Carr *et al.* (1958) as potential purine antimetabolites, but the results were unimpressive in trials against Sarcoma-180. The ethoxycarbonyl derivates of 2-amino-6-selenopurine and their sulfur analogs were found to be virtually uneffective as antitumor agents (Dyer and Minnier, 1968).

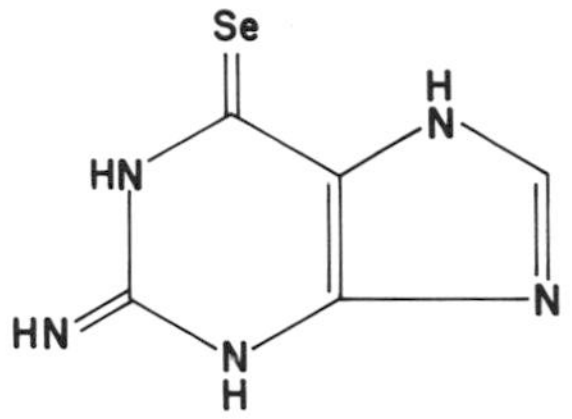

LV 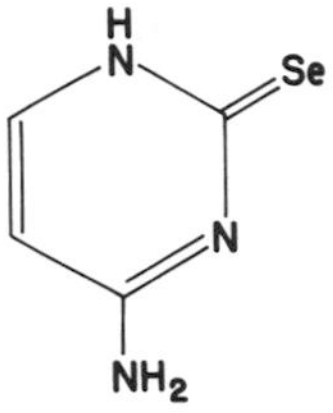   LVI 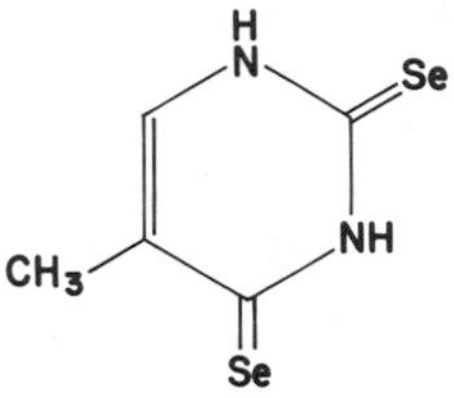   LVII

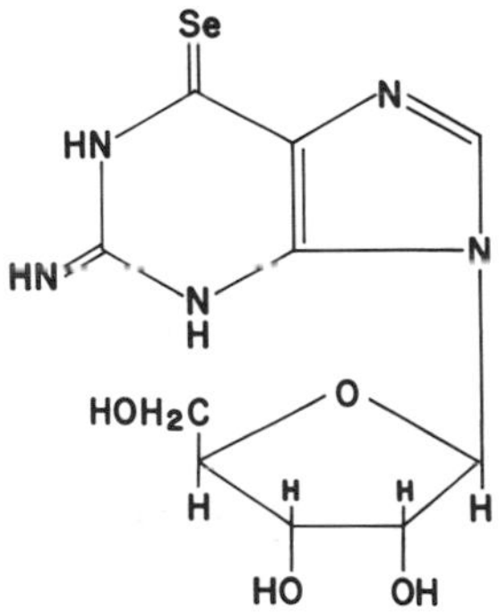

LVIII

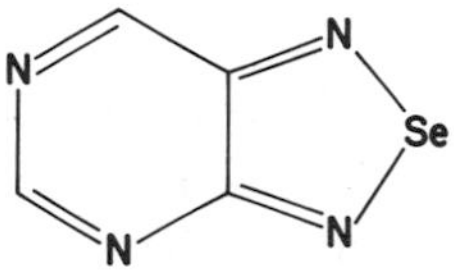

LIX

Milne and Townsend (1974) have synthesized the α (**LX**) and β (**LXI**) anomers of 2-deoxy-6-selenoguanosine. Both of the anomers possessed some antitumor activity against leukemia L-1210. The β anomer was slightly more toxic. Alkylation of (**LXI**) and **LXII** appeared to effect a marked decrease in antitumor activity. The 6-selenoguanosine and the 6-alkylselenoguanosine derivates were found to be the most active compounds in the 6-selenoinosine and 6-selenoguanosine area. In fact, the derivative 6-(1-methyl-4-nitroimidazol-5-yl) selenoguanosine was more active than 6-selenoguanosine. This may be of interest since the reverse appears to be true for the 2-deoxynucleosides. This suggests that the mode of action for this series of nucleosides may be different. The 6-alkylselenopurine is more toxic than the guanosine analogs. Maeda *et al.* (1981) have synthesized the thioxanthine-platinum (II) complexes of selenoguanine and selenoguanosine. These compounds showed medium-strength activity against L-1210 cells in mice and in an *in vitro* system. These platinum complexes also showed very low toxicity.

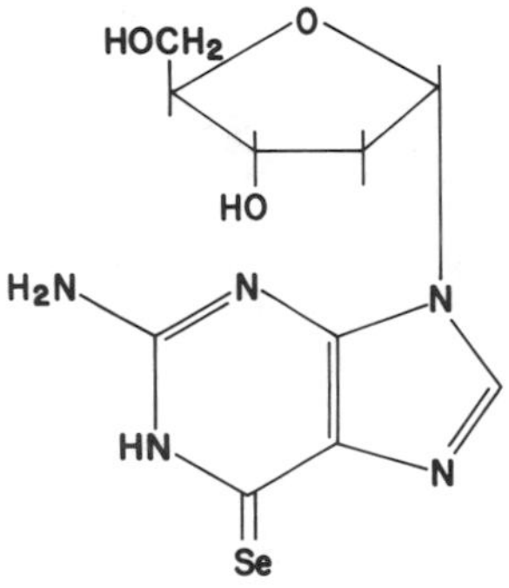

LX 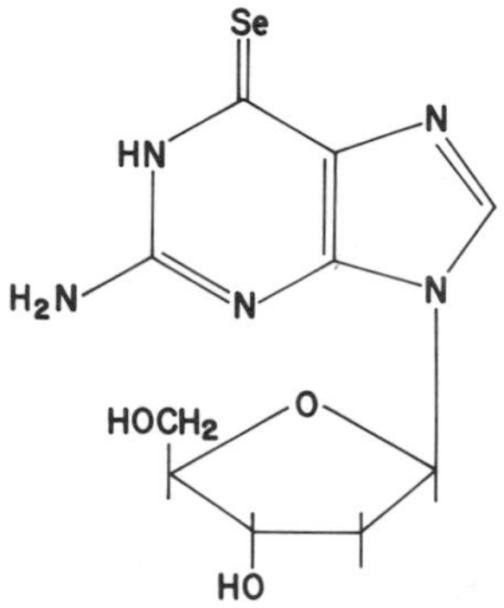   LXI 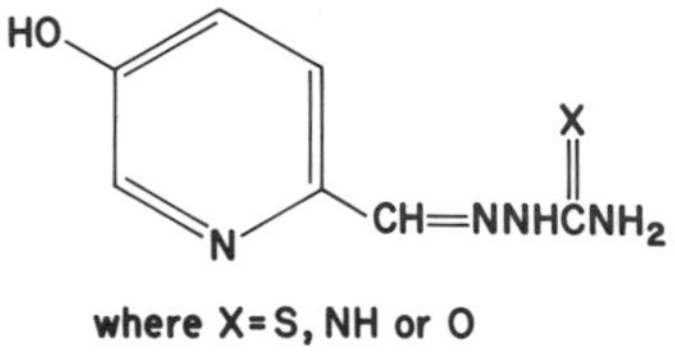   LXII

Several analogs of 5-hydroxy-2-formylpyridine thiosemicarbazone (5-HP) (**LXII**) that contain isosteric replacement of sulfur by Se, NH, or O have been synthesized by Agrawal *et al.* (1974). Measurement of the antineoplastic activity of these compounds in mice with Sarcoma-180 ascites cells have indicated that 5-HP was the most active of the compounds synthesized. The seleno analog was intermediate in activity, but the guanyl hydrazone and the semicarbazone were inactive against this tumor. Moreover, 5-HP and its seleno analog both caused marked inhibition of DNA synthesis *in vitro*, as determined by measuring the incorporated of thymidine-methyl-^{3}H, 5-^{3}H-cytidine, or adenine-8-^{14}C into DNA. These agents, did not affect the syntheses of RNA and protein under the conditions employed. All of the compounds were similar in that none of the compounds prevented the incorporation of 5-^{3}H-cytidine into acid-soluble pyrimidine ribonucleotides, but 5-HP and its seleneosemicarbazone markedly by depressed the incorporation of radioactivity into pyrimidine deoxyribonucleotide pools, which suggested that these two derivatives inhibited the enzyme ribonucleoside diphosphate reductase *in situ*. Both 5-HP and the seleno analog inhibited the isolated enzyme ribonucleoside disphosphate reductase from Novikoff rat hepatoma; in order to achieve 50% inhibition a concentration of 3.5 $\times$ 10^{-6} and 6.8 $\times$ 10^{-6} M were required, respectively.

Some 4-amino-1,2,5-selenadiazoles (**LXIII**) have been reported by

Shealy and Clayton (1967) to be cytotoxic to KB cells in culture. The related derivatives, pyrimido[4,5-c][2,1,3]selenadiazoles were ineffective against mouse leukemia L1210 (Endo *et al.*, 1963). Other selenium-containing heterocycles have been tested. Compound **LXIV** has been reported by Takeda *et al.* (1955) to be the most active against Ehrlich ascites tumor cells of ten selenium compounds screened. Compound **LXIV** was even more effective than nitrogen mustard *N*-oxide. Other selenium derivatives such as 2-phenylbenzoselenazole-4'-arsonic acid (Bogert and Stull, 1927) and 3-methyl-2-benzoselenazolium iodide (Mueller and Phillips, 1967) have been found to lack activity.

Because leukemic leukocytes have been shown to require L-cysteine and L-cystine for continued cell growth, Weisberger and Suhrland (1956a) experimented with selenocystine as an agent which might arrest their multiplication. Selenocystine was orally administered to two patients with acute leukemia and to two patients with chronic myeloid leukemia in a daily average dose of 100 mg. The total leukocyte count showed a rapid decrease and the spleen was reduced in size. In one case, a patient who was resistant to 6-MP appeared to reacquire sensitivity. It was impossible to administer the selenocystine long enough to determine whether a remission could be obtained, because of marked side effects, notable nausea and severe vomiting. The leukocyte count was not decreased in one patient who was given diphenyl diselenide (Weisburger and Suhrland, 1956a). Selenocystine was found in other studies to inhibit the incorporation of ^{35}S-L-cystine by rat Murphy lymphosarcoma tumor cells *in vitro* and *in vivo* (Weisburger and Suhrland, 1956b). Selenocystine also decreased by about one half the incorporation of the ^{35}S-L-cystine into leukemic leukocytes (Weisburger *et al.*, 1956). Inorganic selenium and diphenyl diselenide derivatives were found to be ineffective in the latter system.

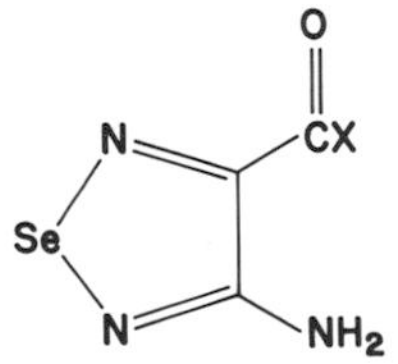

LXIII

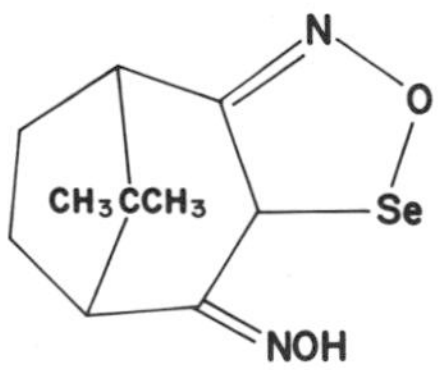

LXIV

Because *N*-hydroxyurea has been found to be antileukemic, Adamson (1965) has tested a series of urea-related compounds, including selenourea, against LIZIO mouse leukemia. Selenourea did not increase the survival time of the animals. Senning and Sorensen (1966) in preliminary tests in mice and rats found that α-w-polymethylenebisselenouronium dibromides (**LXV**) were carcinostatic.

Steroids with various selenium moieties in the 3-position were tested at 10 mg/kg against Dunning leukemia. (Segaloff and Gabbard, 1965). No significant increase was observed in the survival time of the test animals. Some *O*- and *p*-substituted diphyenyl mono- and diselenides have been tested by Matti (1940), but no activity was found.

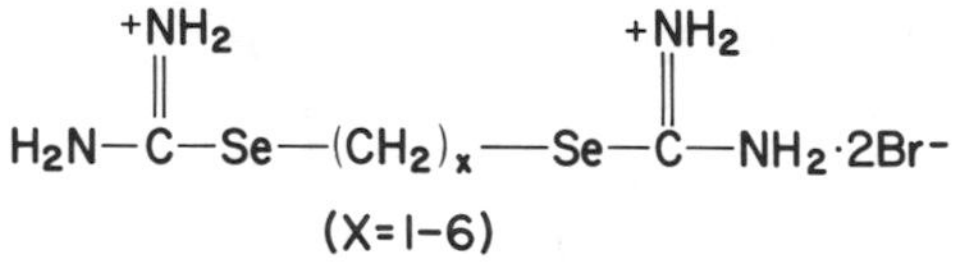

$$\overset{\overset{\displaystyle +NH_2}{\|}}{H_2N-C}-Se-(CH_2)_x-Se-\overset{\overset{\displaystyle +NH_2}{\|}}{C}-NH_2 \cdot 2Br^- \qquad \textbf{LXV}$$

$$(X=1-6)$$

9.10 Antiradiation Agents

The goal of numerous groups throughout the world since the beginning of the atomic age was to develop a drug that, when given before exposure to lethal ionizing radiation, prolongs the life of the exposed organisms. Most all effective antiradiation agents contain sulfur and are chemically related to 2-aminoethanethiol, H_2NCH_2SH.

Selenophenol was the first selenium-containing compound tested for radioprotective activity. When administered to mice ip (0.2 or 0.4 mg/g) in olive oil 10 min before X-irradiation, selenophenol was found by Bacq *et al.* (1963) to be ineffective.

The *in vitro* radioprotective abilities of the selenium-containing amino acids D,L-selenomethionine and D,L-selenocystine have been compared to the known radioprotectants L-cystine, 2 aminoethylthiopseudourea dihydrochloride and 2-aminoethanethiol hydrochloride (Shimazu and Tappel, 1964). The seleno-amino acids were more effective than methionine and the sulfur radioprotectants in their ability to protect the various amino acids and the enzymes yeast alcoholic dehydrogenase and ribonuclease against γ rays. The protective ability of selenomethionine to radiation was attributed to its tendency to form stable radical intermediates which are able to combine with free H atoms or electrons to return the activated selenomethionine to its original state. The ability of seleno-amino acids to protect against radioactive damage may be a clue to the biological antioxidant role of selenium. Schwarz (1965) noted that the selenium levels tested by Shimazu and Tappel (1964) were considerably greater than those found in normal tissue. The radioactive protection of the seleno-amino acids may be related to the experiments of Dickson and Tappel (1969), who have observed that selenomethionine and selenocystine bind reversibly to the sulfhydryl enzymes papain and glyceraldehyde-3-phosphate dehydrogenase to protect them from oxidative inactivation.

Several analogs of well-known mammalian radioprotective agents have been synthesized and screened for their antiradiation activity. The compounds 2-aminoethaneselenol (selenocysteamine) hydrochloride (Klayman, 1965), bis(2-aminoethyl) diselenide (selenocystamine) dihydrochloride (Klayman, 1965), 2-aminoethaneselenosulfuric acid (Klayman, 1965), and 2-aminoethylselenopseudourea dihydrobromide (Chu and Mautner, 1962) are considerably more toxic than their sulfur counterparts and are devoid of activity.

Selenourea has been observed to be superior to cysteine as a radioprotectant for rats irradiated at sublethal (600 R), lethal (750 R), and supralethal (950 R) dosages (Badiello et al., 1967). Selenourea has been shown to protect amino acid from radiation damage in solution by efficiently scavenging free radicals (Badiello and Fielden, 1970). Selenourea, in pulse radiolysis studies, was superior to thiourea and urea in competing for free radicals formed by the radiolysis of water (Badiello and Fielden, 1970). Badiello et al. (1971) found the E. coli B/r irradiated in the presence of 5×10^{-2} M selenourea showed an increased survival when irradiated in air and the opposite effect in nitrogen.

Selenomethionine, colloidal selenium, selenoxanthene (LXVI), selenoxanthone (LXVII), and selenochromone (LXVIII) were evaluated by Breccia et al. (1969) in rats that had been exposed to 600, 750, and 900 R. A protective ability was demonstrated similar to that of cysteine and in some cases activity superior to it.

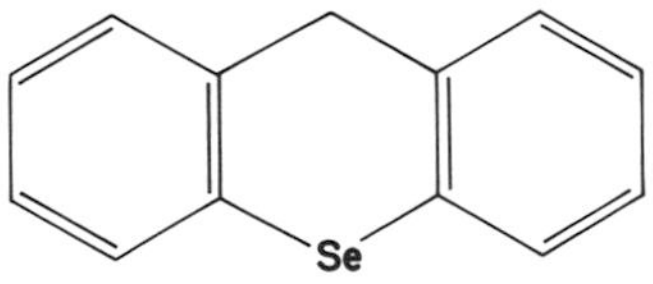

LXVI

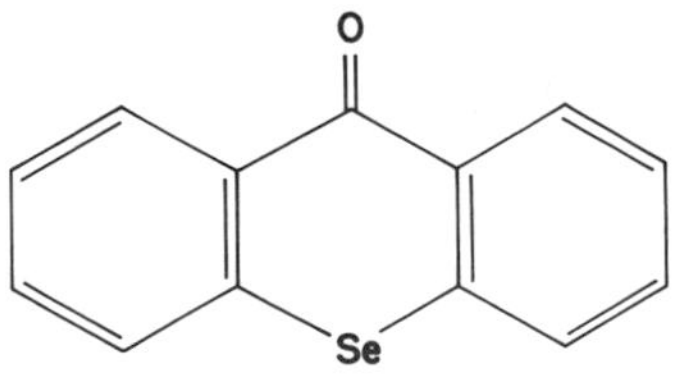

LXVII

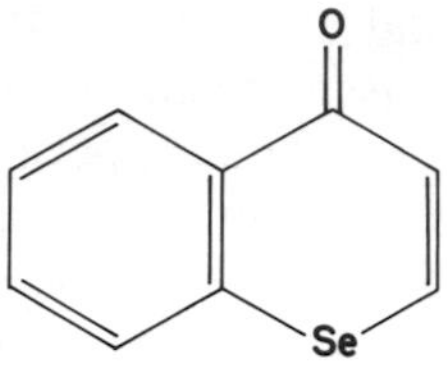

LXVIII

Various selenium heterocycles have been studied for their ability to protect ATP from losing orthophosphate on irradiation (Brucker and Bulka, 1966). Of those heterocycles studied only 2-amino-4,5-dimethyl-selenazole hydrochloride (**LXIX**) showed any radioprotective ability.

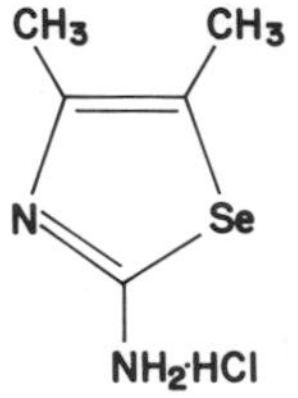

LXIX

Other 2-aminoselenazoles including selenosemicarbazide and ace-toneselenosemicarbazone failed to protect the spores of the mushroom, *Phycomyces blakesleeanus* from the effects of radiation, and in addition showed a strong radiation-sensitizing effect (Brucker and Rohde, 1968). Kozak *et al.* (1976) have demonstrated the protective effect of 2-amino-selenazoline-protected white mice against 500 R/min. However, Se-ami-noethylisoselenouroniumbromide hydrobromide and the selenium analog of mercaptoethylguanidine were ineffective.

When sodium selenate at 4.2–4.6 mg/kg was administered to rats intraperitoneally 55–60 min postirradiation (800 R), followed by an additional 80% subcutaneously, all of the animals survived (Hollo and Zlatarov, 1960).

Radioprotection may result when sulfur-containing antiradiation agents form unsymmetrical disulfides ($RSSR^1$) with the sulfhydryl groups from protein. The oxidation resulting from the radiolysis of water in the tissues would therefore be prevented, and the protein sulfhydryl groups are restored to their normal state, probably due to some enzymatic reaction. On the other hand, if thioselenates (RSSeR) occur in living organisms, it is likely that their great reactivity must severely limit the duration of their existence (Roy *et al.*, 1970).

Another possible mechanism of protective action has been demonstrated by Kuliev *et al.* (1981). They showed that X-irradiation of rats decreased hepatic DNA and RNA. However, pretreatment 3 hr before irradiation with Na_2SeO_3 (3 mg/kg, ip) prevented the radiation-induced decreases in nucleic acids, maintaining DNA and RNA levels at 99.0% and 95.5%, respectively, of control values. Pretreatment with three divided doses of Na_2SeO_3 was slightly more effective than a single dose. The results of these experiments suggest that selenium may be acting as a radioprotection agent by stabilizing RNA and DNA.

9.11 Steroids

A series of steroids related to dehydroepiandrosterone and pregnenolone in which the 3β-hydroxy group is replaced by selenium or sulfur-containing group has been prepared by Segloff and Gabbard (1965). Because the selenols are prone to air oxidation, they were converted to the stable benzoate esters. The diselenide forms were compared with their analogous disulfides. Both types were tested for their androgenicity, antiandrogenicity, progestogenicity, antiprogestogenicity, and the inhibiting effect on ovulation and implantation. The selenium and sulfur isoesters of dehydroepiandrosterone and prognenolone were similar and both were moderately androgenic, but were much less active than the parent steroids.

The diselenide 3β,3β'-diselenobisandrost-5-en-17-one (**LXX**) has been observed to effectively inhibit the endometrial proliferation brought about by progesterone in the rabbit, but this compound was found to be ineffective in the inhibition of ovulation and implantation in the rat. Hiscock et al. (1970) have reported the synthesis of 3α- and 3β-selenyl benzoate derivatives of cholestane and androstanone. Wolff and Zanati (1970) have found the steroids of type **LXXI** to have good androgenic activity when the heteroatom (Z) was sulfur, selenium, or tellurium, but was absent when oxygen was the substitute for (Z). The authors surmised from this observation that the steric rather than the electronic requirements at C-2 and/or C-3 of androgens determine their activity.

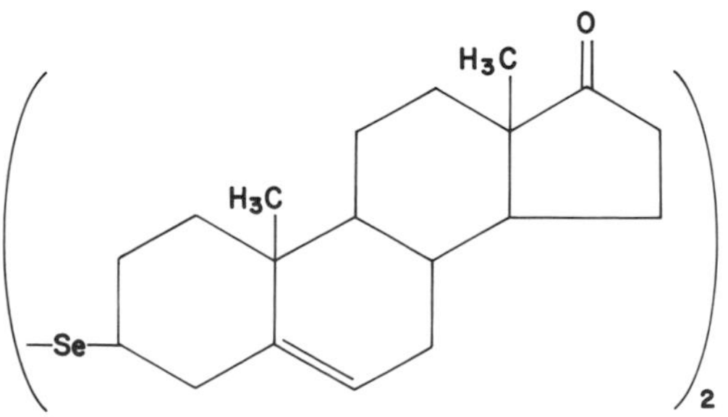

LXX

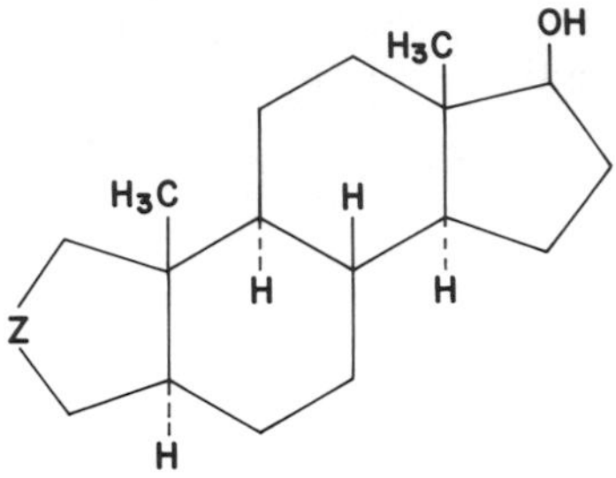

LXXI

Sadek *et al.* (1981) in an effort to produce effective breast tumor imaging agents synthesized and characterized a series of Se-labeled steroids. Starting with natural estrone derivatives nonradioactive selenium at positions 3, 16, and 17 were obtained. Estrogen-receptor assays reveal that 17a[(phenylseleno)methyl]-17B-estradiol (**LXXII**) retains about 12% of the binding activity of 17B-estradiol.

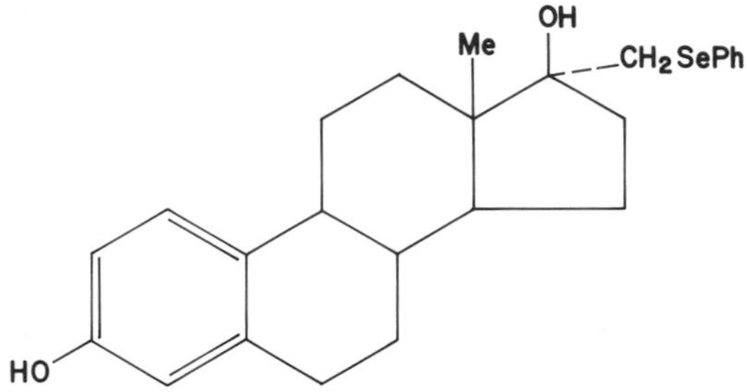

LXXII

The syntheses of phenylselenium-substituted progesterone [21-(1) and 17α-(phenylseleno) progesterone (2)] (**LXXIII**) and testosterone [16β-(3) and 16α-(phenylseleno) testosterone (4)] (**LXXIV**) have been described along with data which help to establish the stereochemistry of the substituents in the testosterone molecules (Konopelski *et al.*, 1980). The progesterone derivatives 1 and 2 and the testosterone derivatives 3 and 4 were assayed for binding affinity to various steroid hormone receptors in a routine screening system (Ojaso and Raynaud, 1978), and the relative binding affinity in each case was determined. Apart from 21-(phenylseleno) progesterone (1), which competes quite effectively for binding to the progestin receptor, the compounds are inactive.

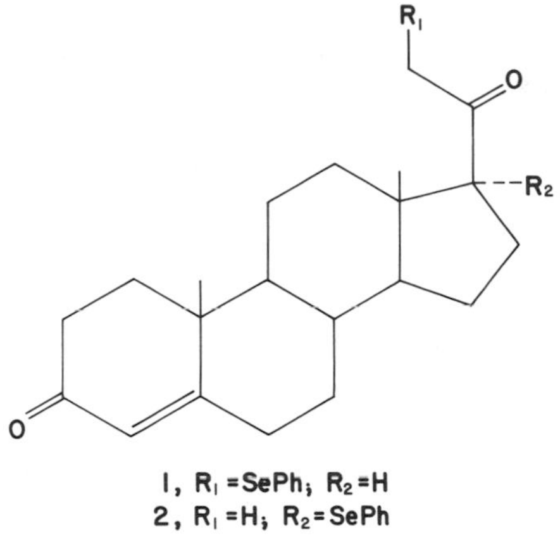

LXXIII

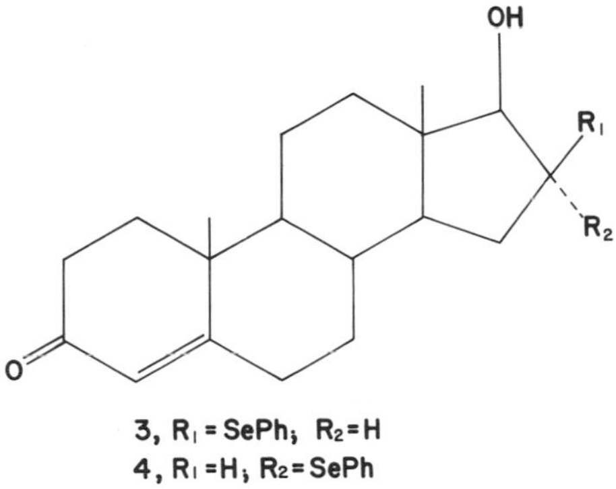

1, R$_1$ =SePh; R$_2$ =H
2, R$_1$ =H; R$_2$ = SePh

LXXIV

3, R$_1$ = SePh; R$_2$=H
4, R$_1$ =H; R$_2$= SePh

9.12 Selenocoenzyme A

Lipmann (1945) discovered the metabolic intermediate coenzyme A (CoA) in the course of investigating the ATP-dependent acetylation of sulfanilamide. From acetyl-CoA ("active acetate") the acetyl group is readily transferred to a suitable acceptor such as an aromatic amine or an amino acid. Coenzyme A has been isolated from pigeon liver extracts as well as many microorganisms. It is composed of adenosine, phosphoric acid, and pantetheine [N-(pantothenyl-β-aminoethanethiol] moieties. Pantetheine is related to pantothenic acid, a vitamin first isolated by Williams et al. (1938) and which is a biogenetic precursor of CoA.

In Lactobacillus helveticu (Gunther and Mautner, 1960) the disulfide form of pantetheine, pantethine, and its selenium isostere, selenopantethine (**LXXV**), were biologically equivalent on a molar basis. This organism has a requirement for pantethine and is believed to be capable of converting it to CoA.

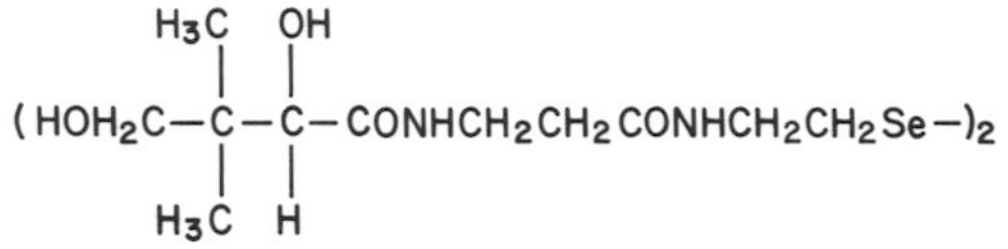

H3C OH
(HOH2C—C—C—CONHCH2CH2CONHCH2CH2Se—)2
H3C H

LXXV

Gunther and Mautner (1965b) have achieved the complete sythesis of selenocoenzyme A (**LXXVI**). Initially the compound was obtained as the diselenide in a mixture with isoselenocoenzyme A, the ribose-2-phosphate. Selenobenzoate esters of the CoA analogs resulted from the reductive acylation with 2-dimethylaminoethyl selenobenzoate. These analogs were then separated by ion exchange chromatography on ECTEOLA–cellulose. Aminolysis of the separated products finally yielded Se-CoA in the selenol form and isoselenocoenzyme A. These latter two substances are subject to rapid air oxidation to the symmetrical diselenides.

NH2
N
N
N
N
O
O
CH3 OH
CH2OP—O—POCH2C—CCONHCH2CH2CONHCH2CH2SeH
OH
OH
CH3 H
O
HO
OPO3H2

LXXVI

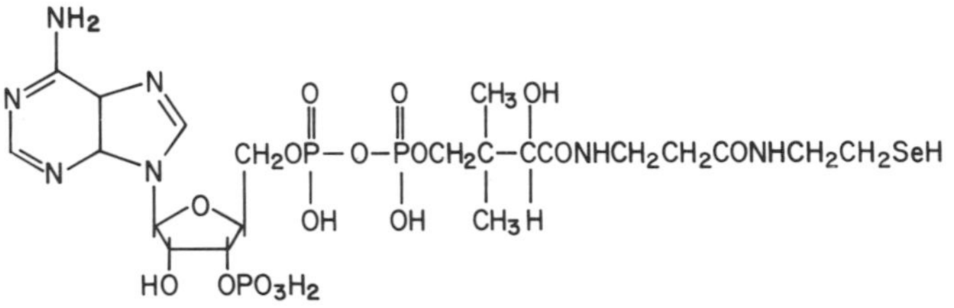

The diselenide form of Se-CoA had no CoA activity in several enzyme systems. In the acetylcoenzyme A synthetase (E.C.y.2.1.1. system Se-CoA was fully functional, however, in the presence of dithiothreitol, a reagent which is able to reduce diselenides to selenols under physiologic conditions (Gunther, 1967). Even though enzyme-binding constants for the sulfur and selenium analogs are identical, under identical kinetic conditions Se-CoA has about one third the rate of acyl transfer. *In vitro* Se-CoA functions as a partial competitive antagonist of coenzyme A.

9.13 Selenium-Containing Carbohydrates

Schneider and Wrede (1917) have reported the preparation of selenoisotrehalose, a disaccharide where a Se atom links two glucose molecules. The taste of the selenium analog was similar to that of the oxygen sugar microorganisms or enzymes did not hydrolyze the new species, but unexpectedly the selenoisotrehalose was nontoxic to mammals (Wrede, 1917). After selenoisotrehalose was injected subcutaneously into rabbits (600 mg/kg), urinary excretion in 24 hr accounted for 85% of the unchanged material.

Several other selenocarbohydrates have been synthesized. These include diglucosyl diselenide (Schneider and Beuther, 1919), selenodigalactose (Schneider and Beuther, 1919), phenyl-seleno-β-D-glucopyranoside (Bonner and Robinson, 1950), 1-seleno-1-deoxy-D-glucose (Kocourek *et al.*, 1963), 5-seleno-5-deoxy-D-xylose (van Es and Whistler, 1967), 2,3,4,6-tetra-*O*-acetyl-1-Se-dimethylarsino 1 seleno-β-D-glucopryranose (Zingaro and Thomson, 1973), 1-Se-dimethylarsino-1-seleno-β-D-glucopyranose (Zingaro and Thomson, 1973), and the selenium derivatives of arylchalcogen-β-D-glucosides (Wagner and Nuhn, 1965). The general relationships found for the acid-, base-, and enzyme-catalyzed scission of the arylcholcogen β-D-glucosides as determined by Wagner and coworkers have been summarized by Klayman *et al.* (1973).

In a single case in which β-glucuronidase was found to be active for the entire series of O-, S-, or Se-substituted arylchalcogen-β-D-glucosides, the enzymatic hydrolytic rates closely followed the rate order observed for acid hydrolysis. Therefore, a protonation step does appear to be implicated for the enzyme–substrate interaction.

The possible role of the unreactive selenoglycoside species of arylchalcogen-β-D-glucosides as well as other seleno sugars as competitive enzyme inhibitors has not been studied.

9.14 Seleno-Amino Acids

Both thialysine (**LXXVII**) and selenalysine (**LXXVIII**) can subtitute for lysine (**LXXIX**) and are used by the protein synthesizing system of *E. coli*, rat liver, and preparations from rabbit reticulocytes (Fabry *et al.*, 1977). Aminoacyl-tRNA synthetase activates both analogs, which are transerred to tRNA-lysine and are incorporated into polypeptides. In addition, both analogs act as competitive inhibitors of lysine in all these reactions. Incorporation of thialysine into protein biosynthesis can be variable. For example, thialysine inhibits the replication of the Mengovirus in mammalian cell cultures, although the growth of the vesicular stomatitis virus was unaffected (Sciosica-Santoro *et al.*, 1977). Efforts to use selenolysine as an antagonist of lysine to depress the growth of cancerous cells, however, have not given promising results (Cavallini *et al.*, 1980).

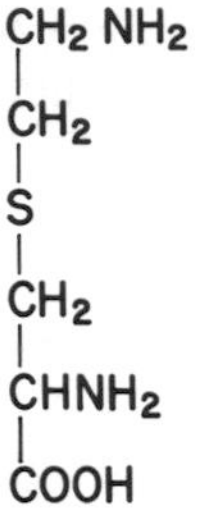

$$
\begin{array}{ccc}
\begin{array}{c}
CH_2\,NH_2 \\
| \\
CH_2 \\
| \\
S \\
| \\
CH_2 \\
| \\
CHNH_2 \\
| \\
COOH
\end{array}
&
\begin{array}{c}
CH_2NH_2 \\
| \\
CH_2 \\
| \\
Se \\
| \\
CH_2 \\
| \\
CHNH_2 \\
| \\
COOH
\end{array}
&
\begin{array}{c}
CH_2\,NH_2 \\
| \\
CH_2 \\
| \\
CH_2 \\
| \\
CH_2 \\
| \\
CHNH_2 \\
| \\
COOH
\end{array} \\[4pt]
\textbf{LXXVII} & \textbf{LXXVIII} & \textbf{LXXIX}
\end{array}
$$

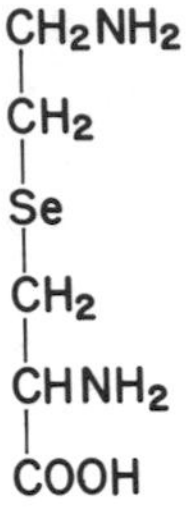

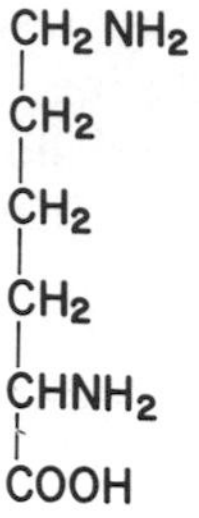

Selenalysine as well as thialysine may be a substrate for lysine de-carboxylase. The reaction rates were about the same for both selenalysine and thialysine, but the rates were about 60% of those rates when lysine was used as the substrate (Blarzino and DeMarco, 1977). On the other hand, selenalysine is more active than lysine in catalyzing the conversion of glyceraldehyde and glyceraldehyde 3-phosphate into methylglyoxal and is more active when reacting with the ketoaldehyde to give rise to a polymeric compound (Leoncini et al., 1977). DeMarco et al. (1976) have comparatively studied the oxidative deamination of Se-carboxymethyl-selenocysteine and S-carboxymethyl-cysteine and that of selenocystam-ine and cystamine. They pointed out that there were no significant dif-ferences between the seleno and sulfo compounds in regard to oxidative deamination.

Thiazolidine carboxylic acid (thiaproline) (**LXXX**) and selenazolidine carboxylic acid (selenaproline (**LXXXI**), which are, respectively, the S-containing and the Se-containing analogs of proline (**LXXXII**), are also used by the protein-synthesizing system of E. coli, rat liver, and prepa-rations from rabbit reticulocytes (Antonucci, 1977). These results as well as those with thialysine indicate that the very precise system of protein biosynthesis is unable to distinguish between an Se or S atom and a methylene carbon in the substrate.

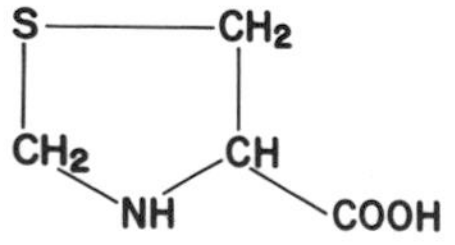

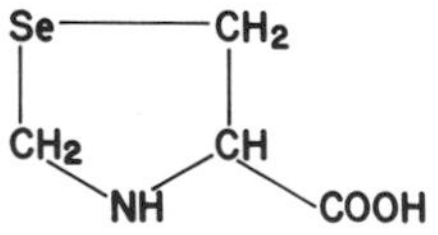

$$
\begin{array}{ccc}
\textbf{LXXX} & \textbf{LXXXI} & \textbf{LXXXII}
\end{array}
$$

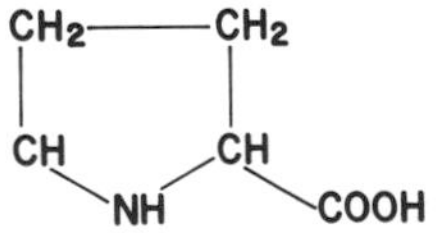

Lesser and Weiss (1912) synthesized selenosaccharin (**LXXXIII**), which was reported to have a weakly astringent rather than a sweet taste. This may result from the greater ionizability of its imide proton in com-parison to that of saccharin.

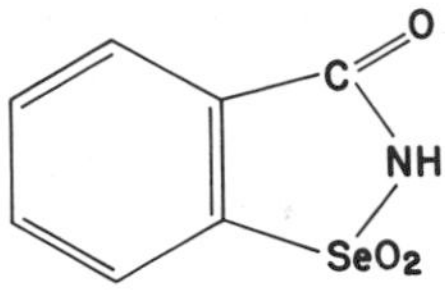

The plant growth-regulating activity of the snythetic compound [benzo(b)selenienyl-3] acetic acid (**LXXXIV**) has been assessed using four biotests (Hofinger *et al.*, 1980) growth of lentil root tips, growth of oat coleoptile segments, ethylene production by lentil root tips, and fresh weight increase and shoot formation in light- and dark-grown tobacco callus after five weeks in culture. Indolyacetic acid (IAA) was used as a standard for comparative purposes. The selenium-containing compound was much more active than IAA in all bioassays.

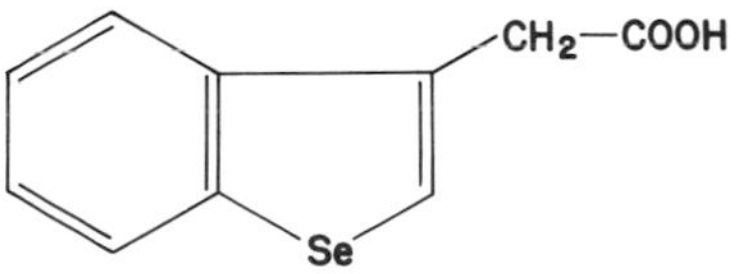

LXXXIV

Walter and duVigneaud (1965) synthesized two isosteres of oxytocin (**LXXXV**), namely, 6-seleno-oxytocin ($Z = NH_2$, X–Y = S–Se) and its deamino analog ($Z = H$, X–Y = S–Se). The replacement of sulfur by selenium in the 6 position of these two compounds yielded highly potent isologs of oxytocin and deamino-oxytocin even though they were some-what lower in potency than the hormone itself and its comparable deamino

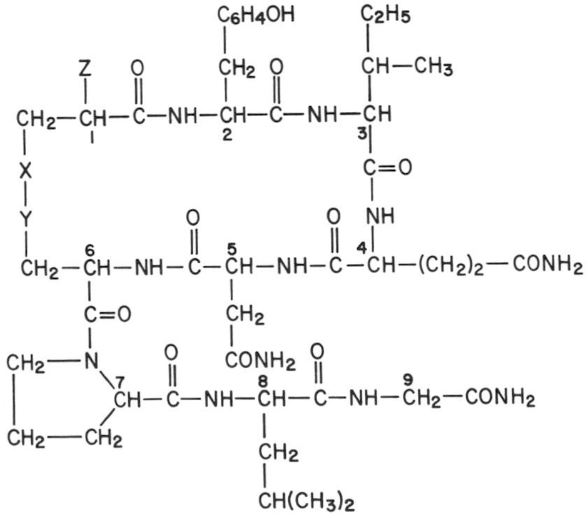

LXXXV

analog. In addition, their results also demonstrate that with 6-seleno-oxytocin, the replacement of the free amino group by hydrogen enhances the oxytocic and avian vasodepressor activities and lowers the pressor activity as it does in the case of oxytocin (Chan and duVigneaud 1972).

References

Adamson, R. H., 1965. Activity of congeners of hydroxyurea against advanced leukemia, *Proc. Soc. Exp. Biol. Med.* 119:456–458.

Aqrawal, K. C., Booth, B. A., Michaud, R. L., Moore, E. C., and Sartorelli, A. C., 1974. Comparative studies of the antineoplastic activity of 5-hydroxy-2-formylpyridine thio-semicarbazone and its seleno-semicarbazone, guanylhydrazone and semicarbazone analogs, *Biochem. Pharmacol.* 23:2421–2429.

Akerfeldt, S., and Fagerlind, L., 1967, Selenophosphorous compounds as powerful cholinesterase inhibitors, *J. Med. Chem.* 10:115–116.

Aktiebolag, A., Antimicrobial seleninic acid compositions, Brit. Pat. 1,174,753; *Chem. Abstr.* 72:59060 (1970).

Antonucci, A., Foppoli, C., DeMarco, C., and Cavallini, D., 1977. Inhibition of protein synthesis in rabbit reticulocytes by selenaproline, *Bull. Mol. Biol. Med.* 2:80–84.

Bacq, Z. M., Onkelinx, C., and Barac, G., 1963. Absence d'activite radioprotectrice du phenylselenol chez la souris, *C.R. Soc. Biol.* 157:899–901.

Badiello, R., and Fielden, E. M., 1970. Pulse radiolysis of selenium-containing radioprotectors: I. selenourea, *Int. J. Radiation Biol.* 17:1–14.

Badiello, R., Trenta, A., Mattii, M., and Moretti, S., 1967. Azione radioprotettiva dei seleno-derivati: effeto della selenourea "in vivo," *Med Nucl. Radiobiol. Lat.* 10:57–68.

Badiello, R., DiMaggio, D. D., Quintiliani, M., and Sapora, O., 1971. The influence of selenurea and of colloidal selenium on the survival of *E. coli* B/r after X-irradiation, *Int. J. Radiat. Biol.* 20:61–68.

Bednarz, K., 1957. Synthesis and biological properties of some new selenosemicarbazides, *Dissertationes Pharm.* 9:249–254. *Chem. Abs.* 52:8083 (1958).

Bednarz, K., 1958. Synthesis and biological properties of some new selenosemicarbazones, Dissertationes Pharm. 10:93–98.

Bekemeir, H., Schmollack, W., Feicht, H. J., and Lemitzer, K. H., 1976. The pharmacology of some 1,3-dioxolanes, 1, 3 oxathiolanes and 1, 3-dithiolanes, *Pharmazie* 31:317–323.

Bergmann, F. and Rashi, M., 1969. Synthesis of new 6-selenopurines by selenohydrolysis of 3-alkyl-6-methylthiopurines, *Israel J. Chem.* 7:63–71.

Billard, W., and Peets, E., 1974. Sulfhydryl reactivity: Mechanism of action of several antiviral compounds—selenocystine, 4-(α-propinyloxy)-β-nitrostyrene and acetylaranotin, *Antimicrob. Agents Chemoth.* 5:19–24.

Blarzino, C., and DeMarco, C., 1977. Selenalysine as subtrate of lysine decarboxylase, *It. J. Biochem.* 26:444–450, 1977.

Bodkov, I. G., 1969. Effect of vetrazine (3,4-dimethoxybenzylhydrazine), selenophene, and their combination on the activity of the heart during its experimental release from blood circulation, *Vop. Farmakol. Regul. Deyatel. Serdsta Mater. Simp.* 1969:106–110.

Bogert, M. T., and Stull, A., 1927. Researches on selenium organic compounds. VII. The synthesis of 2-phenyl-, 2-furyl- and 2-thienylbenzoselenazoles of 2-phenyl-benzoselenazole-4-arsonic acid and of other benzoselenazoles. *J. Amer. Chem. Soc.* 49:2011–2017.

Bonner, W. A., and Robinson, A., 1950a. The preparation and properties of phenyl β-D-selenoglucoside and its tetraacetate, *J. Amer. Chem. Soc.* 72:354–356.

Breccia, A., Badiello, R., Trenta, A., and Mattii, M., 1969. On the chemical radioprotection by organic selenium compounds *in vivo, Radiation Res.* 38:483–492.

Brucker, W., and Bulka, E., 1966. Radiation protection studies in vitro with organoselenium compounds, *Stud. Biophys.* 1:253–255.

Brucker, W., and Rohde, H. G., 1968. Zum einfluss von selenorganischen verbindungen auf den strahleneffekt von Phycomyces blakesleeanus, *Pharmazie* 23:310–315.

Buu-Hoi, N. P., Lambelin, G., Roba, J., Jacques, G., and Gillet, C., 1970. 1-Phenyl-2-aminoethanols and their β-sympatholytic and vasodilating activities, Belg. Pat. 739, 678; *Chem. Abstr.* 73, 109490 (1970).

Carr, A., Sawicki, E., and Ray, F. E., 1958. 8-selenopurines, *J. Org. Chem.* 23:1940–1942.

Cavallini, D., Federici, G., Dupre, S., Cannella, C., and Scandurra, R., 1980. Ambiguities in the enzymology of sulfur-containing compounds, in *Natural Sulfur Compounds*, D. Cavallini, G. E. Gaull, and V. Zappia, (eds.), Plenum Press, New York, pp. 511–523.

Chan, W. Y., and du Vigneaud, 1972. Comparison of the pharmacologic properties of oxytocin and its highly potent analogue, desamino-oxytocin, *Endocrinology* 71:977–982.

Chernysheva, L. F. and Eliseeva, S. V., 1966. Eliminating cardiac effects of histamine with antihistaminic selenophene derivatives, *Farmakol. Toksikol* 29:679–681.

Chernysheva, L. F., and Galbershtam, M. A., 1967. Antihistamine activity of new substances of the selenophene class. *Farmakol. Toksikol. Prep. Selena Mater. Simp.* 1967, 44–50.

Chernysheva, L. F., and Kudrin, A. N., 1967. Pharmacological activity of the selenophene class, *Farmakol. Toksikol. Prep. Selena Mater. Simp.*, 1967:50–53.

Chernysheva, L. F., Kudrin, A. N., and Gigawii, V., 1969. preventative and medical treatment of histaminic and traumatic shock by an antihistamine preparation from the selenophene class, *Vop. Farmkol. Regul. Diyatel. Serdtsa Mater. Simp.*, 1969:110–114.

Chu, S. H., and Davidson, D. D., 1972. Potential antitumor agents. 2 and -2-Deoxy-6-selenoguanosine and related compounds, *J. Med. Chem.* 15:1088–1089.

Chu, S. H., and Mautner, H. G., 1962. Potential antiradiation agents. II. Selenium analogs of 2-aminoethylisothiuronium hydrobromide and related compounds. *J. Org. Chem.* 27:2899–2901.

Chu, S. H., and Mautner, H. G., 1968. Sulfur and selenium compounds related to acetylcholine and choline. VII. Isologs of benzoylthionocholine and of 2-dimethylaminoethyl thionobenzoate, *J. Med. Chem.* 11:446–450.

Chu, Shih-Hsi, Hillman, G. R., and Mautner, H. G., 1972. Sulfur and selenium compounds related to acetylcholine and choline. II. Selenocarbonyl ester and selenocarboxyamide analogs of local anesthetics, *J. Med. Chem.* 15:760–762.

Comrie, A. M., Dingwall, D., and Stenlake, J. B., 1964. Some 2-iminoselenazolidin-4-ones and related compounds, *J. Pharm. Pharmacol.* 16:268–272.

Corbett, C. E., Limongi, J. P., and Ramos, A. O., 1957. Anticholinergic activity of the ortho-carboxybenzeneseleninic acid, *Arch. Int. Pharmacodyn.* 111:245–251.

Craig, P. N., 1962. Substituted phenoselenazine compounds, U.S. Pat. 3,043,839; *Chem. Abst.* 57:16636 (1962).

Dickson, R. C., and Tappel, A. L., 1969. Effects of selenocystine and selenomethionine on activation of sulfhydryl enzymes, *Arch. Biochem. Biophys.* 131:100–110.

DeMarco, C., Rinaldi, A., Dessi, M. R., and Dernini, S., 1976b. Oxidation of S-e-carboxymethyl-selenocysteine by L-aminoacid oxidase and by Daspartate oxidase, *Mol. Cell Biochem.* 12:89–92.

Dyer, E., and Minnier, C. E., 1968. Acylations of some 2-amino-6-halo and 2-amino-5-alkylthiopurines, *J. Med. Chem.* 11:1232–1234.

Ehrlich, P., and Bauer, H., 1915. Uber 3,6-diaminoseleno-pyronin op3,6-diaminoxanthoselenonium, *Ber. Deutsch. Chem. Ges.* 48:502–508.

Elion, G. B., Burgi, E., and Hitchings, G. H., 1952. Studies on condensed pyrimidine systems. X. The synthesis of some 6-substituted purines, *J. Am. Chem. Soc.* 74:411–414.

Endo, H., Sato, K., and Kawasaki, T., 1963. Synthesis of pyrimido (4,5-c) op2,1,3,) selenadiazoles and their antitumor activity on mouse leukemia, L 1210, *Sci. Rep. Res. Inst. Tohoku Univ. Ser. C.* 11:201–202.

van Es, T., and Whistler, R. L., 1967. Derivatives of 5-seleno-D-xylose, *Tetrahedron* 23:2849–2853.

Fabry, M., Hermann, P., and Rychlik, I., 1977. Poly(A)-directed formation of thialysine oligapeptides in *E. coli* cell-free system, *Collec. Czechoslov. Commun.* 42:1077–1081.

Giordano, W. C., 1963. One application treatment for tinea versicolor, *J. Med. Soc. N.J.* 60:186–187.

Glassman, J. M., Begany, A. J., Pless, H. H., Hudyma, G. M., and Seifter, J., 1960. The pharmacology of some phenoselenazines, *Fed. Proc. Fed. Am. Soc. Exp. Biol.* 19:280.

Goble, F. C., Konopka, E. A., and Zoganas, H. C., 1967. Chemotherapeutic activity of certain organic selenium compounds in experimental leptospirosis, *Antimicrob. Agents Chem.* 7:531–533.

Green, H. N., and Bielschowsky, F., 1942. Mode of action of sulphanilamide; relation of chemical structure to bacteriostatic action of aromatic sulphur, selenium and tellurium compounds, *Brit. J. Exp. Path.* 23:13–26.

Gunther, W. H., 1967. Methods in selenium chemistry. III. The reduction of diselenides with dithiothreitol, *J. Org. Chem.* 32:3931.

Gunther, W. H. H., and Mautner, H. G., 1960. Pantethine analogs. The condensation of pantothenic acid with selenocystamine with bis(β-aminoethyl) sulfide and with 1,2-dithia-4-azapane (a new ring system), *J. Am. Chem. Soc.* 82:2762–2765.

Gunther, W. H. H., and Mautner, H. G., 1965a. Analogs of parasympathetic neuroeffectors. 3. Synthesis and study of cholineselenol and related compounds, *J. Med. Chem.* 8:845–847.

Gunther, W. H. H., and Mautner, H. G., 1965b. The synthesis of coenzyme A, *J. Am. Chem. Soc.* 87:2708–2716.

Hannig, E., 1963. Zur C-4-benzoylathylierung des Phenylbutazons. I. "Uber Selenophenolderivate", *Arch. Pharm.* 296:441–445.

Hannig, E., 1964. Zur darstellung and wirkung einiger selenhaltigen beta-ketobasen. 2. Uber selenophenolderivate, *Pharmazie* 19:201–202.

Heilbronn-Wilkstrom, E., 1965. Phosphorylated cholinesterase their formation, reactions and induced hydrolysis, *Svensk Kem. Tidskr.* 77:598–601.

Hiscock, S. M., Swann, D. A., and Turnbull, J. H., 1970. Axial and equatorial selenols: 3α and 3β-selenyl derivatives of cholestane and androstanone, *J. Chem. Soc. D* 1970:1310.

Hofinger, M., Thorpe, T., Bouchet, M., and Gaspar, T., 1980. Auxin-like activity of (benzo(b)selenienyl-3) acetic acid, *Acta Physiol. Plantar* 2:275–280.

Hollo, Z. M., and Zlatarov, S., 1960. The prevention of death from x-rays by selenium salts given after irradiation, *Naturwissenschaften* 47:328.

Hoskin, F. C., Kremzner, L. T., and Rosenberg, P., 1969. Effects of some cholinesterase inhibitors on the squid giant axon. Their permeability, detoxification and effects on conduction and acetylcholinesterase activity, *Biochem. Pharmacol.* 18:1727–1737.

Hunt, D. E., and Pittillo, R. F., 1966. Antimicrobial activity of 1,2,5-selenadiazoles, *Antimicrobial Agents Chemother.* 6:551–4.

Jaffe, J. J., and Mautner, H. G., 1960. A Comparison of the biological properties of 6-selenopurine, 6-selenopurine ribonucleoside, and 6-mercaptopurine in mice, *Cancer Res.* 20:381–386.

Jensen, K. A., and Schmith, K., 1941. Sulfanilyl derivatives of heterocyclic amines. IV. Derivatives of selenazole, *Dansk. Tidskr. Farm.* 15:197–199.

Jilek, J. O., Rajsner, M., Pomykacek, J., and Protiva, M., 1965. Synthetic ataractics. XII. Prothixene and analogs. A new synthesis of chlorprothixene, *Cesk. Farm.* 14:294–303.

Jilek, J., Sindelar, K., Metysova, J., Metys, J., Pomykacek, J., and Protiva, M., 1970. Neurotropic and psychotropic agents. XLIV. 10-(Amino-alkoxy) and 10-piperazino derivatives of dibenzo(b,f)thiepin and related systems, *Collection Czech. Chem. Commun.* 35:3721–3732.

Keimatsu S., and Yokota, K., 1931. Synthesis of organo-selenium compounds, *J. Pharm. Soc. Japan* 51:605–616.

Keimatsu, S., Yokota, K., and Satoda, I., 1933. Organoselenium compounds. VI. *J. Pharm. Soc. Japan* 53:994–1046.

Klayman, D. L., 1965. Synthesis of aminoethyl-substituted selenium compounds, *J. Org. Chem.* 30:2454–2456.

Klayman, D. L., 1973. Selenium compounds as potential chemotherapeutic agents, in *Organic Selenium Compounds Their Chemistry and Biology*, D. L. Klayman and W. H. H. Gunther (eds), John Wiley and Sons, New York, pp. 727–761.

Kocourek, J., Klenha, J., and Jiracek, V., 1963. Preparation of 1-seleno-D-glucose, *Chem. Ind. (London)*, 1397.

Konopelski, J. P., Djerassi, C., and Raynaud, J. P., 1980. Synthesis and biochemical screening of phenylselenium-substituted steroid hormones, *J. Med. Chem.* 23:722–726.

Korzhova, V. V., and Smirnova, E. I., 1969. Pathohistological and histochemical changes in some internal organs of rat fetuses under the influence of experimental late toxicosis and medical treatment of selenophene, *Vop. Farmakol. Regul. Deyatel. Serdtsa Mater. Simp.* 1969:119–122.

Kozak, I., Kronrad, L., and Dienstbier, Z., 1976. Distribution and behaviour of isoselenouronium salts in the body, *Strahlentherapie* 151:84–87.

Kudrin, A. N., and Zaidler, Y. I., 1968. Search for compounds with antiarrhythmic activity among the derivatives of selenophene, amines of the acetylene series, beta-aminoketones, aminopyrazoles, monoamino oxidase inhibitors, *Farmakol Toksikol* 31:41–44.

Kuhn, S. J., and McIntyre, J. S., 1970. Fungicidal and bactericidal 6-methyl-3(3-aryl-3-arylthio (seleno) propionyl)-4-hydroxy-2H-pyran-2-ones, U.S. Pat. 3,493,586; *Chem. Abstr.* 72:90289 (1970).

Kuliev, T. A., Babaev, R. A., and Kuliev, K. A., 1981. Cytofluorimetric data on the dynamics of the nucleic acid content on the liver of radiation - exposed and intact animals treated with sodium selenite, *Radiobiologiya* 21:451–453.

Leonicini, G., Maresca, M., and Bonsignmore, A., 1977. D,L-Glyceraldehyde 3-phosphate breakdown in the presence of selenalysine, *Ital. J. Biochem.* 26:467–472.

LePage, G. A., and Howard, N., 1963. Chemotherapy studies of mammary tumors of C3H mice, *Cancer Res.* 23:622–627.

Lesser, R., and Weiss, R., 1912. Uberden selen-indigo (bis-selenonaphthen-indigo) und selenhaltige aromatische verbindungen, *Ber.* 45:1835–1841.

Limongi, J. P., 1955. Adrenergic blocking action of orthocarboxybenzeno seleninic acid, *Arch. Int. Pharmacodyn.* 103:160–166.

Lipmann, J., 1945. Acetylation of sulfanilamide by liver homogenates and extracts, *J. Biol. Chem.* 160:173–190.

Lotis, V. M., Kudrin, A. N., Korzhova, V. V., and Chernysheva, L. F., 1967. Use of new selenophene preparation for treatment of late pregnancy toxemia, *Formakol. Toksikol. Prep. Selena Mater. Simp.* 1967:64–68.

Maeda, M., Abiko, N., and Sasaki, T., 1981. Synthesis of seleno- and thioguanine-platinum (II) complexes and their antitumor activity in mice, *J. Med. Chem.* 24:167–169.

Markees, D. G., Dewey, V. C., and Kidder, G. S., 1968. The synethesis and biological activity of substituted 2,6-diaminopyridines, *J. Med. Chem.* 11:126–129.

Matti, J., 1940. Organic derivatives of selenium, *Bull Soc. Chim. Fr.* 7:617–621.

Mautner, H. G., 1956. The snythesis and properties of some selenopurines and selenopyrimidines, *J. Am. Chem. Soc.* 78:5292–5294.

Mautner, H. G., 1958. A comparative study of 6-selenopurine and 6-mercaptopurine in the *Lactobacillus casei* and Ehrlich ascites tumor systems. *Biochem. Pharmacol.* 1:169–173.

Mautner, H. G., and Clayton, E. M., 1959. 2-selenobarbiturates. Studies of some analogous oxygen, sulfur and selenium compounds, *J. Am. Chem. Soc.* 81:6270–6273.

Mautner, H. G., and Jaffe, J. J., 1958. The activity of 6-selenopurine and related compounds against some experimental mouse tumors, *Cancer Res.* 18:294–298.

Mautner, H. G., Kumler, W. D., Okano, Y., and Pratt, R., 1956. Antifungal activity of some substituted selenosemicarbazones and related compounds, *Antibiot. Chemother.* 6:51–55.

Mautner, H. G., Chu, S. H., Jaffe, J. J., and Sartorelli, A. C., 1963. The synthesis and antineoplastic properties of selenoguanine, selenocytosine and related compounds, *J. Med. Chem.* 6:36–39.

Mautner, H. G., Bartels, E., and Webb, G. D., 1966. Sulfur and selenium isologs related to acetylcholine and choline. IV. Activity in the electroplax preparation, *Biochem. Pharmacol.* 15:187–193.

Milne, G. H., and Townsend, L. B., 1974. Synthesis and antitumor activity of α and β-2'-deoxyselenoguanosine and certain related derivatives, *J. Med. Chem.* 17:263–268.

Mueller, A. G., and Phillips, J. P., 1967. Heterocycles containing (8-hydroxy-5-methyl-7-quinolyl) vinyl groups, *J. Med. Chem.* 10:110–111.

Muller, P. Buu-Hoi, N. P., and Rips, R., 1959. Preparation and reactions of phenoxazine and phenoselenazine, *J. Org. Chem.* 24:37–39.

Ojasco, T., and Raynaud, J. P., 1978. Unique steroid congeners for receptor studies, *Cancer Res.* 38:4186–4198.

Oxford, J. S., 1973. An inhibitor of the particle-associated RNA-dependent polymerase of influenza A and B virus, *J. Gen. Virol.* 18:11–19.

Phillips, B. M., Sancilio, L. F., and Kurchacova, E., 1967. *In vitro* assessment of anti-inflammatory activity, *J. Pharm. Pharmacol.* 19:696–697.

Pianka, M., and Hall, J. C., 1959. Fungitoxicity. II. Fungitoxicity of certain anilinovinyl quaternaries, *J. Sci. Food Agr.* 10:385–388.

Priestly, H. M., 1972. Dialkyl selenoxide hydronitrates as germicides in detergent compositions, U.S. Pat. 3,642,909; *Chem. Abstr.* 76:112693 (1972).

Protiva, M., Rajsner, M., Adlerova, E., Seidlora, V., and Vejdelek, Z. J., 1964. Neurotropic and psychotropic substances. I. New types of 6, 11-dihydrodibenzo(b,e)thiepine derivatives and analogs, *Collection Czech. Chem. Commun.* 29:2161–2181.

Roberts, M. E., 1963. Antiinflammation studies. II. Antiinflammatory properties of selenium, *Toxicol. Appl. Pharmacol.* 5:500–506.

Robinson, H. M., and Jaffe, S., 1956. Selenium sulfide in the treatment of pityriasis versicolor, *J. Am. Med. Assoc.* 162:113–114.

Romanova, L. A., Gorkin, V. Z., Kudrin, A. N., and Chernysheva, L. F., 1968. Inhibition of monoamine oxidase activity with antihistamine preparations of the selenophene class, *Byul Eksp. Biol. Med.* 66:39–42.

Rosell, S., and Axelrod, J., 1963. Relation between blockage of H3-noradrenaline uptake and pharmacological actions produced by phenothiazine derivatives, *Experientia* 19:318–319.

Rosenberg, P., and Mautner, H. G., 1967. Acetylcholine receptor: Similarity in axons and junctions, *Science* 155:1569–1571.

Ross, A. F., Agarwal, K. C., Chu, S. H., and Parks, R. E., Jr., 1973. Studies on the biochemical actions of 6-selenoguanine and 6-selenoguanosine, *Biochem. Pharmacol.* 22:141–154.

Roy, J., Schwartz, I. L., and Walter, R., 1970. Nucleophilic scission of disulfides by selenolate. Synthesis and some properties of acyclic thiolselenenates, *J. Org. Che.* 35:2840–2842.

Sadek, S. A., Shaw, S. M., Kessler, W. V., and Wolf, G. C., 1981. Selenosteroids as potential estrogen-receptor scanning agents, *J. Org. Chem.* 46:3259–3262.

Schneider, W., and Beuther, A., 1919. Schwefelhaltige disaccharide aus galaktose, *Ber. Deutsch. Chem. Ges.* 52:2135–2149.

Schneider, W., and Wrede, F., 1917. Synthese eines schwefelhaltigen und eines selenhaltigen disaccharides, *Ber. Deutsch. Chem. Ges.* 50:793–804.

Schwarz, K., 1965. Role of vitamin E, selenium and related factors in experimental nutritional liver disease, *Fed. Proc. Fed. Am. Soc. Exp. Biol.* 24:58–67.

Scioscia-Santoro, A., Cavallini, D., Degener, A. M., Perez-Bercoff, R., and Rita, G., 1977. Effect of L-aminoethyl-cysteine, a sulfur analogue of L-lysine, on virus multiplication in mammalian cell cultures, *Experientia* 33:451–453.

Scott, K. A., and Mautner, H. G., 1964. Analogs of parasympathetic neuroeffectors—II. Comparative pharmacological studies of acetylcholine, its thio and seleno analogs, and their hydrolysis products, *Biochem. Pharmacol.* 13:907–920.

Scott, K. A., and Mautner, H. G., 1967. Sulfur and selenium isologs related to acetylcholine and choline. IX. Further comparative studies of the pharmacological effects of acetylcholine and its thio and seleno analogs and their hydrolysis products, *Biochem. Pharmacol.* 16:1903–1918.

Segaloff, A., and Gabbard, R. B., 1965. 3β-Seleno steroids. 3β-Seleno derivatives pregnenolone (3β-hydroxypregn-5-en-20-one) and dehydroepiandrosterone (3β-hydroxyandrost-5-en-17-one and comparison with analogs containing sulfur or oxygen, *Steroids* 5:219–240.

Segaloff, A., and Gabbard, R. B., 1968. Seleno-substituted steroids, U.S. Pat. 3,327,173; *Chem. Abstr.* 69:P77632q, 1968.

Senning, A., and Sorensen, O. N., 1966. The preparation of some selenouronium salts and selenouronium betaines, *Acta Chem. Scand.* 20:1445–1446.

Shealy, Y. F., and Clayton, J. D., 1967. 1,2,5-Selenadiazoles. Synthesis and properties, *J. Heterocycl. Chem.* 4:96–101.

Shimazu, F., and Tappel, A. L., 1964a. Selenoamino acids as radiation protectors *in vitro*, *Radiat. Res.* 23:210–217.

Sindelar, K., Svatek, E., Metysova, J., Metys, J., and Potiva, M., 1969a. Neurotropic and psychotropic agents. XXXVII. Derivatives of selenoxanthene, *Collect. Czech. Chem. Commun.* 34:3792–3800.

Sindelar, K., Metysova, J., and Protiva, M., 1969b. Neurotropic and psychotropic, agents. XXXVIII. Derivatives of 10, 11-dihydrodibenzo(b,f) selenepin, *Collect. Czech. Chem. Commun.* 34:3801–3810.

Sindelar, K., Metysova, J., Metys, J., and Protiva, M., 1969c. 8-chloro-10 (4-methyl-1-piperazinyl)dibenzo(b,f)-thiepin and analogs, new highly active neuroleptics, *Natur-wissenschaften* 56:374.

Supniewski, J., Misztal, S., and Krupinska, J., 1954. Selenium and sulfur derivatives of chloromycetin, *Bull. Acad. Polon. Sci. Classe II* 2:153–159.

Takeda, K., Kitagawa, R., Konaka, T., Noguchi, R., Nishimura, H., Shimoaka, N., Kimura, N., Kaida, K., Sugiyama, K., Ilada, H., and Maki, T., 1952. *Annual Report Shinogi, Res. Lab.* 2:12. Cited by D. L. Kayman, 1973. Selenium compounds as potential chemotherapeutic agents, in D. Klayman and W. H. Gunther (eds.), *Organic Selenium Compounds their Chemistry and Biology*, John Wiley and Sons, New York, pp. 727–761.

Takeda, K., Nishimura, H., Simaoka, N., Noguchi, R., and Nakajima, K., 1955. Anti-tumor substances. I. Effect of some organic selenium compounds on the Ehrlich ascites carcinoma, *Ann. Rept. Shionogi Res. Lab.* 5:1–16.

Townsend, L. B., and Milne, G. H., 1970. Synthesis of the selenium congeners of the naturally occurring nucleoside guanosine, 6-selenoguanosine, *J. Heterocyc. Chem.* 7:753–754.

Wagner, G., and Nuhn, P., 1965. Uber die säurekatalysierte hydrolyse von selenoglucosiden. 7. Uber "selenoglykoside". *Arch. Pharm.* 298:692–695.

Walter, R., and duVigneaud, V., 1965 6-Hemi-L-selenocystine-oxytocin and *1*-deamino-6-hemi-L-selenocystine-oxytocin, highly potent isologs of oxytocin and *1*-deamino-oxytocin, *J. Am. Chem.* 87:4192–4193.

Weaver, W. E., and Whaley, W. M., 1946. Alky seleno cyanates, *J. Am. Chem. Soc.* 68:2115–2116.

Weisburger, A. S., and Suhrland, L. G., 1956a. Studies on analogues of L-cysteine and L-cystine. III. The effect of selenium cystine on leukemia. *Blood* 11:19–30.

Weisburger, A. S., and Suhrland, L. G., 1956b. Studies on analogues of L-cysteine and L-cystine. II. The effect of selenium cystine on Murphy lymphosarcoma tumors cells in the rat, *Blood* 11:11–18.

Weisburger, A. S., Suhrland, L. G., and Seifter, J., 1956. Studies on analogues, of L-cysteine and L-cystine. I. Some structural requirements for inhibiting the incorporation of radioactive L-cystine by leukemic leukocytes, *Blood* 11:1–10.

Williams, R. J., Truesdail, J. H., Weinstock, H. H., Rohramann, E., Lyman, C. M., and McBurney, C. H., 1938. Pantothenic acid. II. Its concentration and purification from liver, *J. Am. Chem. Soc.* 60:2719–2723.

Wolff, M. E., and Zanati, G., 1970. Preparation and androgenic activity of novel heterocyclic steroids, *Experientia* 26:1115–1116.

Wong, A. S., Fasanella, R. M., Haley, L. D., Marshall, C. L., and Krehl, W. A., 1956. Selenium (selsum) in the treatment of marginal blepharitis, *A.M.A. Arch. Ophth.* 55:246–253.

Wrede, F., 1917. Synthesis of two new disaccharides and their biological behavior. *Biochem. Z.* 83:96–102.

Zsolnai, T., 1962. Attempts to find new fungistatics. IV. Organic sulfur compounds, *Biochem. Pharmacol.* 11:271–297.

Zingaro, R. A., and Thomson, J. K., 1973. Thio and seleno sugar esters of dialkylarsinous acids, *Carbohydrate Res.* 29:147–152.

Analytical Methods of Selenium Determination

10

10.1 Introduction

Selenium occurs naturally in biological materials that vary from a few parts per billion to a few per cent. Methods for its determination, therefore, cover a wide range of concentrations. As a result there have been developed a number of procedures for selenium analysis.

Because selenium can occur in a variety of valence states, the methods generally used for analyzing biological materials for selenium are based on the destruction of the organic constituents with the concomitant oxidation of the element to Se or Se^{6+} and its subsequent determination by one of a variety of procedures, either with or without its separation from other elements. Measurement of the element by nondestructive techniques is especially becoming increasingly popular. This chapter reviews some of the principles and techniques involved in both the techniques of destructive and nondestructive testing, and some of the more commonly used methods are outlined in detail.

10.2 Sample Preparation and Storage

Some of the chemical forms of selenium in biological materials are volatile and this must be considered in the preparation and storage of samples for analyses. Early in their experience with "alkali disease" or the "blind staggers" farmers thought they observed a decrease in toxicity when the grains were stored. If these grains were heated to temperatures of 160°C or more, their toxicity decreased and a significant part of the selenium was lost. Farmers also found that when highly seleniferous grains were stored from three to five years without temperature control, much of the selenium content was lost.

311

Beath *et al.* (1935) have found large losses of selenium during the air drying of *Astragalus bisulcatus* plants. This study was later confirmed in other highly seleniferous "indicator" plants, but also found there was no significant loss of the element on air drying ordinary farm crops such as grains and grasses (Beath *et al.*, 1937). When alfalfa was dried at 70°C for 48 hr, Asher *et al.* (1967) have found that only 0.5%–3.0% of the selenium was lost. Ehlig *et al.* (1968) have reported losses of less than 5.0% of selenium on drying a number of species of green plants at 70°C for 30 hr.

Even though the volatization of selenium by animals and molds has extensively been investigated, there has been little study of the loss of the element when animal tissue or fluid is dried. In many cases vacuum-oven drying or lyophilization have extensively been required starting with a dried sample.

Allaway and Cary (1964) have suggested that care be exercised in grinding of plant samples for analysis. Leaves reduce to fine particles much more easily than do stems and it is possible that the two plant parts could separate in the ground sample. Since leaves contain the greatest amounts of selenium, errors could result unless fine grinding and good mixing are done. Olson *et al.* (1973) suggest that for long storage dried tissues should be kept in tight containers at temperatures near or below freezing, especially if they contain volatile selenium compounds. Short-term storage could be done at room temperature if the tissues have been dried. Wet tissues should be stored frozen to reduce the possibility of volatile compounds formed from enzymatic reactions (Ganther, 1966). Falkner *et al.* (1961) have reported that selenium is lost from urine under several conditions of storage, but the loss could be prevented by making the urine 1 M in nitric acid or could be reduced by freezing.

Analyzing samples without drying should give the most reliable re-sults. However, for grasses, grains, and other plants that contain little volatile selenium, drying at 70°C to constant weight should be satisfactory for most analyses. Animal tissues or fluids that need drying before analysis are probably best lyophilized to prevent the possible formation of volatile selenium compounds as the result of biological activity (Ganther, 1966).

10.3 Destructive Analysis

Destruction of the organic matter from biological materials is required to analyze selenium by chemical means. After this step several different methods might be used for measuring the element, some of which require its isolation from interfering substances. Some of the destructive pro-cesses including ashing, closed-system combustion and wet digestion.

10.3.1 Ashing

The large losses of selenium found to occur on open ashing at 700°C led to other methods for the destruction of organic matter. Gleit and Holland (1962) have studied a low-temperature, dry ashing process for decomposing organic matter for trace element analysis. They excited a stream of oxygen using a high-frequency electromagnetic field and then passed the excited oxygen over the sample heated by induction to 100°C. Almost 99% of the radioactivity was recovered by this process when a sample of alfalfa grown on soil enriched with ^{75}Se was ashed. Later, Mulford (1966) used this method and added selenite to powdered cellulose which also contained added mercury, arsenic, and copper. Selenium was almost quantitatively recovered, but when copper alone was added to the cellulose, losses increased as the temperature increased. At present, this process is not being utilized for selenium determinations in biological materials.

10.3.2 Closed-System Combustion

Gutenmann and Lisk (1961) obtained good recoveries of added selenium using the closed-system oxygen flask of the Schoniger flask or the Parr bomb. Johnson (1970) has used the Parr bomb at 20 atm oxygen to combust municipal solid waste, compost, and paper for selenium analysis.

Allaway and Cary (1964) have developed a procedure in which the oxygen flask was used for combustion. Good comparisons were made with neutron activation analysis. Watkinson (1966) has compared the oxygen flask method to the wet digestion using nitric and perchloric acids and found good agreement for selenium analysis.

10.3.3 Wet Digestion

Robinson *et al.* (1934) used a Kjeldahl procedure to destroy organic matter in biological samples before analysis for selenium. A bromine trap for selenium was used, but since the bromine was continuously reduced, repeated addition of bromine was required. Williams and Lakin (1935) have used open digestion with a mixture of sulfuric and nitric acids and low temperatures not exceeding 120°C for plant tissues. After this method was studied for a while, it was modified to include the addition of mercury as a selenium fixative (Osborn, 1939). Using this modification the method was adopted for use in the analysis of foods and plants by the Association of Official Agricultural Chemists. Klein (1941) used a similar procedure and found it satisfactory at even higher temperatures. Good recoveries of added selenium from nitric–sulfuric acid digests of biological samples were observed by Grant *et al.* (1961). Using this digestion mixture Olson (1969) found good agreement with those results obtained by other procedures. However, (Fogg and Wilkinson, 1956) have reported losses of selenium during digestion with nitric–sulfuric acid. Excessive charring after the nitric acid has evaporated may be responsible for the loss of selenium. When charring is avoided, good recoveries can be obtained (Gorsuch, 1959).

A number of methods based on the digestion of samples with mixtures containing perchloric acid have been described. These acids vary as follows: nitric acid and perchloric acids (Watkinson, 1966; Olson, 1969); nitric, sulfuric, and perchloric acids (Hoffman *et al.*, 1968); sulfuric and perchloric acids plus sodium molybdate and water (Cummins *et al.*, 1964; Ewan *et al.*, 1968); nitric and perchloric acids plus ammonium metavandate (McNulty, 1947); and nitric and perchloric acids plus hydrogen peroxide (Hall and Gupta, 1969). Watkinson (1966) has reported very good recoveries of inorganic selenium when added to biological materials. Others have made good comparisons between the results on using wet digestion with acids and the closed-system oxygen combustion (Watkinson, 1966; Olson, 1969) or neutron activation methods (Watkinson, 1966; Olson, 1969).

Perchloric acid helps prevent losses due to charring by maintaining oxidizing conditions after the nitric acid is distilled off. The wet digestion should be carried out to avoid excessive charring, either by using a slow rate of heat, by using a condenser to slow the distillation of nitric acid, or by stopping the digestion and adding more nitric acid. Mixtures containing perchloric acid seem to have decided advantages over those mixtures which do not contain it, and if used properly offer no serious explosion hazard. Selenium is converted to selenite and selenate during digestions with perchloric acid, and for many methods of determination the reagents used require that the selenium be in the selenite form. Treatment with hydrochloric acid in strongly acid solution readily reduces any selenate to selenite (Watkinson, 1966).

10.3.4 Measurement of Selenium

Several different methods are available to measure selenium. These include spectrophotometric, fluorometric, atomic absorption spectrophotometry, gas chromatography, polarography, titration, and gravimetric and mass spectrometry.

10.3.4.1 Spectrophotometric

The most studied methods of selenium analysis are based on the reaction of selenous acid with O-diamines. Hoste and Gilles (1955) first reported such a method for selenium detection. The method was based on the measurement of the yellow color formed when 3,3'-diaminobenzidine (DAB) reacts with Se^{4+}. They believed the yellow compound to be a dipiazselenol, but Parker and Henry (1961) have shown that the colored product is a monopiazselenol. This is supported by experiments showing the effect of pH on the spectrum and extractibility of the compound. The DAB reaction was specific for selenium except that oxidizing agents interfered, and iron(III), copper(II), and vanadium reacted with the reagent. Iron interference was prevented by the addition of fluoride or phosphoric acid and copper interference by oxalic acid, but no method was presented for the prevention of vanadium interference.

Cheng (1956) modified the DAB method which made it more specific for selenium. The use of EDTA was introduced to mask the effect of many of the divalent ions. The piazselenol was extracted into toluene, which removed the interference of vanadium and several other elements as their DAB derivatives were insoluble in toluene. One disadvantage of the procedure was that pH adjustment was necessary, as the formation of the piazselenol required acidic conditions, but the DAB–Se complex was quantitatively soluble in toluene only in neutral or slightly alkaline solution. The yellow piazselenol was measured at 420 nm, and a blank was required because unreacted DAB has a weak absorbance at this wavelength. The method was able to detect about 0.2 mg of selenium. Cummins *et al.* (1964) have modified Cheng's procedure for the analysis of biological samples. Comparisons with neutron activation analysis were in excellent agreement with a lower sensitivity of 1.0 mg, which should be sufficient for studies of selenium toxicosis.

Parker and Harvey (1962) investigated the reaction of selenious acid with a series of aromatic *o*-diamines and observed that 2,3-diaminonaphthalene (DAN) was more sensitive than DAB for selenium detection. Lott *et al.* (1963) have developed a procedure using DAN which was almost identical with that for DAB, except pH adjustment was not necessary before the Se–DAN complex was extracted into the organic solvent. DAN was also found to be a more sensitive reagent in the fluorometric analysis for selenium. Numerous other aromatic *o*-diamines have been used to complex selenium for spectrophotometric analysis of selenium: These include *o*-phenylenediamine (Throop, 1960); 1,2-diaminonaphthalene (Throop, 1960), 4-methylthio-1,2-phenyldiamine (Demeyere and Hoste, 1962), 4-dimethylamino-1,2-phenyldiamine (Demeyere and

Hoste, 1962), 4,5-diamino-6-thiopyrimidine (Chan, 1964), and 1-methyl-4-chloro- and 4,5-dichloro-*o*-phenylenediamine (Cukor and Lott, 1964).

Another type of spectrophotometric method involves the formation of complexes between sulfur and selenium. Both 2-mercaptobenzothiazole (Bera and Chakrabartty, 1968) and 2-mercaptobenzoic acid (Cresser and West, 1968) have given sensitivity and selectivity comparable to those obtained with DAB.

10.3.4.2 Fluorometry

Even though spectrophotometric methods are suitable for toxic concentrations of selenium, they are generally not sensitive enough for biological specimens. In a search for more sensitive methods Cousins (1960) found that the Se–DAB complex fluoresced in hydrocarbon solvents. The maximum excitation for transmission was found to be 450 nm and the fluorescent emission was measured at 580 nm. The blank fluorescence was low and the method was about 20 times more sensitive than the DAB spectrophotometric procedures. Parker and Harvey (1962) found Se–DAN to be strongly fluorescent and more sensitive than the Se–DAB complex. Lott *et al.* (1963) also found the method to be more sensitive using Se–DAB, but also more subject to interference from foreign ions. Allaway and Cary (1964) also used DAN to determine selenium, but they separated the element by coprecipitation with arsenic. However, they made an important advance in the procedure by adding hydroxylamine to decrease the oxidation of DAN during the formation of Se–DAN complex.

Watkinson (1966) digested samples in a nitric–perchloric acid mixture and then heated the digest with HCl to convert all of the selenium to the 4+ state. The pH was then adjusted to 1 and the selenium was complexed with DAN in a water bath. After extraction with cyclohexane, the fluorescence was measured with a 606-nm filter. Olson (1969) conducted a collaborative study comparing a slightly modified method of Watkinson with several spectrophotometric, fluorometric procedures and with neutron activation analysis. The results from the slightly modified fluorescent procedure compared well with the other techniques.

Although mesitylene was found to be the most efficient in extracting the piazselenol (Dye *et al.*, 1963) toluene was chosen primarily because of cost, but Parker and Harvey (1962) have found that the Se–DAN complex was not stable in toluene. They observed that cyclohexane was a good solvent, but it was quite volatile and they recommended decalin. Watkinson (1966) nonetheless recommended cyclohexane because the volatility could be controlled and large numbers of samples could be processed with a detection minimum of about 0.02 mg of selenium. Nazarenko *et al.* (1975) have assayed selenium in food with DAN at a sensitivity of 0.001 μg/g. Brown and Watkinson (1977) have proposed an automatic method for estimating nanogram amounts of selenium directly in whole blood. The method is based on the measurement of the fluorescence of 4,5-benzopiazselenol in cyclohexane and is capable of measuring 40 samples per hour with a detection limit of 0.44 ng/g. Spallholz *et al.* (1978) have described a single-test-tube method for the microdetermination of selenium.

10.3.4.3 Atomic Absorption Spectrometry

Allen (1962) first recognized that selenium could be measured in the flame at 196 nm. The various forms of the element still needed to be converted to the inorganic state before measurement. The early lamps used for measuring selenium could detect only 1.5–5 ppm. Rann and Hambly (1965) have measured the absorption of selenium at 196 nm in an air–acetylene flame and achieved a sensitivity of about 1 ppm.

The most intense resonance line of selenium (196.03) corresponds to a range near the vacuum ultraviolet. In addition, the most frequently used air–acetylene flame absorbs about 55% of the radiation intensity of the light source. If electrodeless discharge lamps are used with an air–acetylene flame, a lower detectibility level of 0.2 μg/ml can be achieved. These levels can be extended down to 0.1 μg/g by using a deuterium lamp for background correction. The argon–hydrogen flame can also increase sensitivity, but it increases interferences, too (Kahn and Schallis, 1968). Although the flame methods lack sensitivity for submicrogram quantities of selenium, advances in flameless atomic absorption are able to increase the sensitivity of the selenium method. Improvement in sensitivity has been obtained by use of the sampling boat technique (Kahn *et al.*, 1968), the Delves sampling cup technique (Kerber and Fernandez, 1971), the heated graphite atomizer (Baird *et al.*, 1972) and the hydride generator, which converts selenium to a gaseous hydride before sweeping it into the flame (Fernandez and Manning, 1971).

Flameless atomic absorption techniques offer a high sensitivity (5 $\times$ 10^{-11} g Se) but owing to the high volatility of selenium, the methods are not simple nor are they free from interference in the graphite cell. Addition of nickel (Ishizaki, 1978) or cobalt salts (Brodie, 1977) enhances significantly the sensitivity for selenium (by almost 30%) and allows higher ashing temperature (1000°C) without losses. Montaser and Mehrabzadeh (1978) reached a detectability level of 1 $\times$ 10^{-12} g of selenium by using a graphite electrothermal furnace and background correction with a deuterium lamp.

The hydride generation method is becoming more popular, as the graphite furnace method is often not reliable owing to the strong and variable matrix effect and signal splitting. The selenium hydride is formed as a result of the reduction of selenium compounds [Se (IV)] with different reducing mixtures $Zn-SnCl_2-KI$ or usually $NaBH_4$. The detection limit is usually around 0.2 ng Se/g. The method has an additional advantage of separating selenium from the matrix before atomization, thus avoiding interferences inherent in conventional techniques. Practical analytical working ranges for selenium are 3–250, 0.03–3, and up to 0.12 $\mu g/ml$, respectively, for flame atomic absorption, furnace atomic absorption, and vapor generation methods (Brodie, 1979).

10.3.4.4 Gas Chromatography

Estimation of selenium by gas–liquid chromatography (GLC) is almost always based on the measurement of the amount of piazselenol formed during the reaction of SE(IV) with an appropriate reagent in acid media. After extraction with toluene, the piazselenols are usually estimated with an electron capture detector. In addition to its superior sensitivity and selectivity the GLC method allows for elimination of matrix interference.

Nakashima and Toei (1968) first used the GLC method for estimation of selenium with 4-chloro-o-phenylenediamine and reported a sensitivity of 4×10^{-8} g. Shimoishi (1977), looking for increased sensitivity in the determination of selenium, studied 13 derivatives of 1,2-diaminibenzene and found 1,2-diamino-3,5-dibromobenzene capable of detecting a level of 1 ng. Kurahasi et al. (1980) has detected Se(VI) in blood and plasma with a detection limit of 2 ng. Other solvents that have been used for the GLC assay of selenium include 2,3-diaminonaphathalene (Young and Christian, 1973), 1,2-diamino-4,5-dichlorobenzene (Stijve and Cardinale, 1975), 4-chloro-1,2-diaminobenzene (Bycroft and Clegg, 1978), 4-bromo-1,2-diaminobenzene (Bycroft and Clegg, 1978), bis(trimethylsilyl)trifluoroacetamide (Burgett, 1974), and 4-nitro-o-phenylenediamine (McCarthy et al., 1981). The limit of detection for the latter compound using electron capture detection was about 0.5 pg (McCarthy et al., 1981). The nitro group is more sensitive to electron capture detection than the chlorine group.

Determination of selenium in biological materials by GLC requires previous decomposition of organic matter usually with HNO_3, $HNO_3-H_2SO_4-HClO_4$, or HNO_3 and Mg. Utilization of nitric acid for digestion prevents ashing but results in disturbing peaks on chromatograms due to decomposition reactions of the nitric acid with diaminobenzene derivatives. However, in the wet digestion method with HNO_3 and $HClO_4$, the HNO_3 is distilled off. This step should eliminate the HNO_3 problem, but other more cumbersome procedures could be utilized to eliminate the HNO_3.

The combination of GLC and AAS makes it possible to assay both dimethyl selenide and dimethyl diselenide (Chau et al., 1975). In conjunction with a microwave emission spectrometric detection system GLC gave a detection limit of 40×10^{-12} g using 4-nitrodiaminobenzene (Talmi and Andren, 1974). A very sensitive photometric detector for selenium analysis enables determination of 2×10^{-12} g Se/sec (Flinn and Aue, 1978). Buchtela et al. (1975) have developed a radio–gas chromatographic separation of metal chelates which allows detection of 10^{-15} g of selenium.

10.3.4.5 Polarography

Polarography seems to be similar to spectrometry in sensitivity and rapidity. Schwaer and Suchy (1935) first applied polarography to selenium analysis, but Faulkner et al. (1961) first applied this technique to the analysis of selenium in biological materials. Christian et al. (1965) further improved the analysis for selenium. Interference by copper may be removed by extraction with dithizone in chloroform (Christian et al., 1965). Extraction of DAB complex into a mixture of chloroform and ethylene dichloride followed by back extraction into perchloric acid to obtain the polarogram was found to be best for samples with less than 4 μg of selenium. As little as 0.2 μg of selenium could be determined in 1- to 2-g samples. Griffin (1969) has reported that, by using single-sweep polarography instead of conventional polarography to determine the Se–DAB complex, sensitivity equal or greater than that with the spectroflurometric method can be achieved (1 ppb). Diamandis and Hadjiioannou (1981) have described a potentiometric method for selenium determination of selenium with a picrate-sensitive electrode. Amounts of selenium in the range 3–30 μg in a sample volume of 15.00 ml can be determined with good accuracy and precision.

10.3.4.6 Electrochemical Methods

Selenium has also been measured by electrochemical methods using anodic (Andrews and Johnson, 1975) and cathodic stripping voltammetry (Dennis et al., 1976). Andrews and Johnson (1975) have reported a procedure for the estimation of Se(IV) by anodic stripping voltammetry with a detection limit of 0.04 ppb. Posey and Andrews (1981) have applied a gold-plated glassey carbon disk electrode to selenium(IV) determination by anodic stripping voltammetry and were about to detect 2×10^{-9} M.

Dennis *et al.* (1976) have determined selenium by cathodic stripping volammetry with a hanging drop electrode. They prepared their sample by complexing other metals and reducing selenium by $NaBH_4$ to a hydride which was absorbed in alkaline solution. Graphite electrodes have also been used for the estimation of Se(IV) in water solutions. Lunev *et al.*, (1976) have in this way assayed selenium in food dissolved in HNO_3 and $Mg(NO_3)_2$ with a detection limit of 0.1 ppm. Grier and Andrews (1981), using a hanging mercury drop electrodes, have determined selenocystine, cystine, and cysteine by cathodic stripping voltammetry in 0.1 M $HClO_4$ or 0.1 M H_2SO_4. The detection limit for selenocystine is 5×10^{-10} M in the presence of 100-fold amounts of cystine and cysteine.

10.3.4.7 Mass Spectrometry

Gorman *et al.* (1951) described a method for the analysis of solids using a spark source mass spectrometer (SSMS). Evans and Morrison (1968) reported on the analysis of lung tissue for 30 elements which also included selenium. Yurachek *et al.* (1969) analyzed human hair for 23 elements, including selenium. Harrison *et al.* (1971) have used SSMS for the forensic analysis of hair and fingernails for selenium and a number of other elements.

The accuracy of SSMS should also be improved when it is combined with the isotopic dilution method. Reamer and Veillon (1981) have spiked samples with ^{82}Se, and the isotopic ratio of ^{80}Se to ^{82}Se was measured by combined gas chromatography–mass spectrometry using dual ion monitoring. Precise determination at the ppb level is possible.

Mass spectrometry also requires destruction of the sample by dry or wet ashing, which is a disadvantage when this method is compared to nondestructive neutron activation analysis, (NAA), but as with (NAA) the greatest disadvantage is the cost of the required equipment. However, SSMS has considerable advantage when a number of other elements must be surveyed.

10.3.4.8 Isotopic Dilution

Holynska and Lipinska-Kalita (1977) have used Se-75, sodium selenite tracers in an isotopic dilution procedure for selenium determination in blood and tissue. They used a wet digestion procedure and reduced selenium to its elemental form and then coprecipitated it on a filter with tellurium and a reducing agent and then counted the filter. About 95% was recovered with a precision of 4.2%. Parts per billion could be detected.

10.4 Nondestructive Analysis

Two important nondestructive techniques that are used are neutron activation analysis and X-ray fluorescence analysis. The main advantage of these two techniques is the small amount of required sample preparation compared to destructive analysis.

10.4.1 Neutron Activation Analysis

Because of its high detectability (10^{-8}–10^{-9} g Se) neutron activation analysis (NAA) is widely used for determination of selenium in biological materials. Thermal neutron activation is performed most frequently, as neutrons are a source of radioactive selenium isotopes. Only two ^{77M}Se and ^{75}Se are presently used in activation analysis. ^{77M}Se has a very short half-life (17.5 sec), which helps provide the greatest sensitivity and is employed only in instrumatic neutron activation analysis (INAA). ^{75}Se is used most frequently, as its long half-life time (120.4 days) allows for chemical separation even though long activation times are required.

The material to be tested is usually irradiated in a nuclear reactor with a flux of 10^{13}–10^{15} N/cm^2 sec for 7–14 days in the case of ^{75}Se or several seconds for ^{77M}Se. The activities of the irradiated samples is measured with a γ-ray multichannel analyzer and high-resolution Ge/Li or sometimes NaI/Tl detectors.

Activation analysis for selenium has been accomplished in most instances by use of the comparator technique. The unknown sample and a known weight of the pure element are irradiated together, so that each is exposed to the same neutron flux at the same time. After irradiation, both the standard and the unknown samples are counted under conditions in which, ideally the disintegration rate of the product radionuclide/unit weight of the element (specific activity) would be the same. The most popular standard reference samples in biological studies are bovine liver, Bowen's kale, and Orchard leaves, and in environmental studies coal and water (NBS).

Detection limits using either ^{75}Se or ^{75M}Se range from 0.005 to 0.10 μg Se. The range of values for a given selenium radionuclide reflects the sample size, the neutron energies and flux used to achieve activation, the irradiation period, and whether or not chemical separations were employed to remove interferences. The sensitivity is dependent on the abundance and the activation cross section of the nuclide producing the radioisotope whose photopeak(s) has(ve) been chosen for measurement after activation. The sensitivities achieved with the short-lived radioisotopes can be markedly affected by appreciable transportation time or long chemical separations after irradiation.

Both the radiochemical (NAA) and instrumental (INAA) variants have been used. Radiochemical separation methods have been based on the precipitation of selenium (Raie and Smith, 1977), separation on ion exchangers, and more often, distillation (Zmijewska and Semkow, 1978) and extraction (Zmijewska and Semkow, 1978).

Zmijewska and Semkow (1978) have extracted selenium with benzene containing 1% phenol or distilled it from a HBr–HCl medium following a wet digestion with H_2SO_4–$HClO_4$. Distillation was found to be more time consuming than extraction. With the distillation method 98.4% was recovered and the detection limit was 2 ng. In many cases several elements can be determined on the same sample.

Another variant of instrumental neutron activation analysis (INAA) is used more frequently. Large numbers of elements can be analyzed at the same time. Brotter *et al.* (1977) measured the content of 25 elements in a human skeleton and Cornelis *et al.* (1975) have determined 16 elements in human urine.

Human hair is frequently used as an index of human exposure to environmental metals or for identification in criminology. Usually single hairs are analyzed by INAA. Dybczynski and Boboli (1976) have measured 14 elements in a single human hair. Pillay and Juis (1978) found an average selenium content of 1.3 μg/g in human hair and also measured 20 other elements. INAA has also been used for water and air analysis, whereas in studies of soil, fertilizers, and rocks, radiochemical variants of NAA ar usually used.

10.4.2 X-Ray Fluorescence Analysis

X-Ray-induced fluorescence (XFA) is another nondestructive technique which is possible to analyze a number of elements simultaneously, often without a necessity of previous sample preparation. Selenium is determined by measurement of its K-α line (11.2 keV) with a semiconductor (e.g., Si/Li) and computer-coupled multichannel analyzer. Using this method selenium has been estimated in geological samples, fly ash, airborne particles, and fresh water. Strausz *et al.* (1975) and Holynska and Morkowicz (1977) have used the XFA method for measuring selenium biological samples which had been wet digested, and Strausz *et al.* (1975) reached a detection limit of 0.2 μg in a 5-g sample but preconcentration was necessary. Without preconcentration only 2–3 ppm of selenium can be detected.

10.4.3 Proton-Induced X-Ray Emission

Berti *et al.* (1977) have determined selenium in blood serum on thick dry-ashed samples using X-ray emission spectra induced by 1.8–4-MeV proton beams. The sensitivity was about 10 ppb for 100-min counts with 1.8-MeV protons and for 30-min counts with 4-MeV protons.

References

Allan, J. E., 1962. Atomic absorption spectrophotometry absorption lines and detection limits in the air-acetylene flame, *Spectrochim Acta* 18:259–263.

Allaway, W. H., and Cary, E. E., 1964. Determination of submicrograms of selenium in biological materials, *Anal. Chem.* 36:1359–1362.

Andrews, R. W., and Johnson, D. C., 1975. Voltammetric deposition and stripping of selenium (IV) as a rotating gold-disc electrode in decimolar perchloric acid, *Anal. Chem.* 47:294–299.

Asher, C. J., Evans, C. S., and Johnson, C. M., 1967. Collection and partial characterization of volatile selenium compounds from *Medicago sativa*, *Aust. J. Biol. Sci.* 20:737–748.

Baird, R. B., Pourian, S., and Gabrielian, S. M., 1972. Determination of trace amounts of selenium in wastewaters by carbon rod atomization, *Anal. Chem.* 44:1887–1889.

Beath, O. A., Eppson, H. F., and Gilbert, C. S., 1935. Selenium and other toxic minerals in soils and vegetation, *Wyoming Agr. Expt. Sta. Bull.* 206.

Beath, O. A., Eppson, H. F., and Gilbert, C. S., 1937. Selenium distribution in and seasonal variation of type vegetation occurring on seleniferous soils, *J. Am. Pharm. Assoc.* 26:394–405.

Bera, B. C., and Chakrabartty, 1968. Spectrophotometric determination of selenium with 2-mercaptobenzothiazole, *Analyst* 93:50–55.

Berti, M., Buso, G., Colautti, P., Moschini, G., Stievano, B. M., and Tregnaghi, C., 1977. Determination of selenium in blood serum by proton-induced X-ray emission, *Anal. Chem.* 49:1313–1315.

Brodie, K. G., 1977. Determining arsenic and selenium by atomic absorption spectroscopy— a comparative study, *Intern. Lab.* (September/October) pp. 65–74.

Brodie, K. G., 1979. Analysis of arsenic and other trace elements by vapour generation, *Intern. Lab.* (July/August), pp. 40–47.

Brotter, P., Gawlik, S., Lausch, J., and Rosick, V., 1977. On the distribution of trace elements in human skeletons, *J. Radioanal. Chem.* 37:393–397.

Brown, M. M., and Watkinson, J. H., 1977. An automated fluorometric method for the determination of nanogram quantities of selenium, *Anal. Chim. Acta* 89:29–35.

Buchtela, K., Grass, P., and Mueller, G., 1975. Radio–gas chromatography of metal chelates, *J. Chromatog.* 103:141–151.

Burgett, C. A., 1974. The gas chromatography of selenium as the trimethylsilyl derivative, *Anal. Lett.* 7:799–806.

Bycroft, B. M., and Clegg, E. D., 1978. Gas–liquid chromatographic determination of selenium in biological materials using 4-bromo and 4-chloro-1,2-diaminobenzene as derivatizing agent, *J. Assoc. Off. Anal. Chem.* 61:923–926.

Chan, F. L. 1964. 4-5-diamino-6-thiopyruimidine as an analytical reagent, I. Spectrophotometric determination of selenium, *Talanta* 11:1019–1029.

Chau, Y. K., Wong, P. T. S., and Goulden, P. D., 1975. Gas chromatography–atomic absorption method for the determination of dimethyl diselenide, *Anal. Chem.* 47:2279–2281.

Cheng, K. L., 1956. Determination of traces of selenium. 3,3-Diaminobenzidine as selenium (IV) organic reagent, *Anal. Chem.* 28:1937–1742.

Christian, G. D., Knoblock, E. C., and Purdy, W. C., 1965. Use of highly acid supporting electrolytes in polarography. Observed changes in polarographic waves of selenium (IV) upon standing, *Anal. Chem.* 37:425–427.

Cornelis, R., Speecke, A., and Hoste, J., 1975. Neutron activation analysis for bulk and trace elements in urine, *Anal. Chim. Acta.* 78:317–327.

Cousins, F. B., 1960. A fluorometric microdetermination of selenium in biological material, *Aust. J. Exp. Biol. Med. Sci.* 38:11–16.

Cresser, M. S., and West, T. S., 1968. Analytical chemistry of selenium: absorptiometric determination with 2-mercaptobenzoic acid, *Analyst* 93:595–600.

Cukor, P., Walzcyk, J., and Lott, P. F., 1964. The application of isotopic dilution analysis to the fluorometric determination of selenium in plant material, *Anal. Chim. Acta* 30:473–482.

Cummins, L. M., Martin, J. L., Maag, G. W., and Maag, D. D., 1964. A rapid method for the determination of selenium in biological material, *Anal. Chem.* 36:382–384.

Demeyere, D., and Hoste, J., 1962. o-Diamines as reagents for selenium. I. 4-Dimethylamino-o-phenylenediamine, *Anal. Chim. Acta* 27:288–294.

Dennis, B. L., Morris, J. L., and Wilson, G. S., 1976. Determination of selenium as selenide by differential pulse cathodic stripping voltammetry, *Anal. Chem.* 48:1611–1616.

Diamandis, E. P., and Hadjiioannau, T. P., 1981. Catalytic determination of selenium with a picrate-selective electrode, *Anal. Chim. Acta* 123:143–150.

Dybczynski, R., and Boboli, K., 1976. Forensic and environmental aspects of neutron activation analysis of single human hairs, *J. Radioanal. Chem.* 31:267–289.

Dye, W. B., Bretthauer, E., Seim, H. J., and Blincoe, C., 1963. Fluorometric determination of selenium in plants and animals with 3,3'-diaminobenzidine, *Anal. Chem.* 35:1687–1695.

Ehlig, C. F., Allaway, W. H., Cary, E. E., and Kubota, J., 1968. Differences among plant species in selenium accumulation from soils low in available selenium, *Agron, J.* 60:43–47.

Evans, C. A., and Morrison, G. H., 1968. Trace element survey analysis of biological materials by spank source mass spectrometry, *Anal. Chem.* 40:869–875.

Ewan, R. C., Baumann, C. A., and Pope, A. L., 1968. Determination of selenium in biological materials, *J. Agr. Food Chem.* 16:212–215.

Falkner, A. G., Knoblock, E. C., and Purdy, W. C., 1961. The polarographic determination of selenium in urine, *Clin. Chem.* 7:22–29.

Fernandez, F. J., and Manning, D. C., 1971. Determination of arsenic at submicrogram levels by atomic absorption spectrophotometry, *At. Abs. Newslett.* 10:86–88.

Flinn, C. G., and Aue, W. A., 1978. Photometric detection of selenium compounds for gas chromatography, *J. Chromatog.* 153:49–55.

Fogg, D. N., and Wilkinson, N. T., 1956. Determination of selenium, *Analyst* 81:525–531.

Ganther, H. E., 1966. Enzymic synthesis of dimethyl selenide from sodium selenite in mouse liver extracts, *Biochemistry* 5:1089–1098.

Gleit, C. E., and Holland, W. D., 1962. Use of electrically excited oxygen for the low temperature decomposition of organic substances, *Anal. Chem.* 34:1545–1547.

Gorman, J. G., Jones, E. J., and Hipple, J. A., 1951. Analysis of solids with a mass spectrometer, *Anal. Chem.* 23:438–440.

Grant, C. A., Thafvelin, B., and Christella, R., 1961. Retention of selenium by pig tissues, *Acta Pharmacol. Toxicol.* 18:285–297.

Grier, R. A., and Andrews, R. W., 1981. Cathodic stripping voltammetry of selenocystine, cystine, and cysteine, in dilute aqueous acid, *Anal. Chim. Acta* 124:333–339.

Griffin, D. A., 1969. Optimizations of parameters for single sweep polarographic analysis of the selenium-diaminobenzidine complex, *Anal. Chem.* 41:462–466.

Gutenmann, W. H., and Lisk, D. J., 1961. Determination of selenium in oats by oxygen flask combustion, *J. Agr. Food. Chem.* 9:488–489.

Hall, R. J., and Gupta, P. L., 1969. The determination of very small amounts of selenium in plant samples, *Analyst* 94:292–299.

Harrison, W. W., Clemena, G. G., and Magee, C. W., 1971. Forensic applications of spark source mass spectrometry, *J. Assoc. Offic. Agr. Chemists* 54:929–936.

Hoffman, I., Westerby, R. J., and Hirdiroglou, M., 1968. Precise fluorometric microdeterminations of selenium in agricultural materials, *J. Assoc. Offic. Anal. Chemists* 51:1039–1042.

Hoste, J., and Gillis, J., 1955. Spectrophotometric determination of traces of selenium with diaminobenzidine, *Anal. Chim. Acta* 12:158–161.

Holynska, B., and Lipinska-Kalita, K., 1977. Optimization of procedure for wet digestion procedure of blood and tissue for selenium determination by means of ^{75}Se tracer, Report INT 120/1, Institute of Physics and Nuclear Techniques, Krakow, Poland, pp. 1–7.

Holynska, B., and Markowicz, A., 1977. Application of energy dispensive x-ray fluorescence for the determination of selenium and blood and tissue, *Radiochem. Radioanal. Lett.* 31:165–170.

Ishizaki, M., 1978. Simple method for determination of selenium in biological materials by flameless atomic absorption spectrometry using carbon-tube atomizer, *Talanta* 25:167–169.

Johnson, H., 1970. Determination of selenium in solid waste, *Environ. Sci. Technol.* 4:850–853.

Kahn, H. L., and Schallis, J. E., 1968. Improvement of detection limits for arsenic, and other elements with an argon-hydrogen flame, *At. Abs. Newslett.* 7:5–9.

Kahn, H. L., Peterson, G. E., and Schallis, J. E., 1968. Atomic absorption microsampling with the sampling boat technique, *At. Abs. Newslett.* 7:35–39.

Kerber, J. D., and Fernandez, F. J., 1971. Determination of trace metals in aqueous solution with the Delves sampling cup technique, *At. Abs. Newslett.* 10:78–80.

Klein, A. K., 1941. Report on the determination of selenium in foods, *J. Assoc. Offic. Agr. Chemists* 24:363–380.

Kurahashi, K., Inoue, S., Yonekura, S., Shimoishi, Y., and Toei, K., 1980. Determination of selenium in human blood by gas chromatography with electron-capture detection, *Analyst* 105:690–695.

Lott, P. F., Cukor, P., Moriber, G., and Solga, J., 1963. 2,3-Diaminonaphthalene as a reagent for the determination of milligram to submicrogram amounts of selenium, *Anal. Chem.* 35:1159–1163.

Lunev, M. L., Kamenev, A. I., and Agasyan, P. K., 1976. Voltammetric determination of selenium (IV) at graphite electrodes, *Zh. Anal. Khim.* 31:1476–1481.

McCarthy, T. P., Brodie, B., Milner, J. A., and Bevill, R. F., 1981. Improved method for selenium determination in biological samples by gas chromatography, *J. Chromatog.* 225:9–16.

McNulty, J. S., 1947. Routine method for determining selenium in horticultural materials, *Anal. Chem.* 19:809–810.

Montaser, A., and Mehrabzadeh, A. A., 1978. Atomic absorption spectrometry with an electrothermal graphite braid atomizer, *Anal. Chem.* 50:1697–1699.

Mulford, C. E., 1966. Low-temperature ashing for the determination of volatile metals by atomic absorption spectroscopy, *At. Abs. Newslett.* 5:135–139.

Nakashima, S., and Toei, K., 1968. Determination of ultramicro amounts of selenium by gas-chromatography, *Talanta* 15:1475–1476.

Nazarenko, I. I., Kislova, I. V., Guseinov, T. M., Mkrtchyan, M. A., and Kiselov, A. M., 1975. Fluorometric determination of selenium in biological materials (e.g., foods) with 2,3-naphthalenediamine, *Zh. Anal. Khim.* 30:733–737.

Olson, O. E., 1969. Fluorometric analysis of selenium in plants, *J. Assoc. Offic. Anal. Chemists* 52:627–634.

Olson, O. E., Plamer, I. S., and Whitehead, E. I., 1973. Determination of selenium in biological materials, *Meth. Biochem. Anal.* 21:39–78.

Osborn, R. A., 1939. Report on the determination of selenium in foods, *J. Assoc. Offic. Agr. Chemists* 22:346–349.

Parker, C. A., and Harvey, L. G., 1961. Fluorometric determination of sub-microgram amounts of selenium, *Analyst* 86:54–62.

Parker, C. A., and Harvey, L. G., 1962. Luminescence of some piazselenols, *Analyst* 87:558–565.

Pillay, K. K. S., and Kuis, R. L., 1978. The potentials and limitations of using neutron activation analysis data on human hair as a forensic evidence, *J. Radioanal. Chem.* 43:461–478.

Posey, R. S., and Andrews, R. W., 1981. Determination of selenium (IV) by anodic stripping voltammetry with an in situ goldplated rotating glassy carbon disk electrode, *Anal. Chim. Acta* 124:107–112.

Raie, R. M., and Smith, H., 1977. The determination of selenium in biological material by thermal neutron activation analysis, *Radiochem. Radioanal. Lett.* 28:215–219.

Rann, C. S., and Hambly, A. N., 1965. Determination of selenium by atomic absorption spectrophotometry, *Anal. Chim. Acta* 32:346–354.

Reamer, D. C., and Veillon, C., 1981. Determination of selenium in biological materials by stable isotope dilution gas chromatography–mass spectrometry, *Anal. Chem.* 53:2166–2169.

Robinson, W. O., Dudley, H. C., Williams, K. T., and Byers, H. G., 1934. Determination of selenium and arsenic by distillation in pyrites, shales, soils and agricultural products, *Ind. Eng. Chem. Anal. Ed.* 6:274–276.

Schwaer, L., and Suchy, K., 1935. Dropping-mercury cathode. XLV. Electroreduction of selenites and tellurites *Collect. Czech. Chem. Commun.* 7:25–32.

Shimoishi, Y., 1977. Some 1,2-diaminobenzene derivatives as reagents for gas chromatographic determination of selenium with an electron capture detector, *J. Chromatogr.* 136–85:93.

Spallholz, J. E., Collins, G. F., and Schwarz, K., 1978. A single-test-tube method for the microdetermination of selenium, *Bioinorg. Chem.* 9:453–459.

Stijve, T., and Cardinale, E., 1975. Rapid determination of selenium in various substrates by electron capture gas–liquid chromatography, *J. Chromatogr.* 109:239–245.

Strausz, K. I., Purdham, J. T., and Strausz, O. P., 1975. X-Ray fluorescence spectrometric determination of selenium in biological materials, *Anal. Chem.* 47:2032–2034.

Talmi, Y., and Andrew, A. W., 1974. Determination of selenium in environmental samples using gas chromatography with a microwave emission spectrometric detection system, *Anal. Chem.* 46:2122–2126.

Throop, L. J., 1960. Spectrophotometric determination of selenium in steroids, *Anal. Chem.* 32:1807–1809.

Watkinson, J. H., 1966. Fluorometric determination of selenium in biological material with 2,3-diaminonaphthalene, *Anal. Chem.* 38:92–97.

Williams, K. T., and Lakin, H. W., 1935. Determination of selenium in organic matter, *Ind. Eng. Chem. Anal. Ed.* 7:409–410.

Young, J., and Christian, G. D., 1973. Gas-chromatographic determination for selenium, *Anal. Chim. Acta* 65:127–138.

Yurachek, J. P., Clemena, G. G., and Harrison, W. W., 1969. Analysis of human hair by spark source mass spectrometry, *Anal. Chem.* 41:1666–1668.

Zmijewska, W., and Semkow, T., 1978. Determination of selenium in biological materials, furnaces waste dumps and soil by neutron activation analysis, *Chem. Anal. (Warsaw)* 23:583–592.

Index